Proceedings of the Workshop on

Applications of High Intensity Proton Accelerators

Proceedings of the Workshop on

Applications of High Intensity Proton Accelerators

Fermilab, Chicago 19 – 21 October 2009

editors

Rajendran Raja

Fermi National Accelerator Laboratory, USA

Shekhar Mishra

Fermi National Accelerator Laboratory, USA

World Scientific

NEW JERSEY • LONDON • SINGAPORE • BEIJING • SHANGHAI • HONG KONG • TAIPEI • CHENNAI

Published by

World Scientific Publishing Co. Pte. Ltd.
5 Toh Tuck Link, Singapore 596224
USA office: 27 Warren Street, Suite 401-402, Hackensack, NJ 07601
UK office: 57 Shelton Street, Covent Garden, London WC2H 9HE

British Library Cataloguing-in-Publication Data
A catalogue record for this book is available from the British Library.

APPLICATIONS OF HIGH INTENSITY PROTON ACCELERATORS
Proceedings of the Workshop

ISBN-13 978-981-4317-28-3
ISBN-10 981-4317-28-4

Printed in Singapore by Mainland Press Pte Ltd.

Program Organizing Committee:

Chris Adolphsen (SLAC)
Swapan Chattopadhyay (Cockcroft Inst.)
Rodney Gerig (ANL) Co-Chair
Yousry Gohar (ANL)
Stuart Henderson (SNS)
Kenji Kikuchi, Ibaraki University (Japan)
K.Y.Kim (PEFP, Korea)
Stuart Maloy (LANL)
Shekhar Mishra (FNAL) Co-Chair
Kenneth Olsen (SPAFOA)
Peter Ostroumov (ANL)
Steve Peggs (BNL)
Rajendran Raja (FNAL)
Jean-Pierre Revol (CERN)
Robert Rimmer (Jlab)
Vinod Sahni (RRCAT, India)
Vladmir Shiltsev (FNAL)
Maury Tigner (Cornell)
JiuQing Wang (IHEP, China)
Mike Zisman (LBNL)

Local Organizing Committee:

Jeffrey Appel (FNAL)
Shilpee Arora (FNAL)
Elizabeth Clements (FNAL)
Jeanmarie Guyer (FNAL)
Ted Liu (FNAL)
Shekhar Mishra (FNAL)
Philippe Piot (NIU)
Kevin Pitts (UIUC)
Cynthia Sazama (FNAL) Co-chair
Yagmur Torun (IIT)
Suzanne Weber (FNAL)
Harry Weerts (ANL)

Preface

The workshop on Applications of High Intensity Proton Accelerators AHIPA09 was held at Fermilab, October 19–21, 2009. The workshop brought together over 170 participants from many countries and disciplines to discuss the various applications of high intensity proton machines. The workshop explored the use of a high intensity proton source to study long baseline neutrino oscillations and rare kaon and muon decays in working group 1. The proton source can be used to serve as the front end of a neutrino factory and muon collider and this parameter space was explored in working group 2. High intensity proton sources in the multi-megawatt range (10 MW or higher, 1 GeV protons) can be used to produce fast neutrons in a new class of nuclear reactors called accelerator driven sub-critical systems (ADS). Such devices have been studied since the 1990's but await their realization upon the demonstration of such an intense proton source. Reactors based on the ADS principle are safer than conventional reactors since they operate away from criticality. They can transmute elements such as Uranium 238 and Thorium 232 into elements that are fissile as well as use existing nuclear waste as fuel. Given the current drive towards greener energy technologies, ADS systems are generating considerable interest worldwide as was evidenced by the talks given at the workshop. The superconducting rf technology envisaged for Fermilab Project-X also holds great promise to realize the proton driver for the first ADS system. Working groups 3 and 4 investigated the ADS systems from the accelerator and the reactor ends, respectively.

High intensity proton sources make demands on the materials they interact with. Working group 5 investigated the material science as well as the production of nuclear isotopes using such a source. Each of the working groups submitted a summary of their findings as a whitepaper at the Department of Energy sponsored "Accelerators for America's future" workshop held in Washington D.C the following week. These proceedings document in greater detail the talks given at AHIPA09 and tell a coherent tale on the promise held by high intensity proton sources of the future for both a variety of physics investigations as well as for a safer form of nuclear energy.

AHIPA09 was an enjoyable experience for the participants and brought together a number of experts from several disciplines.

Rajendran Raja
Shekhar Mishra

Contents

Plenary Talks

SUPERCONDUCTING RADIOFREQUENCY LINAC DEVELOPMENT AT FERMILAB*

STEPHEN D. HOLMES

Fermi National Accelerator Laboratory, P.O Box 500
Batavia, IL, 60510, USA

As the Fermilab Tevatron Collider program draws to a close, a strategy has emerged of an experimental program built around the high intensity frontier. The centerpiece of this program is a superconducting H^- linac that will support world leading programs in long baseline neutrino experimentation and the study of rare processes. Based on technology shared with the International Linear Collider, Project X will provide multi-MW beams at 60-120 GeV from the Main Injector, simultaneous with very high intensity beams at lower energies. Project X also supports development of a Muon Collider as a future facility at the energy frontier.

1. Strategic Context and Evolution of the Fermilab Complex

Fermilab currently operates both the highest energy particle collider and the highest intensity accelerator-based neutrino beam in the world. This situation will change by the end of 2009 when the Large Hadron Collider (LHC) at CERN in Geneva, Switzerland will initiate operations at energies substantially higher than the Tevatron, and the J-PARC facility in Tokai, Japan, will initiate a long baseline neutrino program in strong competition with the Fermilab program. In anticipation of these developments the U.S. elementary particle physics (EPP) community has adopted a strategic plan for the U.S. over the coming decade [1] emphasizing forefront research on three frontiers: the energy frontier, the intensity frontier, and the cosmic frontier. This strategy includes accelerator-based initiatives on the energy and intensity, based both in the U.S. and abroad. The strategy recognizes that over the coming decade Fermilab will be the sole site for accelerator-based EPP in the U.S. The Fermilab strategy [2] is fully aligned with the U.S. plan, and features the development of a multi-Megawatt proton source as the centerpiece of the long term U.S. program.

* Work supported by the Fermi Research Alliance, under contract to the U.S. Department of Energy

1.1. *Evolution of the Fermilab Accelerator Complex*

Project X is the name applied to a multi-MW proton facility proposed for construction at Fermilab. Project X is based on an H- linear accelerator (linac) employing superconducting rf technology. Project X is the linchpin for future development of the Fermilab accelerator complex, providing long-term opportunities on both the energy and intensity frontiers. The technology development is closely aligned with the technologies required for the International Linear Collider (ILC), and this development preserves Fermilab as a possible host, or major contributor, to such a future facility. The development of multi-MW capabilities also could provide a basis for a future Muon Collider or Neutrino Factory. In parallel Project X provides great flexibility on the intensity frontier, offering the opportunity for a long-term world-leading program in neutrino physics and other beyond-the-standard-model phenomena. Finally, the technology developed for Project X opens opportunities beyond traditional EPP applications, for example in cold-neutron physics and accelerator driven (subcritical) systems (ADS) for energy generation and/or transmutation of waste.

2. Project X Performance Criteria and Initial Configurations

The initial design criteria for a multi-MW proton source have been established based on requirements outlined in the P5 report [1]:

- A neutrino beam for long baseline neutrino oscillation experiments based on a proton source capable of delivering at least 2 MW of beam power at any energy over the range 60-120 GeV;
- High intensity 8 GeV protons supporting muon and kaon precision experiments, simultaneous with the neutrino program; and
- A path toward a muon source for a possible future Neutrino Factory or Muon Collider. This requires that the initial facility retain the potential for upgrading to 2-4 MW at 8 GeV.

Two configurations that meet these requirements have been developed. The first is based on an 8 GeV pulsed linac, and the second on a 2 GeV continuous wave (CW) linac followed by an 8 GeV rapid cycling synchrotron (RCS). Both configurations utilize the existing Recycler and Main Injector rings for the accumulation of beam and acceleration to 60-120 GeV.

2.1. *Project X Initial Configuration-1*

An intial configuration, identified as Initial Configuration-1 (IC-1), based on an 8 GeV pulsed linac is shown schematically in Figure 1. This schematic corresponds to operations at 120 GeV; operations at 60 GeV rely on a cycle time of roughly half that of 120 GeV, in order to preserve the total beam power without changing the number of protons accelerated per pulse.

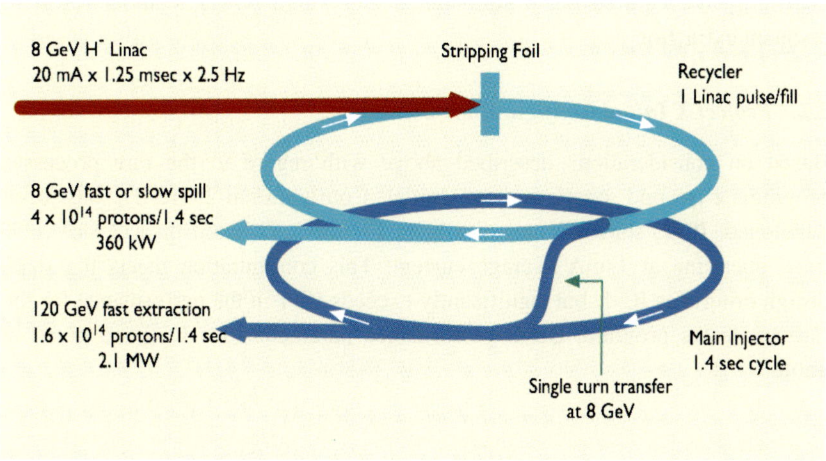

Figure 1. Project X Initial Configuration-1.

IC-1 is based on an 8 GeV, superconducting, pulsed H^- linac. The linac beam is accumulated in the Recycler Ring over ~100 turns, and is then transferred in a single turn to the Main Injector ring for acceleration and delivery to a neutrino target. The linac delivers 1.6×10^{14} H^- per pulse at a 2.5 Hz repetition rate: the total beam power available at 8 GeV is thus 500 kW. The Main Injector operates on a 1.4 second cycle time at 120 GeV, corresponding to 2.1 MW of beam power delivered to a neutrino program. In this scenario the Main Injector utilizes only two out of every seven linac pulses, leaving 5/7 of the linac pulses (360 kW at 8 GeV) available for other programs. In short, IC-1 is capable of delivering 2.1 MW at 120 GeV, simultaneous with 360 kW at 8 GeV.

Unfortunately IC-1 has limitations in providing a flexible platform for mounting a rare processes program – these programs (muons and kaons) typically require high duty factor beams that are traditionally provided by resonant extraction from a circular ring. The Recycler, because of its large ratio of circumference to energy is not well suited to resonant extraction – space charge effects and the need to create a substantial momentum spread to stabilize the beam compromise the overall efficiency of the process. Other smaller rings at Fermilab may be better suited, most notably the Debuncher Ring with its larger aperture and smaller circumference. However, even in this ring simulations show that there is an upper limit to the time averaged beam power that can be resonantly extracted at about 150 kW. We believe that a limit in this range is fundamental to the resonant extraction process, due to losses generated on a finite width extraction septum. Hence, we have explored options for providing the required beam structure, at very high power, without resort to resonant extraction.

2.2. *Project X Initial Configuration-2*

Based on considerations described above with regard to the rare processes program a second configuration, Initial Configuration-2 (IC-2), has been developed. IC-2, shown schematically in Figure 2, is based on a 2 GeV CW linac operating at 1 mA average current. This configuration meets the same design criteria as IC-1, but significantly exceeds IC-1 in the performance for the rare processes program. Beam performance parameters for IC-2 are given in Table 1.

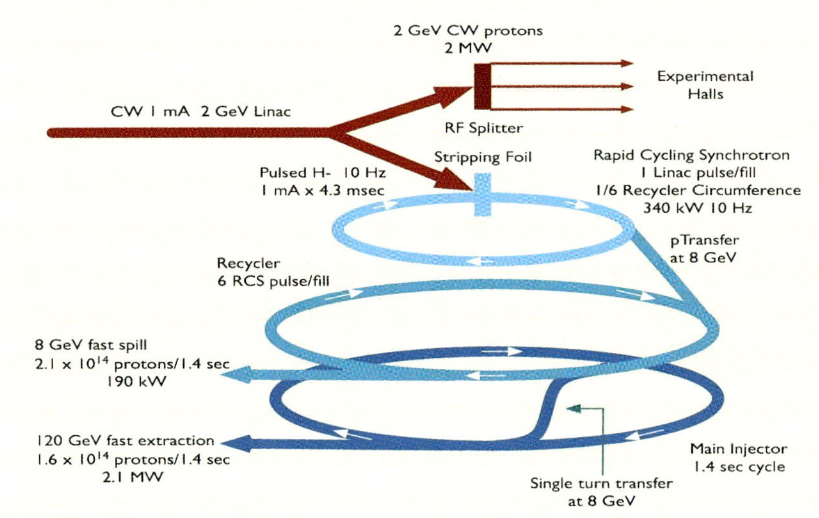

Figure 2. Project X Initial Configuration-2.

Table 1. Beam performance parameters for Project X Initial Configuration-2.

Linac		
Particle Type	H^-	
Kinetic Energy	2.0	GeV
Average Beam Current	1.0	mA
Pulse Rate	CW	Hz
Beam Power	2000	kW
Beam Power to 2 GeV Program	1920	kW
RCS		
Particle Type	protons	
Kinetic Energy	8	GeV
Cycle Time	0.1	sec
Particle per Pulse to Recycler	2.6×10^{13}	
Beam Power to 8 GeV Program	200	kW
Main Injector/Recycler		
Kinetic Energy (maximum)	120	GeV
Cycle Time	1.4	sec
Particles per Pulse	1.6×10^{14}	
Beam Power to 120 GeV	2100	kW

In IC-2 the neutrino program is supported by the acceleration of beam in a rapid cycling synchrotron (RCS) that injects into the Recycler. The RCS is one sixth the circumference of the Recycler (and the Main Injector) and operates at 10 Hz. The Main Injector operates at 120 GeV on the same 1.4 second cycle as in IC-1. The RCS is filled every 0.1 seconds by diverting 4.3 msec of linac beam and injecting over ~2000 turns. Six such cycles are then accumulated in the Recycler Ring and transferred to the Main Injector in a single turn every 1.4 seconds. With RCS operations at 10 Hz this leaves 8 of every 14 cycles available for supplying a fast extracted beam at 8 GeV for other programs. In this mode of operation only 4% of the linac beam is diverted to the RCS, leaving nearly the full 2 MW available for a low energy rare processes program. Thus, IC-2 provides simultaneously, 1.9 MW at 2 GeV, 200 kW at 8 GeV, and 2.1 MW at 120 GeV.

The rare processes program is supported directly from the CW linac with a bunch structure that can be individually tailored to the needs of the experimental program. The bunch format is created by a chopper at the front end with a bandwidth of ~300 MHz. The ultimate destination of the bunches is determined by their phase with respect to a subharmonic (1/4) transverse deflector at the end of the linac that can send individual bunches to one of three different experimental areas, depending upon the phase of the bunch. The chopper must create bunch patterns that average to 1 mA over the ~1 μsec memory time of the superconducting cavities, but otherwise the disposition of current within the bunch stream is arbitrary.

An example is shown in Figure 3. Rf buckets are available at a frequency of 162.5 MHz for acceptance of beam particles. Bunches colored blue have a phase that will transmit them undeflected through the transverse deflector at the end of the linac. The red bunches carry a phase that will deflect them in one direction, and the green bunches in the opposite direction. In this particular example the blue bunches could be used for a muon-to-electron conversion experiment requiring beam on for 100 nsec followed by ~1 μsec of no beam. The red and green beams are both continuous. The average current over the cycle shown in the figure is 1 mA.

Both IC-1 and IC-2 have been documented and are available online [3].

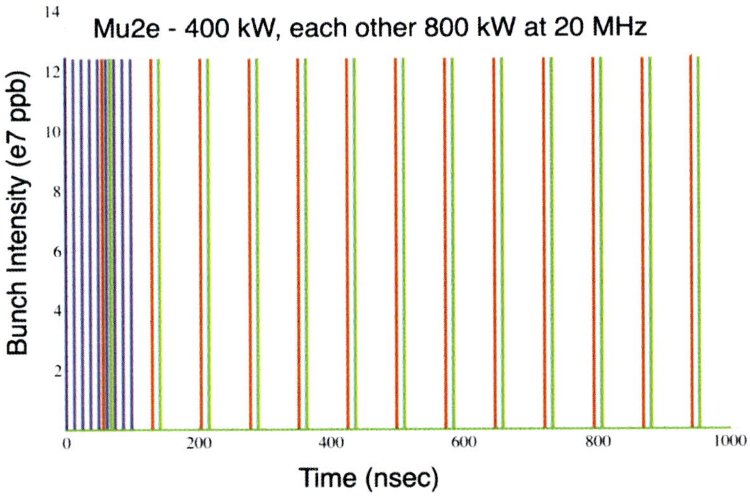

Figure 3. Possible bunch pattern averaging to 1 mA current, and capable of supplying 400 kW to one experiment, simultaneous with 800 kW to each of two additional experiments.

3. Project X Research, Design, and Development Plan

The goal of the Project X Research, Design, and Development (RD&D) Plan is to complete a fully developed technical description, construction schedule, and cost estimate by 2013. This effort is currently underway and is being undertaken by a multi-institutional U.S. collaboration composed of Argonne National Laboratory, Brookhaven National Laboratory, Cornell University, Fermilab, Lawrence Berkeley National Laboratory, Michigan State University, Oak Ridge National Laboratory, Thomas Jefferson National Accelerator Facility, SLAC

National Accelerator Laboratory, and the Americas Regional Team of the ILC Global Design Effort. Assuming successful completion of the RD&D phase, physical construction could start in 2015, with construction completed over a roughly five year construction period.

3.1. *Technology Development*

Technology deployment for the IC-2 linac is shown in Figure 4. This configuration is not yet optimized and hence only represents an initial concept.

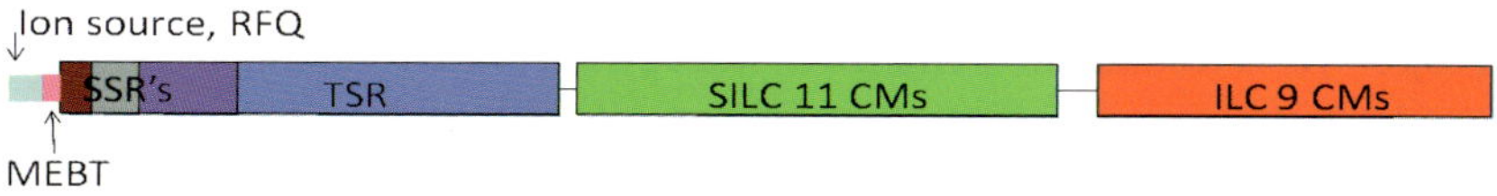

Figure 4. Initial configuration of the cw linac in IC-2.

A 10 mA dc ion source and 162.5 MHz RFQ are followed by a MEBT operating at 2.5 MeV. Development of the flexible chopper described above, which resides in the MEBT, will be one of the most challenging goals of the RD&D phase of the program. The upstream acceleration section is based on superconducting spoke resonator cavities designed to various betas and utilizing triple spoke resonators (TSR's) at the end. This section operates at 325 MHz and is envisioned as extending up to roughly 450 MeV. The following section is based on squeezed elliptical cavities at roughly beta=0.8, and operating at 1.3 GHz. The technology in this section is "ILC like". Eleven cryomodules are envisioned in this section, taking the beam up to about 1.2 GeV. These cavities are operated with a gradient of 14 MV/m. The linac is completed with nine beta=1.0 cyromodules, again operating at 1.3 GHz. This section takes the energy to 2.0 GeV, or possibly beyond. The cavity operating gradient in this section is 17 MV/m with a Q0 of 1.5×1010. The cavity technology is very similar to that required for the ILC, although the couplers and rf distribution system will be different. We know that the deployment outlined here is not optimal and expect to modify the various beta and frequency transition points over the next several months. Prototypes of both the lowest beta single spoke resonator and the beta=1.0 elliptical cavities have been built and successfully tested.

4. Other Applications

The technology being developed for Project X is broadly applicable in areas beyond support of research in elementary particle physics. Examples of additional applications include:

- Accelerator Driven Energy Systems
- Rare isotope production for nuclear physics
- Neutron Sources
- X-ray FELs
- Energy recovery linacs
- Muon facilities for materials research…

Several of these applications are discussed in detail in this workshop.

5. Summary

Project X is central to Fermilab's strategy for future development of the accelerator complex. Project X technology development is aligned with the needs of future Energy Frontier facilities such as the ILC and a Muon Collider. Construction of Project X would provide a world leading Intensity Frontier facility based on programs in neutrinos and rare processes.

Two intial configurations have been established that meet the general design criteria established for the facility. Both configurations support the delivery of 2 MW to a neutrino target, simultaneous with high power beam deliver to a low energy rare processes program. However the configuration based on a CW linac appears to offer significant advantages in terms of supporting the rare processes program, with an opportunity for multi-MW of beam power devoted to this program in parallel with the neutrino program.

Strong synergies exist with applications beyond elementary particle physics, either through the technology development or through the development of experimental programs based at the Project X facility. Such applications include, but are not limited to, Accelerator Driven Energy Systems, cold neutron sources, and other nuclear physics applications.

With sufficient support, Project X could be constructed over the time period 2015-1019.

References

1. Particle Physics Project Prioritization Panel, "U.S. Particle Physics: Scientific Opportunities, A Strategic Plan for the Next Ten Years", May 2008, http://www.science.doe.gov/hep/files/pdfs/P5_Report%2006022008.pdf
2. Fermilab Steering Group Report, September 2007, http://www.fnal.gov/pub/directorate/steering/index.shtml
3. Project X website, http://projectx.fnal.gov/

RARE MUON DECAY EXPERIMENTS

Y. KUNO

Department of Physics, Osaka University,
Osaka, Japan 560-0043
E-mail: kuno@phys.sci.osaka-u.ac.jp

Rare muon decay experiments to search for charged lepton flavor violation are described. Future prospects of rare muon decay experiments with highly intense proton accelerators are discussed.

Keywords: Muons, Charged lepton flavor violation

1. Introduction

Elementary particle physics has played a key role in expanding our understanding of the fundamental components of matter and the interactions between them. Now, the next important steps toward answering long-standing questions such as "What is the nature of the Universe?", "What is the origin of mass?", "How did the Universe begin?", and "How did the predominance of matter over antimatters initially appear?" are about to be taken in the field of elementary particle physics.

Our understanding of elementary particle physics is based on the Standard Model (SM), which is a gauge theory of the strong and electroweak interactions. The SM has been scrutinized by numerous experimental tests, and it is still consistent with most precision measurements to date. However, the SM cannot answer any of the long-standing questions mentioned above. Therefore, a more complete theory of particle physics has been demanded.

1.1. *Intensity Frontier and Energy Frontier*

There are two experimental approaches that can be used to tackle fundamental questions in elementary particle physics. They are the high-energy frontier approach and the high-precision frontier approach. In the high-energy frontier approach, studies are performed at very high energies, so particle accelerators capable of generating the very highest energies are needed. One example of such an accelerator is the Large Hadron Collider (LHC) at the European Center for Nuclear Research (CERN) in Switzerland.

In the high-precision frontier approach, measurements are made at low energies, but with high precision, and searches for very rare or so-called forbidden processes are performed in order to study phenomena at very high energy scale. Such works are largely based on quantum effects. In the Feynman diagrams entering some rare processes, one occasionally encounters intermediate virtual particles with heavy mass. In some instances, the mass of these particles is so large that they cannot be directly observed in high energy accelerators. It is thus possible in rare processes to study physics phenomena at an energy scale unattainable by using existing particle accelerators.

The physics subject described in this article is about the high precision frontier using muons.

2. Physics Motivation of cLFV

In the minimal version of the SM where massless neutrinos are assumed, lepton flavor conservation is a natural consequence of the gauge invariance. Therefore, it has been considered to be a naive explanation on why lepton flavor violation (LFV) in charged leptons in highly suppressed. Charged lepton flavor violation will be referred to "cLFV" from now on.

It have been confirmed that neutrinos are massive and mixed among different neutrino flavor species, by the observation of phenomena of neutrino oscillation. Therefore, lepton flavor for neutrinos is known to be violated. However, cLFV has been yet to be observed, and a discovery of cLFV is considered to be one of the important subjects in elementary particle physics.[1]

In the framework of the Standard Model (SM) with massive neutrinos and their mixing, the branching ratio of $\mu \to e\gamma$ decay ($B(\mu \to e\gamma)$) can be estimated by

$$B(\mu \to e\gamma) = \frac{3\alpha}{32\pi} |\sum_l (V_{\rm MNS})^*_{\mu l}(V_{\rm MNS})_{el}\frac{m^2_{\nu_l}}{m^2_W}|^2, \tag{1}$$

where $V_{\rm MNS}$ is the Maki-Nakagawa-Sakata (MNS) matrix elements, m_{ν_l} and m_W are masses of the neutrino ν_l and the W boson, respectively. This process is suppressed by the GIM mechanism and the estimated branching ratio is about 10^{-54}. Therefore, the contribution from the SM with massive neutrinos turns out to be very tiny. As a result, it can be concluded that

"**(1) observation of cLFV would indicate a clear signal of new physics beyond the SM with massive neutrino.**".

The precision measurement of the muon anomalous (g-2) magnetic moment has shown a possible deviation from the SM prediction, up to about 3 sigma. If this is true, it suggests new physics contributions to the dipole interaction of muons with photons. The cLFV processes, such as $\mu \to e\gamma$ has similar dipole interaction with photons. And it is conceivable that similar new physics contributions would also appear in cLFV. As a result, it can be concluded that "**(2) new physics contribution to the muon g-2 would also contributes to cLFV**".

Let us consider a general effective Lagrangian for cLFV. It can be given by

$$\mathcal{L}_{cLFV} = y\frac{em_\mu}{\Lambda^2}\bar{\mu}_R\sigma^{\mu\nu}e_L F_{\mu\nu} + h.c., \tag{2}$$

where y is a coupling constant of the new interaction, Λ is an energy scale of the new interaction, e and m_μ are an electric charge and the muon mass, respectively. μ_R and e_L are wave functions for a right-handed muon and a left-handed electron, respectively. $F_{\mu\nu}$ is an electromagnetic field tensor. By using Eq.(2), the branching ration of $\mu \to e\gamma$ at a tree level can be given by

$$B(\mu \to e\gamma) = y^2\frac{3(4\pi)^3\alpha}{G_F^2\Lambda^4} \tag{3}$$

$$= 1 \times 10^{-11} \times \left(\frac{400\ \text{TeV}}{\Lambda}\right)^4\left(\frac{y}{1}\right)^2, \tag{4}$$

where G_F and α are the Fermi coupling constant and the fine structure constant, respectively. Equation (4) indicates that if $y \sim 1$ the search for $\mu^+ \to e^+\gamma$ at a sensitivity of 10^{-11} has a sensitivity to reach an energy scale of 400 TeV, which cannot be directly accessed by high energy accelerators. At one loop level, by using the effective Lagrangian in Eq.(2), $B(\mu \to e\gamma)$ can be written by

$$B(\mu \to e\gamma) = 1 \times 10^{-11} \times \left(\frac{2\ \text{TeV}}{\Lambda}\right)^4\left(\frac{\theta_{\mu e}}{10^{-2}}\right)^2, \tag{5}$$

where $y = g^2\theta_{\mu e}/(16\pi^2)$ and $\theta_{\mu e}$ is a mixing angle between the muon and electron. Equation (5) indicates when the energy scale of new physics is $O(1\text{TeV})$, the mixing angles are suggested to be in an order of 10^{-2}, which is almost the same magnitudes for those of quarks and neutrinos. Therefore, cLFV is sensitive to new physics at TeV, as the LHC does, and flavor physics in the neutrino and quark sectors. As a result, it can be concluded that "**(3) the physics of cLFV is very complementary to that of the LHC.**".

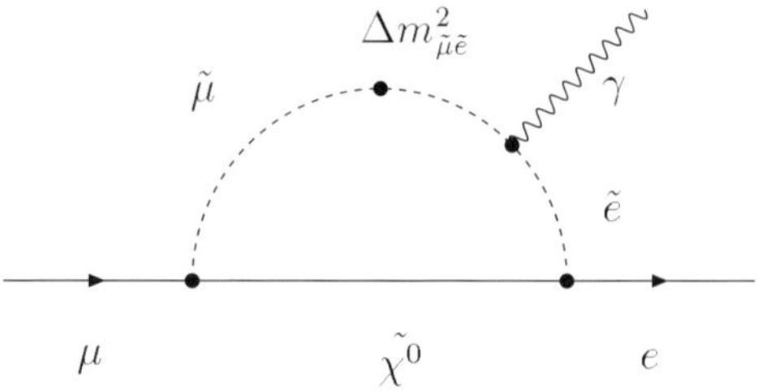

Fig. 1. A typical diagram for $\mu \to e\gamma$.

In fact, cLFV is known to be sensitive to various extension of new physics beyond the SM. Among them, a well-motivated physics model is a supersymmetric (SUSY) model. FIgure 1 shows one typical diagram of the SUSY contribution to cLFV. In Fig.1 it can be seen that the slepton mixing, which is shown as $\Delta m^2_{\tilde{\mu}\tilde{e}}$ (an off-diagonal element of the slepton mass matrix), would introduce cLFV. In the minimum SUSY scenario, the slepton mass matrix is assumed to be diagonal at the Planck scale. At a low energy scale, new physics phenomena such as grand-unification (GUT) or neutrino seesaw would introduce off-diagonal matrix elements such as $\Delta m_{\tilde{\mu}\tilde{e}}$ through quantum corrections (renormalization). Therefore, the slepon mixing is sensitive to GUT (at 10^{16} GeV) or neutrino seesaw mechanism (at 10^{13-14} GeV). As a result, it can be concluded that "**(4) cLFV has potential to study physics at very high energy scale.**".

Figure 2 shows predicted branching ratios[2] for muon to electron conversion, which is one of the cLFV processes with muons. In Fig.2, green and red points indicate the predictions with optimistic and pessimistic parameters respectively, and also the present experimental upper limit is presented. It is seen that if experimental bounds are improved in several orders of magnitude, there is a chance to make discovery of cLFV. As a result, it can be concluded that "**(5) theoretical predictions are just below the present experimental bounds**".

In a short summary, the five key conclusions, which are mentioned above from (1) to (5), would provide a very strong physics case.

3. cLFV Experiments

The experimental upper limits for the cLFV muon processes are summarized in Talbe 1. In the following, two typical examples of cLFV processes with muons, namely $\mu^+ \to e^+\gamma$ and $\mu^- N \to e^- N$, are discussed.

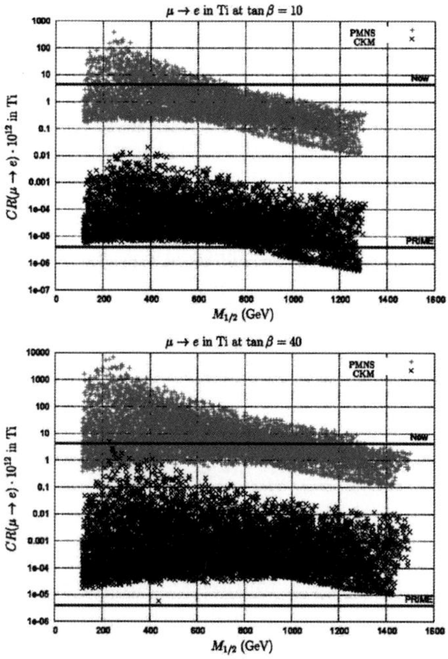

Fig. 2. A typical diagram for $\mu \to e\gamma$.

3.1. $\mu^+ \to e^+\gamma$ *Decay*

The event signature of $\mu^+ \to e^+\gamma$ decay at rest is a positron and a photon are moving back-to-back in coincidence with their energies equal to half that of the muon mass ($m_\mu/2 = 52.5$ MeV). There are two major backgrounds to the search for $\mu^+ \to e^+\gamma$ decay. One of them is a physics (prompt) background from radiative muon decay, $\mu \to e\nu\overline{\nu}\gamma$, when e^+ and photon are emitted back-to-back with the two neutrinos carrying off a small amount

Table 1. Experimental limits for the cLFV muon decays.

Reaction	Present limit
$\mu^+ \to e^+\gamma$	$< 1.2 \times 10^{-11}$
$\mu^+ \to e^* e^* e^-$	$< 1.0 \times 10^{-12}$
$\mu^- Ti \to e^- Ti$	$< 4.3 \times 10^{-12}$
$\mu^- Au \to e^- Au$	$< 7 \times 10^{-13}$
$\mu^- Pb \to e^- Pb$	$< 4.6 \times 10^{-11}$
$\mu^+ e^- \to \mu^- e^+$	$< 8.3 \times 10^{-11}$

of energy. The other background is an accidental coincidence of an e^+ in a normal muon decay, $\mu \to e\nu\overline{\nu}$, accompanied by a high energy photon. Possible sources of the latter would be either $\mu \to e\nu\overline{\nu}\gamma$ decay, annihilation-in-flight or external bremsstrahlung of e^+s from a normal muon decay.

The present experimental upper limit for $\mu^+ \to e^+\gamma$ is 1.2×10^{-11}, which was obtained by the MEGA experiment at Los Alamos National Laboratory (LANL) in the US. A new experiment at PSI called MEG, which aims to achieve a single event sensitivity of 2×10^{-13} was built. An improvement factor in a $\mu^+ \to e^+\gamma$ sensitivity will be expected to utilize a continuous muon beam of 100% duty factor at PSI. Physics data taking has already started in 2008, obtaining a preliminary result from the 2008 data of $B(\mu \to e\gamma) < 3.0 \times 10^{-11}$.

3.2. $\mu^- N \to e^- N$ *Conversion in a Muonic Atom*

Another prominent muon cLFV processes is coherent neutrino-less conversion of a negative muon to an electron ($\mu^- N \to e^- N$ conversion) in a muonic atom. When a negative muon is stopped in some material, it is trapped by an atom, and a muonic atom is formed. After it cascades down energy levels in the muonic atom, the muon is bound in its $1s$ ground state. The fate of the muon is then either decay in orbit (DIO) ($\mu^- \to e^-\nu_\mu\overline{\nu}_e$) or nuclear muon capture by a nucleus $N(A, Z)$ of mass number A and atomic number Z, namely, $\mu^- + N(A, Z) \to \nu_\mu + N(A, Z - 1)$. However, in the context of lepton flavor violation in physics beyond the Standard Model, the exotic process of neutrino-less muon capture, such as

$$\mu^- + N(A, Z) \to e^- + N(A, Z), \tag{6}$$

is also expected. This process is called $\mu^- N \to e^- N$ conversion in a muonic atom. This process violates the conservation of lepton flavor numbers, L_e and L_μ, by one unit, but the total lepton number, L, is conserved.

The final state of the nucleus (A, Z) could be either the ground state or one of the excited states. In general, the transition to the ground state, which is called coherent capture, is dominant. The rate of the coherent capture over non-coherent capture is enhanced by a factor approximately equal to the number of nucleons in the nucleus, since all of the nucleons participate in the process.

The event signature of coherent $\mu^- N \to e^- N$ conversion in a muonic atom is a mono-energetic single electron emitted from the conversion with an energy ($E_{\mu e}$) of

$$E_{\mu e} = m_\mu - B_\mu - E_{recoil}, \tag{7}$$

where m_μ is the muon mass, and B_μ is the binding energy of the $1s$ muonic atom. E_{recoil} is the nuclear recoil energy which is small and can be ignored. Since B_μ varies for various nuclei, $E_{\mu e}$ could be different. For instance, $E_{\mu e} = 104.3$ MeV for titanium (Ti) and $E_{\mu e} = 94.9$ MeV for lead (Pb).

From an experimental point of view, $\mu^- N \to e^- N$ conversion is a very attractive process in the following reasons.

- The energy of the signal electron of about 105 MeV is far above the end-point energy of the normal muon decay spectrum ($\sim$ 52.8 MeV).
- Since the event signature is a mono-energetic electron, no coincidence measurement is required. The search for this process has the potential to improve sensitivity by using a high muon rate without suffering from accidental background events, which would be serious for other processes, such as $\mu \to e\gamma$ and $\mu \to eee$ decays.

There are several potential sources of electron background events in the energy region around 100 MeV, which can be grouped into three categories as follows. The first group is intrinsic physics backgrounds which come from muons stopped in the muon-stopping target. The second is beam-related backgrounds which are caused by beam particles of muons and other contaminated particles in a muon beam. The third is other backgrounds which are, for instance, cosmic-ray backgrounds, and fake tracking events, and so on.

The previous search for $\mu^- N \to e^- N$ conversion was performed by the SINDRUM II collaboration at PSI. The SINDRUM II spectrometer consisted of a set of concentric cylindrical drift chambers inside a superconducting solenoid magnet of 1.2 Tesla. They set an upper limit of $\mu^- N \to e^- N$ in Au of $B(\mu^- + Au \to e^- + Au) < 7 \times 10^{-13}$.

4. Future Experimental Prospects

Considering its marked importance to physics, it is highly desirable to consider a next-generation experiment to search for cLFV with muons. There are three muon cLFV processes to be considered; namely, $\mu^+ \to e^+\gamma$, $\mu \to eee$ decays and $\mu^- N \to e^- N$ conversion.

The three muon LFV processes have different experimental issues that need to be solved to realize improved experimental sensitivities. They are summarized in Table 2. The processes of $\mu^+ \to e^+\gamma$ and $\mu \to eee$ decays are limited by accidental backgrounds. If the incident muon beam rate is increased by a factor N, background suppression has to be improved by a factor of N^2. To achieve this, the detector resolutions have to be significantly

Table 2. A list of major backgrounds, beam requirement and issues for various cLFV processes with muons.

Process	Backgrounds	Beam Requirement	Issue
$\mu^+ \to e^+\gamma$	accidentals	continuos beam	detector resolutions
$\mu \to eee$	accidentals	continuous beam	detector resolutions
$\mu^- N \to e^- N$	beam-associated	pulsed beam	beam qualities

improved, which is in general very challenging. In particular, improving the photon energy resolution for $\mu^+ \to e^+\gamma$ is difficult. On the other hand, for $\mu^- N \to e^- N$ conversion, there are no accidental background events, and thus an experiment with higher rates can be performed. If a new muon source with a higher beam intensity and a better beam quality for suppressing beam-associated background events can be constructed, measurements of higher sensitivity can be performed.

Furthermore, it is known that there are more physics processes contributing to $\mu^- N \to e^- N$ conversion and a $\mu \to eee$ decay than a $\mu^+ \to e^+\gamma$ decay. Namely, the dipole interaction of photon-mediation can contribute to all the three processes, but the box diagrams and four-fermion contact interaction can contribute to only $\mu^- N \to e^- N$ conversion and $\mu \to eee$ decay. In summary, in consideration of the experimental and theoretical aspects, a search for $\mu^- N \to e^- N$ conversion would be a natural next choice to accomplish significant improvements in future.

4.1. *Next Searches for $\mu^- N \to e^- N$ at 10^{-16}*

Next experimental projects to search for $\mu^- N \to e^- N$ conversion with a higher sensitivity are being pursued in the USA and Japan. To suppress background events, in particular beam-related backgrounds, the following key elements have been proposed. They are based on the ideas developed in the MELC proposal at the Moscow Meson Factory.[3]

- *Beam pulsing* :
  Since muonic atoms have lifetimes of the order of 1 μsec, a pulsed beam with its width that is short compared with these lifetimes would allow one to remove prompt background events by performing measurements in a delayed time window. To eliminate prompt beam-related backgrounds, proton beam extinction is required during the measurement interval.
- *High Field Solenoids for Pion Capture* :
  Superconducting solenoid magnets of a high magnetic field surrounds a proton target to capture pions in a large solid angle. It leads a dramatic increase of muon yields by several orders of magnitude.

- *Curved Solenoids for Muon Transport* :
  The solenoid system for muon transport has high transmission efficiency, resulting a significant increase of muon flux. The curved solenoids select charges and momenta of muons as well as removing neutral particles in a beam. The principle is as follows. In a curved solenoidal magnetic field, a center of the helical trajectory of a charged particle is shifted perpendicular to the curved plane. The shift, whose amount is given as a function of momentum and its charge, makes a dispersive beam. By placing appropriate collimators, charges and momenta of muons can be selected.

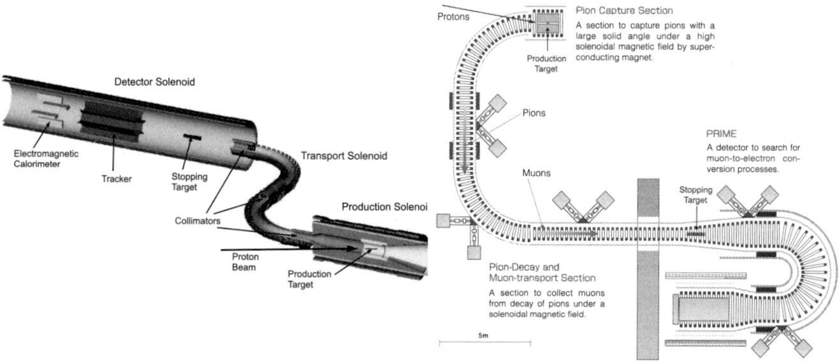

Fig. 3. Schematic layout of the Mu2e experiment at FNAL (left) and the COMET experiment at J-PARC (right).

One proposal in the USA was the MECO experiment at BNL.[4] It was mostly based on the MELC design and aimed to search for $\mu^- N \to e^- N$ conversion at a sensitivity of better than 10^{-16}. It consists of the production solenoid system, the transport solenoid system and the detector solenoid system. Unfortunately, the MECO proposal was cancelled in 2005, owing to its funding problems. In 2008 a new initiative at Fermi National Accelerator Laboratory (FNAL), which is called the Mu2e experiment, has been made to perform a MECO-type experiment.[5] The Mu2e experiment is planned to combat beam-related background events with the help of a 8 GeV/c proton beam of 25 kW in beam power from the Booster machine at FNAL. Figure 4.1(left) shows the proposed layout of the Mu2e experiment.

The other experimental proposal to search for $\mu^- N \to e^- N$ conversion, which is called COMET (COherent Muon to Electron Transition), is being prepared at the Japan Proton Accelerator Research Complex (J-PARC), Tokai, Japan.[6] The COMET uses a proton beam of 56 kW from the J-PARC main ring. The aimed sensitivity at COMET is similar to Mu2e, better than 10^{-16}. A schematic layout of the COMET experiment is presented in Figure 4.1(right). The differences of the designs between Mu2e and COMET exist in the adoption of C-shape curved solenoid magnets for both a muon beamline and a e^{+} spectrometer in COMET. First of all, in Mu2e, after the first 90-degree bending, the muons of their momenta of interest are necessarily shifted back to the median plane in the second 90-degree bending with opposite bending direction (therefore a S-shape), whereas in COMET, by applying a vertical correction magnetic field, the muons of interest can be kept on the median curved plane. From this fact, any opposite bending direction is not needed and a 180-degree bending in COMET would provide larger dispersion to give a better momentum selection. Secondary, a curved solenoid spectrometer in COMET is useful to eliminate low-energy events from muon decays in orbit before going into the detector, resulting in lower single counting rates in the detectors. To eliminate beam-related backgrounds at this sensitivity, the both experiments, Mu2e and COMET, require stringent requirement on proton beam extinction of 10^{-9} during the measurement interval. To meet the requirement, additional kicker magnets in the accelerator ring as well as in the extracted proton beam line is being considered.

4.2. *Future Searches for $\mu^{-}N \to e^{-}N$ at* 10^{-19}

In the long-term future, significant improvements to aim at an experiment with a 2×10^{-19} single-event sensitivity should be considered to make a full coverage of the SUSY parameter space, as seen in Fig.2. The sensitivity of 2×10^{-19} can be achieved only together with *multi MW proton beam power*, which can be provided by future proton accelerators. The other potential key requirements for the improvement are the following.

- Muon beam purity
  A low-momentum ($<$ 70 MeV/c) μ^{-} beam with no pion contamination ($< 10^{-20}$) would keep prompt background events at a negligible level. This could be achieved by adopting a muon storage ring, where pions decay out during their flight of many turns in the ring. An additional advantage of the method is that heavy muon-stopping targets such as gold, whose muonic-atom lifetimes is around 100 nsec, can be studied.

- Narrow energy spread of a muon beam
  The e^- energy resolution is determined by multiple scattering and energy straggling in the muon-stopping target. To improve the resolution. a thiner muon-stopping target is required. To keep a good muon-stopping efficiency, a narrow energy spread of a muon beam is needed.
- Extinction of a muon bcam
  As discussed, requirements on beam extinction is very stringent. In addition to the proton beam extinction, the extinction of a muon beam in low energy, which might be easier than high-energy protons, would be needed. To achieve this, fast kicker magnets are needed.

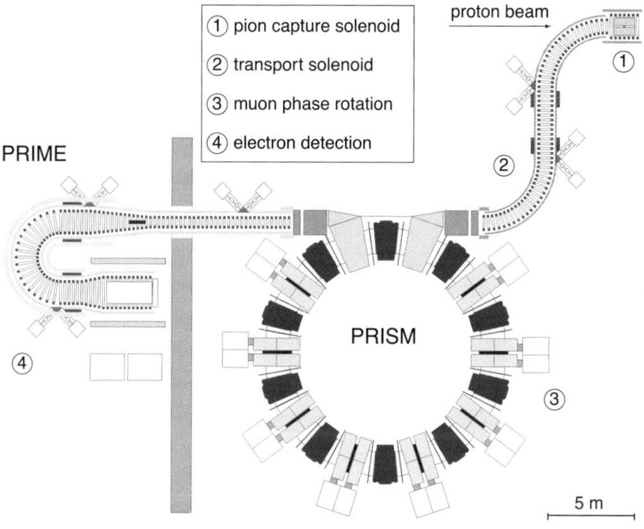

Fig. 4. Schematic layout of the PRISM/PRIME detector.

In consideration of these requirements, the PRISM (Phase Rotated Intense Slow Muon source) project is being developed in Japan.[7] In the PRISM project, a muon storage ring, which comprises a fixed field alternating gradient (FFAG) ring, is considered. The FFAG ring has large aperture to accept a muon beam of a large size and allows fast acceleration due to a fixed magnetic field. To achieve narrow energy spread, phase rotation, where fast muons are decelerated and slow muons are accelerated by RF fields in the muon storage ring, is adopted. Furthermore, the kicker magnets for injection and extraction to the muon storage ring would serve the muon-beam extinction, although they might be running at a low repetition such as 100−1000 Hz. The time structure of a proton beam should coincide with their repetition. A schematic layout of PRISM and its PRIME detector is shown in Figure 4.

5. Summary

The search for cLFV with muons has large potential to obtain hints for new physics beyond the Standard Model. The physics of cLFV has received glowing attention from both theorists and experimentalists. In this article, the phenomenology, the experimental status and future experimental prospects for various cLFV processes with muons are described. Ultimate searches for $\mu^- N \to e^- N$ conversion at a single event sensitivity of 2×10^{-19} have been considered. Here, high intensity proton accelerator would play a crucial role, since these searches would not be achieved without high intensity proton accelerators of multi MW beam power, such as Project-X. These attempts would offer extraordinary opportunities for exploring new phenomena which would otherwise be directly inaccessible at future high energy colliders.

References

1. Y. Kuno and Y. Okada, *Rev. Mod. Phys.* **73**, 151 (2001)
2. L. Calibbi, A. Faccia, A. Masierro, and S.K. Vempati, Phys. Rev. **D74** 116002 (2006).
3. R.M. Dzhilkibaev and V.M. Lobashev, *Sov. J. Nucl. Phys.* **49** 384 (1989) [*Yad. Fiz.* **49** 622 (1989)]
4. M. Bachman *et al.*, MECO, BNL Proposal E940 (1997).
5. R.M. Carry *et al.* (Mu2e collaboration), "Proposal to Search for $\mu^- N \to e^- N$ with a Single Event Sensitivity Below 10^{-16}, FNAL proposal (2008).
6. Y. Kuno *et al.* (COMET collaboration), J-PARC 50 GeV Proton Synchrotron Proposal P21 "A Experimental Search for Lepton Flavor Violating $\mu^- N \to e^- N$ Conversion at Sensitivity of 10^{-16} with A Slow-Extracted Bunched Proton Beam", unpublished, 2007. see http://j-parc.jp/NuclPart/pac_0801/pdf/Kuno.pdf
7. Y. Kuno *et al.* (PRISM collaboration), J-PARC 50 GeV Proton Synchrotron LOI P20 " An Experimental Search for a $\mu^- N \to e^- N$ Conversion at Sensitivity of the Order of 10^{-18} with a Highly Intense Muon Source: PRISM", unpublished, 2006. see http://j-parc.jp/NuclPart/pac_0606/pdf/p20-Kuno.pdf

RARE KAON DECAYS

DOUGLAS BRYMAN

Department of Physics and Astronomy, University of British Columbia
Vancouver, BC, Canada V6T 1Z1

The prospects for measuring the ultra-rare decays $K^+ \rightarrow \pi^+ \nu\bar{\nu}$ and $K_L^0 \rightarrow \pi^0 \nu\bar{\nu}$ are discussed. Several new experiments are being constructed or have been proposed at existing facilities and ideas for reaching very high precision experiments at a future high intensity proton source like Project X ICD2 are under discussion.

1. Introduction

The ultra-rare decay modes $K^+ \rightarrow \pi^+ \nu\bar{\nu}$ and $K_L^0 \rightarrow \pi^0 \nu\bar{\nu}$ are prime targets in the search for new physics offering incisive probes of quark flavor dynamics. Both reactions have been calculated to high precision in the framework of the Standard Model (SM) and hypothetical new phenomena coming from a wide range of physics beyond the SM would lead to measurable departures from the predictions.

The predictions for the $K^+ \rightarrow \pi^+ \nu\bar{\nu}$ and $K_L^0 \rightarrow \pi^0 \nu\bar{\nu}$ branching ratios are theoretically robust to the 2-5% level. No other loop-dominated quark processes can be predicted with this level of certainty. Since these reactions are highly suppressed in the SM to the 10^{-11}-10^{-10} level, new physics can compete and be observed with large enhancement factors, e.g. 5-10, over the SM values. However, the certainty with which the SM contributions are known permits 5σ discovery potential for new physics even for enhancements of the branching fractions as small as 30%. This sensitivity is unique in quark flavor physics and probes essentially all models of new physics that couple to quarks within the reach of the LHC. Further, high precision measurements of $K^+ \rightarrow \pi^+ \nu\bar{\nu}$ and $K_L^0 \rightarrow \pi^0 \nu\bar{\nu}$ are sensitive to many models of new physics with mass scales beyond the reach of the LHC.

The SM predictions for the $K^+ \rightarrow \pi^+ \nu\bar{\nu}$ and $K_L^0 \rightarrow \pi^0 \nu\bar{\nu}$ branching ratios [1] are

$B(K^+ \rightarrow \pi^+ \nu\bar{\nu}) = (8.51 \pm 0.75) x 10^{-11}$ and

$B(K_L^0 \rightarrow \pi^0 \nu\bar{\nu}) = (2.54 \pm 0.41) x 10^{-11}$

where the uncertainties arise from the CKM matrix, input parameters, electroweak and QCD perturbation theory, and the (isospin-corrected) $K \to \pi e \nu$ normalization. In general, non-SM particles could alter the branching fraction. New particles with flavor-changing neutral currents will change the short-distance coefficient of the effective interaction.

In the following, the prospects for measuring $K^+ \to \pi^+ \nu\bar{\nu}$ and $K_L^0 \to \pi^0 \nu\bar{\nu}$ at current and planned facilities will be discussed.

2. $K^+ \to \pi^+ \nu\bar{\nu}$

2.1. *Measurement Issues*

Measuring $K^+ \to \pi^+ \nu\bar{\nu}$ decay at the 10^{-10} branching ratio level represents a significant experimental challenge. The poorly defined signal consists of a charged kaon followed by a charged pion, $K^+ \to \pi^+$, with no other observed particles. Potential backgrounds, primarily from other K decays, at branching ratios 10 orders of magnitude higher, have similar signatures. Therefore, the experimental strategy involves proving that candidate events have low probabilities of being due to background. To be successful at detecting $K^+ \to \pi^+ \nu\bar{\nu}$ and separating it from background, the detector must have powerful π^+ particle identification so that $K^+ \to \mu^+ \nu$ and $K^+ \to \mu^+ \nu\gamma$ decays can be rejected, highly efficient 4-π solid angle photon detection coverage for vetoing $K^+ \to \pi^+ \pi^0$ events and other decays, and an efficient K^+ identification system for eliminating beam-related backgrounds.

2.2. *BNL E787/E949*

BNL E787/E949 [2] was the culmination of long series experiments searching for $K^+ \to \pi^+ \nu\bar{\nu}$ spanning 50 years. Like its predecessors, it employed a low momentum beam of stopping kaons. E949 used a 710 MeV/c K^+ beam which was slowed in a degrader and stopped in a scintillating fiber detector. Measurement of the $K^+ \to \pi^+ \nu\bar{\nu}$ decay involved observation of the daughter π^+ in the absence of other coincident activity in two momentum regions: 1) "PNN1", above the $K^+ \to \pi^+ \pi^0$ peak and 2) "PNN2", below the peak. The π^+ was identified by its kinematic features obtained from energy in a plastic scintillator calorimeter, momentum, and range measurements. In addition, high-speed 500 MHz digitizers were used on all fast scintillation detectors to make precise observations of the complete $K^+ \to \pi^+ \to \mu^+ \to e^+$ decay sequence. The entire spectrometer was immersed in a 1 T solenoidal magnetic field along the beam direction. In addition to the use of scintillating fibers and the large systems of 500 MHz digitizers, the challenges of E949 spurred the development of an exceptionally efficient detector of radiation and the invention and development of blind analysis methodology to avoid bias in background predictions and analysis of data. The numerous sources of potential background were extensively studied in E787 (the predecessor experiment) and E949 resulting in reliable and testable background predictions and a likelihood analysis method for evaluating potential candidate events for the probability of being due to $K^+ \rightarrow \pi^+ \nu\bar{\nu}$ decay or background. For the entire PNN1 high momentum region data set from the E787/E949 experiments the number of background events expected was considerably less than 1 event.

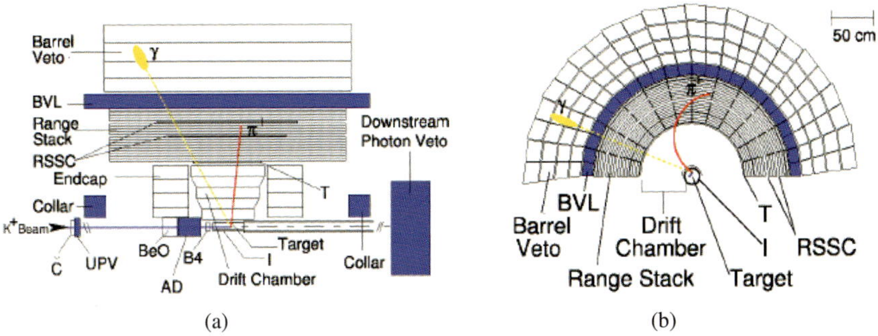

Fig. 1. Schematic side (a) and end (b) views of the upper half of the E949 detector. Illustrated in this figure, an incoming K+ that traverses all the beam instruments, stops in the target and undergoes the decay $K^+ \rightarrow \pi^0\pi^+$. The outgoing π+ and one photon from $\pi^0 \rightarrow \gamma\gamma$ are also shown.

Three events were observed by E787 and E949 in the momentum region 211 ≤ P ≤ 229 MeV/c above the $K^+ \rightarrow \pi^+\pi^0$ peak. Analysis of E787 and E949 data in the phase space region below the peak gave 4 additional events resulting in $\mathrm{B}(K^+ \rightarrow \pi^+ \nu\bar{\nu}) = (1.73^{+1.15}_{-1.05})x10^{-10}$. The estimated probability that all the $K^+ \rightarrow \pi^+ \nu\bar{\nu}$ candidates observed by E787 and E949 were due to background was <0.001. The measured branching ratio although twice as large as the SM prediction was consistent with it within the statistical uncertainty.

2.3. *NA62*

NA62 [3] at CERN is a new experiment dealing with $K^+ \rightarrow \pi^+ \nu\bar{\nu}$ seeking to reach an order of magnitude beyond the sensitivity obtained by E787/E949. The aim is to collect 100 events at the SM level in two years of data taking with a 15% background. The experiment uses a 400 GeV/c proton beam from the CERN-SPS exploiting a decay-in-flight technique with 75 GeV/c kaons. The detector requires sophisticated technology for which an intense R&D program has been in progress since 2006. The detection scheme shown in fig. 2 includes tracking detectors, particle ID detectors and veto detectors. Tracking will be performed by a kaon spectrometer and a pion spectrometer. Two new tracking detectors, the Gigatracker (three planes of pixel detectors) in the kaon spectrometer and the Straw Tracker in the pion spectrometer are under development. Particle ID is done with Cerenkov detectors. The hermetic photon system covers the forward small and large angles and includes the liquid krypton calorimeter inherited from NA48.

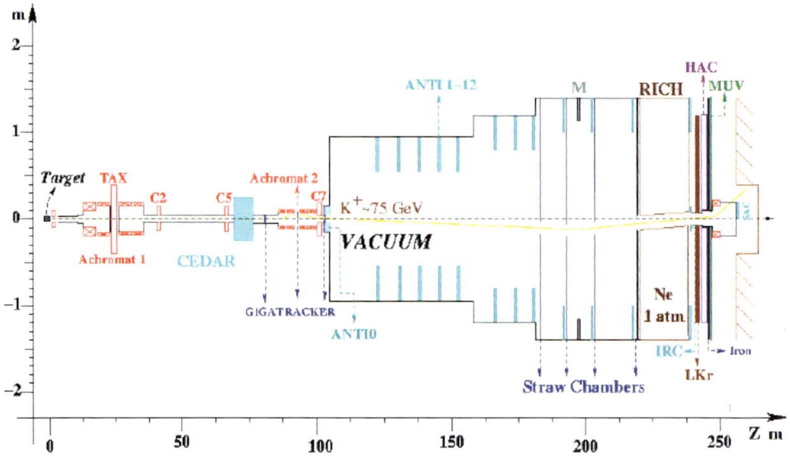

Fig. 2. Layout of NA62 at CERN[3].

2.4. *Fermilab Tevatron Stretcher*

A unique opportunity for producing extremely high rates of low energy charged kaons under ideal conditions for a measurement of the rare process $K^+ \to \pi^+ \nu\bar{\nu}$ exists at Fermilab once Run-II of the collider is completed. The "Tevatron Strecher" scheme involves operating the Tevatron as a 150 GeV high-duty factor (95%) synchrotron "stretcher" of the beam supplied by 10% of the available Main Injector flux. With kaon production from the Tevatron Stretcher, the demonstrated performance of the BNL E949 technique can be extrapolated to a Fermilab experiment capable of observing O(1000) $K^+ \to \pi^+ \nu\bar{\nu}$ events at the SM level.

In order to evaluate the potential of such an experiment at Fermilab the experience of E949 can be used taking advantage of the potential for incremental improvements in a well established technique. Sensitivity and background estimates can be reliably made relative to the measurements for E949.* For the new proposed Fermilab experiment, kaon beams of lower momentum, 550 MeV/c, would be used, in order to substantially improve the kaon stopping efficiency. Low energy K production cross sections at 0 degrees for 150 GeV protons were estimated using simulation resulting in favorable yields compared to BNL. A new design for a shorter higher acceptance charged K+ beam has been made leading to higher flux with good pion separation.

An acceptance gain of >10 can be attained compared with the BNL experiment with reduced backgrounds by making several incremental improvements to the E949 technique. An existing magnet, such at the CDF solenoid, run with a 1.25 T magnetic field may be used in comparison to E949 which used a 1 T field; the detector could also be made longer to increase the solid angle acceptance and momentum resolution. Several other improvements are anticipated including finer segmentation of the pion stopping region detectors (RS) which will substantially reduce muon backgrounds allowing a significant improvement in acceptance. Finer segmentation of the RS will also facilitate improvements in π^+ tracking prior to stopping. Improved momentum resolution will make primary 2-body background rejection easier and result in acceptance gains. In addition, the photon veto detector can be improved by using more radiation lengths than in E949 and reducing inactive material. This would result in better background rejection and higher acceptance. In spite of considerably higher stopping kaon rates, accidental losses (killing of valid events) due to extraneous hits in the photon veto system will be comparable to the E949 situation due to the higher duty factor.

Based on the assumptions above and five years of operation, the numbers of events to be expected for a Fermilab experiment using the Main Injector/Tevatron Stretcher combination would be approximately 200/yr.; precisions, including projected background subtraction, of approximately 5% would be anticipated in a five year run if the branching ratio is consistent with the SM. This would represent an improvement of two orders of magnitude over the work done at BNL. Further improvements in precision of $K^+ \to \pi^+ \nu\bar{\nu}$ measurements would be anticipated at Project X beam intensities.

* Operation of E949 was prematurely stopped after one short run.

3. $K_L^0 \to \pi^0 \nu\nu$

3.1 Measuring $K_L^0 \to \pi^0 \nu\bar{\nu}$ decay at the few × 10^{-11} branching ratio level represents, perhaps, a more difficult experimental challenge than for the charged mode. The poorly defined signal consists of a neutral kaon followed by a neutral pion, $K_L \to \pi^0$, with the pion immediately decaying into two γ's with no other observed particles. Potential backgrounds, from other *K* decays, at branching ratios many orders of magnitude higher, have similar signatures. In addition, neutrons, which inevitably dominate a neutral beam, can create π^0s from material in or near the beam. The experimental strategy involves proving that candidate events have low probabilities of being due to background. The principal intrinsic source of background is $K_L \to \pi^0\pi^0$ decay with a branching ratio 8.64 × 10^{-4}. This process can fake the signal either when one of the two π^0s is missed entirely, or when one γ from each of the π^0s is missed and the two odd γ's happen to reconstruct to a π^0 to within the resolution of the detector. There are many other possible background processes.

3.1. *KOTO*

The current experimental limit B($K_L^0 \to \pi^0 \nu\bar{\nu}$) < 2.6×$10^{-8}$ (90% C.L.) was reported by the KEK-PS E391a experiment [4]. The new J-PARC E14 *KOTO* experiment [5] aims at sensitivity sufficient for observation of $K_L^0 \to \pi^0 \nu\bar{\nu}$ at the SM level with an upgrade of the E391apparatus shown in fig. 3. Energy and position of the gammas from $\pi^0 \to \gamma\gamma$ are measured by the main pure CsI calorimeter at the downstream end of the decay region. To ensure that nothing else was present, the decay region is hermetically covered by photon veto counters with high efficiency. In addition, KOTO has a well-collimated "*pencil*" neutral beam. In the experiment, $\pi^0 \to \gamma\gamma$ decay will be reconstructed

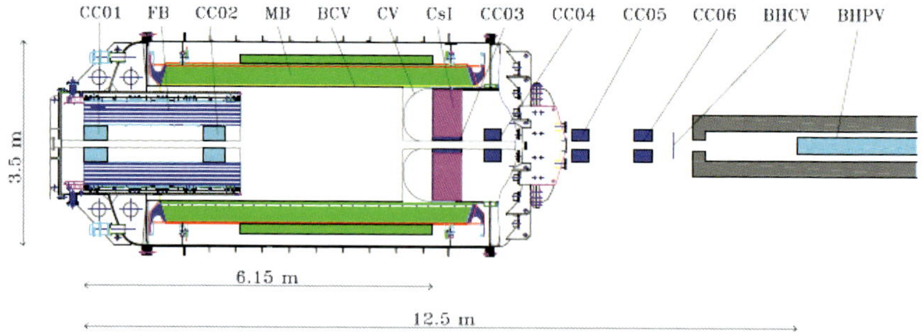

Fig. 3. Diagram of the *KOTO experiment at JPARC[5]*.

assuming that the decay vertex is on the beam-axis. Other background issues revolve around the scattering of neutrons in the halo of the neutral beam in beam line producing π^0 and η which can contaminate the signal region due to the assumption of π^0 reconstruction.

3.2. *Fermilab Project X ICD2*

To definitively measure and study the decay $K_L^0 \to \pi^0 \nu\bar{\nu}$ a low energy technique originally proposed for operation at BNL [6] may be highly effective and suitable for implementation at Project X ICD2 [7]. This approach focuses on obtaining the maximum possible information about each event, i.e. the direction, energy, production time and decay position of the K_L, and the directions, energies and times of the γ's from π^0 decay. In addition, the technique requires highly efficient hermetic rejection of events (vetoing) with extra particles. A low energy neutral beam is created by protons tightly bunched in time, so that the production time of the kaons is known. Combined with direction and timing measurements on the final state γ's [8], this gives the time-of-flight (TOF) and therefore the energy of the incident kaon. The directional measurement of the γ's gives the kaon decay position, assuring that the photons originated in a $\pi^0 \to \gamma\gamma$ decay from the beam. Finally, the energy measurement of the γ's allows powerful kinematic constraints to be imposed on candidate events in the K_L^0 center-of-mass system enabling rejection of the background from $K_L \to \pi^0 \pi^0$ decays.

The tight time bunch structure and high duty factor of the primary beam for the Project X ICD2 accelerator [7] could make the ideal beam for the low energy TOF approach to $K_L^0 \to \pi^0 \nu\bar{\nu}$. Rough estimates indicate that combining an intense pencil beam approach with elements of the technique proposed in [6] could lead to an experiment capable of recording 200 events per year at the SM level. Such an experiment would cover nearly all non-SM effects accessible to $K_L^0 \to \pi^0 \nu\bar{\nu}$ [9] or lead to severe constraints on theoretical hypotheses.

References

1. A. Kronfeld, private communication; A. J. Buras, M. Gorbahn, U. Haisch, and U. Nierste, JHEP 11 002 (2006); F. Mescia and C. Smith, Phys. Rev. D 76, 034017(2007).
2. A. V. Artamonov *et al.*, Phys. Rev. D 79:092004 (2009), and references therein.
3. G. Ruggiero, PoS KAON2009:043 (2009).
4. J. K. Ahn *et al.*, Phys. Rev. Lett. 100:201802 (2008); J. K. Ahn *et al.*, arxiv.org/abs/0911.4789 (2009).

5. See T. Shimogawa, PoS KAON09: 020 (2009).
6. See D. A. Bryman and L. Littenberg, Nucl. Phys. Proc. Suppl. 99B:61-69, 2001.
7. See discussions of Project X ICD2 in these proceedings.
8. D. Bryman *et al.*, Proc. Como 2005, Astroparticle, particle and space physics, detectors and medical physics applications 323 (2005).
9. D. Bryman, A. Buras, G. Isidori, and L. Littenberg. Int. J. Mod. Phys. A21, 487(2006).

MUON COLLIDER*

R. B. PALMER
Brookhaven National laboratory,
Upton, New York, 11786, USA
E-mail: ab_palmer@bnl.gov

Parameters are given of muon colliders with center of mass energies of 1.5 and 3 TeV. Pion production is from protons on a mercury target. Capture, decay, and phase rotation yields bunch trains of both muon signs. Six dimensional cooling reduces the emittances until the trains are merged into single bunches, one of each sign. Further cooling in 6 dimensions is then applied, followed by final transverse cooling in 50 T solenoids. After acceleration the muons enter the collider ring. Ongoing R&D is discussed.

Keywords: Muon Collider; Ionization Cooling; Acceleration

1. Introduction

This work is in collaboration with: The Neutrino Factory and Muon Collider Collaboration[1] (NFMCC) and the FNAL Muon Collider Task Force[2] (MCTF).

Muon colliders were first proposed by Budker in 1969.[3] A more detailed study was done for Snowmass 96.[4] This report will address the current approach to such a scheme.

Muon colliders would allow the high energy study of point-like collisions but without so much synchrotron radiation that requires the electron acceleration to be linear and long. Muons can be accelerated in smaller rings and offer other advantages, but the disadvantages are that muons are produced diffusely and they decay rapidly. The proposed scheme outlined here includes the pion production, capture, pion decay to muons, acceleration, and collider rings will be outlined.

Table 1 gives parameters for muon colliders at two energies. Both use the same muon intensities and emittances, although the repetition rates for

*Work supported by US Department of Energy under contracts AC02-98CH10886 and DE-AC02-76CH03000

the higher energy machines are reduced to control neutrino radiation that could be a radiation hazard.

Fig. 1a shows a schematic of the components of the system.

2. Components

2.1. *Introduction*

The muons are generated by the decay of pions produced by proton interacting in a mercury jet target. The pions are captured by a 20 T solenoid surrounding the target, followed by an adiabatic lowering of the magnetic field to a constant value in the decay channel.

Table 1. Parameters of two muon colliders

C of m Energy	TeV	1.5	3
Luminosity	10^{34} cm^{2}sec^{-1}	1	4
Beam-beam Tune Shift		0.087	0.087
Muons/bunch	10^{12}	2	2
Ring circumference	km	2.6	4.5
Beta at IP $= \sigma_z$	mm	10	5
rms mom. spread	%	0.1(0.3)	0.12
Required depth for ν rad	m	≈20	≈ 200
Mu per 8GeV p		0.16	0.16
Muon survival		0.07	0.06
Repetition Rate	Hz	15	12
Proton Driver power	MW	≈4	≈ 4
Trans Emittance	mm	25	25
Long Emittance	μm	72,000	72,000

Following the target, the pions drift and decay into muons in a solenoid focused channel. The captured muons, with mean momenta of $\approx$ 200 MeV/c, and longitudinal emittance ($\beta\ \gamma\ dp/p\ \sigma_z$) of approximately 0.5 m and transverse emittance ($\beta\ \gamma\ \sigma_\theta\ \sigma_r$) of approximately 20 mm. These must be cooled to the specified emittances of 72 mm and 25 μm respectively, as shown in Tb.1. This represents a cooling of the 6 dimensional phase space by approximately 10^7. This is achieved by Ionization Cooling in which the muons lose forward and transverse momentum in absorbers, followed by re-acceleration that restores only the forward loss.

Fig. 1b shows a plot of the longitudinal vs. transverse emittances of the muons as they progress from production at the target to the end of the cooling systems, using subsystems numbered 1-5 in the figures.

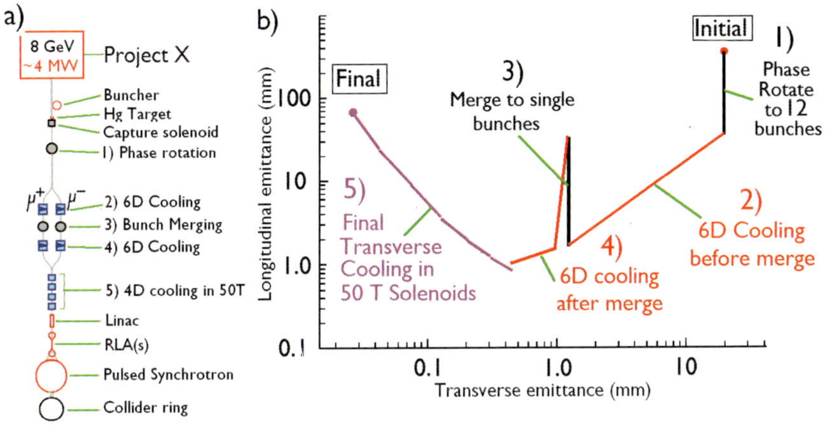

Fig. 1. a) shows a schematic of the components of the system; b) shows the longitudinal vs. transverse emittances through the five beam manipulations and cooling.

1) Bunching and Phase Rotation

The phase rotation[8] converts the single muon burst (with short length but huge momentum spread) into strings of bunches over a longer time but with lower momentum spreads.

In a 80 m drift (including the initial taper), the muon burst is allowed to lengthen and develop a time energy correlation. It is then, over a distance of 33 m, bunched by rf (with frequencies from 333 to ≈ 220 MHz, dependent on location) into a train without reducing the time energy correlation. Then, over 42 m, by phase and frequency control, the rf accelerates the low energy bunches and decelerates the high energy ones, to form a mono-energetic train. Muons of both signs are captured into interleaved bunches.

2) 6D cooling before merge

The next stages cool simultaneously in all 6 dimensions until the longitudinal emittance is small enough to allow the bunches to be merged into just 2 bunches, one of each sign. The cooling parameters would vary along the length of the cooling systems: starting with lower fields, lower frequency rf, and longer cells, and progressing to higher magnetic fields and rf frequencies. A section of cooling in only transverse cooling might be employed initially.

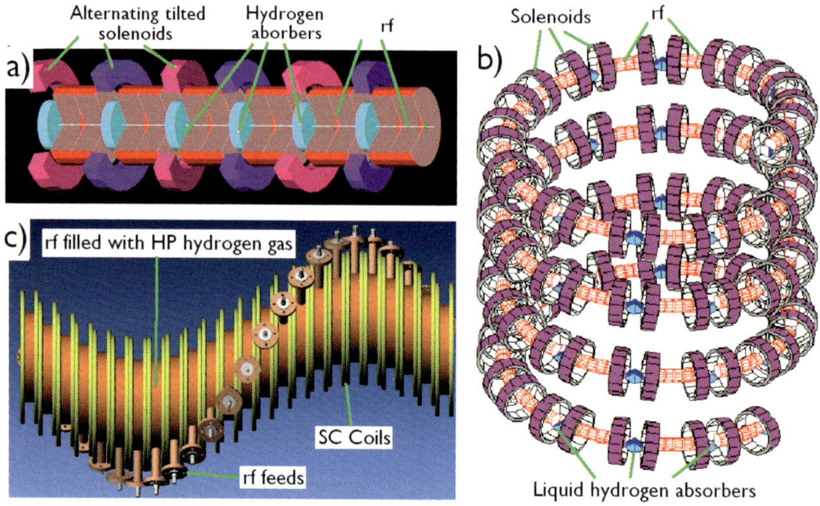

Fig. 2. 6D cooling systems: a) Snake; b) Guggenheim; c) Helical Cooling Channel.

There are broadly 3 different methods illustrated in fig.2 to accomplish this 6D cooling. Each has a somewhat different mechanism to reduce momentum spreads to achieve the cooling in all dimensions.

(a) In the Snake[9] system the lattice is approximately straight and consists of alternating thin solenoids with small (7 mrad) tilt angles that repeat after 6 solenoids. Liquid hydrogen absorbers with thin windows are located under each solenoid and the rest of the space is full of vacuum rf. A great advantage of this solution is that it cools both signs simultaneously. Simulations are in an early stage, but appear to meet requirements. Longitudinal cooling (momentum spread reduction) arises because the higher momentum tracks pass through the absorbers at larger angles, see greater absorption lengths, and lose more energy.
(b) The Guggenheim lattice[10] also uses vacuum rf and tilted alternating field solenoids, but the tilts are such as to bend the lattice into a slowly rising or falling helix. The absorbers are also of liquid hydrogen, but they are wedge shaped, rather than flat as in the Snake. Longitudinal cooling arises because dispersion brings tracks through the thicker parts of the absorbers.

Both methods a) and b) employ vacuum rf in significant magnetic fields and experiments have shown breakdown problems with such cavities. Two solutions are being tested: magnetic insulation and the use of beryllium surfaces. A third approach is to fill the cavities with high pressure hydrogen gas. A method using this approach is:

(c) 6D cooling is achieved in a high pressure gas filled Helical Cooling Channel (HCC).[11] Momentum cooling is achieved by having longer path lengths for higher momentum particles. RF breakdown in high pressure gas is not affected by the magnetic fields. But it is not yet known whether the gas will breakdown, or become excessively lossy in the presence of the ionizing muon beam. Integrating the rf and waveguides into the helical magnets will be more difficult than in a) or b).

The choice of which method to pursue is not yet decided. Significant R&D is still needed to determine the best approach.

3) Bunch merge

Collider luminosity is proportional to the square of the number of muons per bunch. It is thus desirable to use few bunches with many muons per bunch. The trains of bunches are thus merged into one prior to their use in the collider. This is done as soon as longitudinal cooling is sufficient. There are two stages: a) using a drift followed by the bunch spacing frequency rf, with harmonics, the individual bunches are phase rotated to minimize the spaces between them and lower their energy spread; and b) very low frequency ($\approx$5 MHz) rf, plus harmonics, interspersed along a long drift to phase rotate the entire train into a single bunch that can be captured using the bunch spacing rf.

4) 6D cooling after merge

After the bunch merging, the longitudinal emittance of the single bunch is now similar to that at the start of cooling, and it can be taken through the same, or similar, cooling systems as used before merging. But now cooling is continued to the least possible emittances. Simulations of all three method appear to achieve about the same minimum emittances: 0.4 mm transverse and 1 mm longitudinal, although the engineering practicality of the final 6D systems is yet to be determined.

After this stage, the transverse emittance is a factor of 16 greater than that specified, but the longitudinal emittance is a factor of 72 less. Thus in the final cooling of the longitudinal emittance can be allowed to grow.

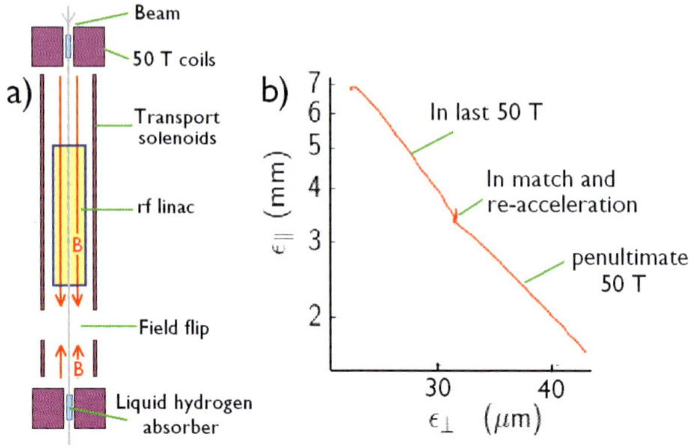

Fig. 3. a) Schematic of a part of the 50 T final cooling; b) Long. vs. trans. emittances through the last two stages including matching and re-acceleration.

5) Final cooling in high field solenoids

To attain the required final transverse emittance, the cooling needs stronger focusing than is achievable in the 6D cooling lattices. But, if the momentum is allowed to fall below 60 MeV/c, it can be obtained in liquid hydrogen in long 50T solenoids. At this low momentum the energy spread, and thus longitudinal emittance, rises, but this is acceptable.

Fig. 3a shows a schematic of one cell of cooling in a 50 T solenoid. Each cell includes liquid hydrogen absorber in the high field, transport between them, re-acceleration, and a field flip. Cooling from 300 to 25 μm is achieved in 8-16 cells. Matching and re-acceleration has been simulated only through the last two stages. Fig. 3b shows the longitudinal vs. transverse emittances of an ICOOL[6] simulation through these last two cells. Very little emittance dilution is observed, either longitudinally or transversely.

The calculated space charge tune shifts are moderate, but space charge is not yet in the simulations.

2.2. *Acceleration*

Initial acceleration after cooling would be in a sequence of linacs with frequencies increasing as the energy rises and bunch lengths decrease. Then there would be one or more Recycling Linear Accelerators (RLAs). RLAs could be used for the remaining acceleration, but a lower cost solution would use rapidly field ramped synchrotons.[7] In order to avoid high field pulsed bending magnets or excessive diameters, fixed super conducting magnets are alternated with pulsed magnets swinging from -1.8 T to +1.8T.

2.3. *Collider Rings*

The 1.5 TeV center of mass collider parameters are based on a recently designed lattice[5] that gives a 4.5 sigma acceptance for the 25 μm transverse emittance, an momentum acceptance of almost 10 sigma. The parameters for the 3 TeV center of mass collider are based on the 4 TeV Oide[4] lattice designed for the 1996 Snowmass study.[4]

3. R&D

(i) The use of a free mercury jet has been demonstrated in a recently run experiment MERIT[12] experiment at CERN. In this experiment 24 GeV protons with intensities up to 30 10^{12} protons intersected a mercury jet inside a pulsed 15 T solenoid magnet. The velocities of the observed splash, which could be damaging at a few 100 m/s, are suppressed by the damping effects of the magnetic field. Extrapolation of this data suggests that with 20 T, the splash should be acceptable.

(ii) The experiment MICE,[13] at RAL, UK, is under construction and should demonstrate ionization cooling of a muon beam using equipment similar to those in the Snake, or Guggenheim 6D cooling schemes.

(iii) A BNL/PBL collaboration is winding two nested YBCO high temperature super-conducting solenoids that when tested in a 19 T resistive magnet at the NHMFL should give a field approaching 40 T: a step towards 50 T.

(iv) An FNAL/Muons Inc.[15] collaboration has built and tested a few superconducting coils of a helical cooling channel.

(v) The MuCool[16] Collaboration based at Fermilab has tested rf vacuum cavities, at both 805 and 201 MHz, with and without external magnetic fields. These experiments have shown that copper surfaces in the cavities get damaged, and the maximum rf gradients are reduced,[17] when operated in magnetic fields. It is conjectured[18] that field emitted electrons focused by the magnetic field are responsible. Damage is not observed on beryllium windows in the cavities, and a test comparing copper with beryllium is planned. Magnetic insulation will also soon be tested.

In collaboration with Muons Inc.[15] cavities filled with high pressure hydrogen[19] have also been tested using differing materials, both with and without external magnetic fields. A test soon will expose such a cavity to a proton beam to investigate gas breakdown and possible reduction of the cavity Q.

In collaboration with Japan, the collaboration has tested a liquid hydrogen absorber.

4. Conclusion

If a solution is found for the rf breakdown in magnetic fields, then the system described here is a plausible scenario for capturing, manipulating, cooling, accelerating, and colliding muons with useful luminosities and energies even up to 3 or more TeV in the center of mass. But much work remains to be done.

References

1. NFMCC, http://www.cap.bnl.gov/mumu/
2. MCTF, https://mctf.fnal.gov/
3. G. I. Budker, 1969 Yerevan Conference, AIP Conf. Proc.352 (1996) 4, G. I. Budker, 1970 Kiev Conference, AIP Conf. Proc.352 (1996) 5
4. *$\mu^+\mu^-$ Collider. A Feasibility Study*, BNL-52503, Fermi Lab-Conf. 96/092, LBNL-38946, Snowmass (1996).
5. Y. Alexahin and E. Gianfelice-Wendt, http://www.cap.bnl.gov/mumu/conf/collider-091201/talks/YAlexahin-2-091201.ppt
6. http://pubweb.bnl.gov/users/fernow/www/icool/readme.html
7. Don Summers, http://www.cap.bnl.gov/mumu/conf/collider-091201/talks/DSummers1-091201.pdf
8. D. Neuffer, *Exploration of the high-frequency buncher concept*, MUC-NOTE-269 (2003). This series of technical notes can be found at http://www-mucool.fnal.gov/mcnotes/
9. Y. Alexahin, http://www.cap.bnl.gov/mumu/conf/collider-091201/talks/YAlexahin-1-091201.ppt
10. J. S. Berg, R. C. Fernow, and R. B. Palmer, *An Alternating Solenoid Focused Ionization Cooling Ring,* MUC-NOTE-THEORY-239 (2002), Pavel Snopoc, http://www.cap.bnl.gov/mumu/conf/MUTAC-090406/talks/PSnopok1-090406.pdf
11. Y. Derbenev and R. P. Johnson, PRSTAB 8, 041002 (2005). http://www.muonsinc.com/reports/PRSTAB-HCCtheory.pdf
12. http://proj-hiptarget.web.cern.ch/proj-hiptarget/
13. Muon Ionization Cooling Experiment (MICE), http://mice.iit.edu/
14. Particles Beams Lasers, http://www.bnl.gov/magnets/Staff/gupta/talks/ltsw09/gupta-ltsw09-hts.pdf
15. Muons Inc., http://www.muonsinc.com/
16. MuCool Collaboration, http://mice.iit.edu/mta/
17. A. Moretti et al, *Effects of high solenoidal fields on rf accelerating cavities*, MUC Note 318 (2005)
18. R. B. Palmer et al, Phys. Rev. ST Accel. Beams 12, 031002 (2009)
19. P. Hanlet, et al, *High pressure rf cavities in magnetif fields*, Proc EPAC 2006, Edinburgh, TUPCH147, and R. Johnson, private

NEUTRINO FACTORIES

STEVE GEER*

Fermi National Accelerator Laboratory, PO Box 500, Batavia, IL 60510, U.S.A.

Over the last decade there has been significant progress in developing the concepts and technologies needed to produce, capture and accelerate $O(10^{21})$ muons/year. This development prepares the way for a new type of neutrino source : a Neutrino Factory. This article reviews the motivation, design and R&D for a Neutrino Factory.

1. Introduction

Over the last decade there has been significant progress in developing the concepts and technologies needed to produce, capture and accelerate $O(10^{21})$ muons/year. This prepares the way for a Neutrino Factory (NF) in which high energy muons decay within the straight sections of a storage ring to produce a beam of neutrinos and anti-neutrinos. The NF concept was proposed[1] in 1997 at a time when the discovery that the three known types (flavor) of neutrino (ν_e,ν_μ,ν_τ) can change their flavor as they propagate through space (neutrino oscillations[2]) was providing us with a first glimpse of physics beyond the Standard Model.

2. Concept, Beam Properties, and Physics Reach

Conventional neutrino beams are produced from charged pions ($\pi^\pm$) decaying in a long channel. The resulting beam consists of ν_μ ($\bar{\nu}_\mu$) from π^+ (π^-) decays. To fully explore neutrino oscillations it is desirable to also have ν_e ($\bar{\nu}_e$) beams by, for example, exploiting the decays $\mu^+ \to e^+ \nu_e \bar{\nu}_\mu$ ($\mu^- \to e^- \nu_\mu \bar{\nu}_e$). Since $\mu^\pm$ live 100 times longer than $\pi^\pm$, a linear muon decay channel would need to be tens of kilometers long. To overcome this difficulty, in a NF the muons are injected into a storage ring with long straight sections. The fraction f of muons

* This work is supported at the Fermi National Accelerator Laboratory, which is operated by the Fermi Research Association, under contract No. DE-AC02-76CH03000 with the U.S. Department of Energy.

that decay in the straight section is given by the ratio of the straight section length to ring circumference. In NF designs typically $f \sim 0.3$.

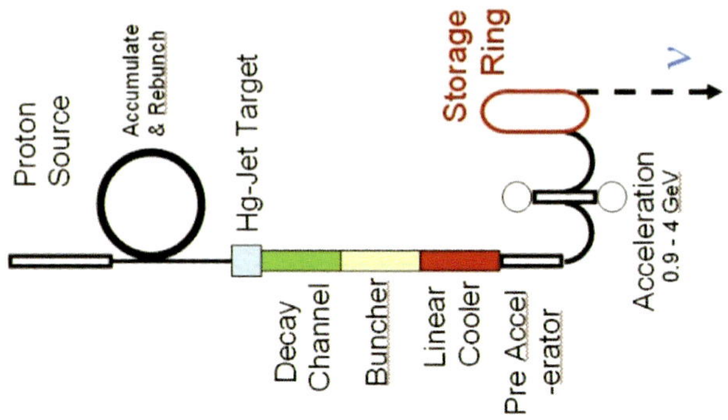

Figure 1: Neutrino Factory schematic.

A NF is shown schematically in Fig. 1. It consists of:

(i) A high-power multi-GeV proton source.
(ii) A target within a high-field solenoid followed by a lower field larger radius solenoid $\pi^{\pm}$ decay channel.
(iii) A system of rf cavities to capture the daughter muons into a bunch train, and then apply a time-dependent acceleration to increase the energy of the slower (low energy) bunches and decrease the energy of the faster (high energy) bunches (phase rotation).
(iv) A cooling channel that reduces the beams transverse phase space so that it fits within the acceptance of the first acceleration stages.
(v) An acceleration scheme that accelerates the muons.
(vi) A storage ring with at least one long straight section that points through the earth to a distant detector.

End-to-end NF simulations show that if the primary proton beam power is ~4MW there will be up to $O(10^{21})$ muons per year decaying in the beam-forming straight section. In a near detector the resulting event rates would be very large. For example, a few tens of meters from the end of a 50 GeV NF, 10^{21} muon decays would result in $O(10^{7})$ neutrino events/g/cm^2. Perhaps even more exciting, in a very distant detector the event rates are sufficient to probe neutrino flavor transition probabilities down to $O(10^{-4} - 10^{-5})$.

Since muon decays are well understood, a NF offers low systematic uncertainties on beam fluxes and spectra. This provides an advantage compared to conventional neutrino beams, but the real NF advantage comes from the ν_e

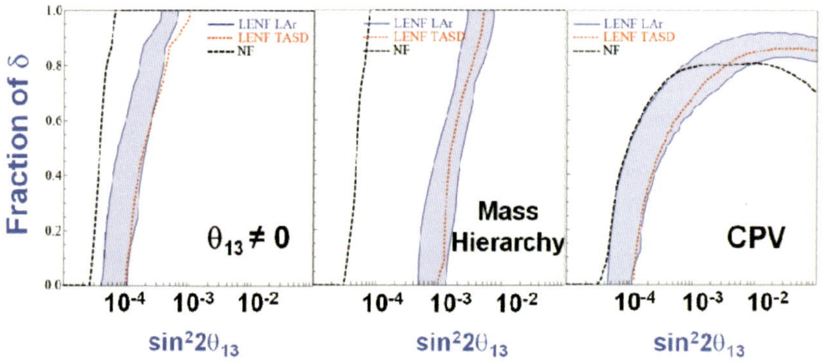

Figure 2: Fraction of all possible values of CP phase δ for which a discovery would be made at >3σ, shown as a function of $\sin^2 2\theta_{13}$ for a low energy 4 GeV NF (labeled LENF) and high energy 25 GeV NF (labeled NF). A discovery would be made in the regions to the right of the curves. Figure based on Ref. 3.

($\bar{\nu}_e$) in the beams since the transitions $\nu_e \leftrightarrow \nu_\mu$ ($\bar{\nu}_e \leftrightarrow \bar{\nu}_\mu$) are expected to play a special role in future measurements. In a conventional beam the initial flavor is ν_μ and experiments must search for $\nu_\mu \rightarrow \nu_e$ transitions. The experimental sensitivity is eventually limited by a small ν_e component in the initial beam and by backgrounds in which π^0s from ν_μ interactions are misidentified as electrons, faking ν_e interactions. This makes it difficult to probe oscillation probabilities below $O(10^{-2})$. However, a NF experiment can search for ν_μ appearance from $\nu_e \rightarrow \nu_\mu$ transitions by looking for a "wrong-sign muon[1]", i.e. a muon of opposite sign to the muons stored in the NF. It is straightforward to suppress backgrounds to $\leq O(10^{-4})$ of the total event rate.

A series of studies[4,5] have explored the ability of NF experiments to establish a non-zero value for the mixing angle θ_{13}, determine the neutrino mass hierarchy, and observe CP violation (CPV). The sensitivity depends upon $\sin^2 2\theta_{13}$ and the CP phase δ. In Fig. 2 the simulated sensitivity for a 25 GeV NF is compared with corresponding sensitivity for a lower energy 4 GeV NF. The NF would extend the "physics reach" in $\sin^2 2\theta_{13}$–space down to $O(10^{-3} - 10^{-4})$.

3. Design and R&D

Over the last decade NF R&D has been advanced by a series of design and simulation studies complemented by component development and experimental tests. Studies[6] in the U.S. established viability, defined an initial R&D program, and produced a first NF cost estimate. The later studies improved the design to increase performance and reduce cost. During this period NF R&D also became "internationalized". In 2006 the International Scoping Study (ISS), hosted by RAL in the U.K., produced physics[5], accelerator[7], and detector[8] reports.

3.1. *Proton Beam and Target*

At the NF front-end the proton source must deliver short high-intensity multi-GeV bunches onto a target. The ISS baseline parameters are:

(i) Proton energy $5 < E_P < 10$ GeV, chosen to maximize $\pi^{\pm}$ production at fixed beam power.
(ii) Bunch length $\sigma_t \leq 3$ns, needed because the downstream phase rotation channel requires initially short muon bunches.
(iii) Beam power = 4MW.
(iv) A liquid Hg-jet target injected into a 20T solenoid. A high-Z target is chosen to maximize $\pi^{\pm}$ production. The 20T solenoid radially confines essentially all $\pi^{\pm}$ coming from the target. The Hg-jet choice avoids the shock and radiation damage related target-lifetime issues that arise in a solid target. A system of solenoids downstream of the target matches the 20T solenoid into a larger bore 2T decay channel.

Multi-GeV multi-MW proton source designs are now being developed at a number of laboratories around the World, motivated by a broad range of interests which include, for example, high-intensity convention neutrino beams. Proton source R&D is being pursued within this broader context. We can anticipate that at least one viable design for a NF-class proton source will exist within a few years. Target R&D has also advanced in recent years, and has culminated in the Mercury Intense Target experiment (MERIT[9]) which has successfully demonstrated a Hg-jet injected into a 15T solenoid and hit by a suitably intense beam from the CERN PS. The jet was viewed by high speed cameras which enabled measurement of (i) the time before the jet was disrupted, (ii) the velocity of jet fragments after disruption, (iii) the length of the segment disrupted, and (iv) the time before the jet re-established itself. Preliminary results suggest this technology could support beam powers > 4MW.

3.2. *RF in Magnetic Channels*

The bunching, phase rotation, and cooling channel designs require high gradient normal conducting rf cavities operating in a magnetic channel. The initially preferred design exploited the penetrating nature of muons by using cavities in which the normally open rf cells are closed with thin conducting windows. At fixed peak power this doubles the effective accelerating gradient, and hence halves the required number of rf power sources. Thin beryllium windows for this purpose have been demonstrated in an 805 MHz test cavity. However, tests show that when this type of cavity is operated within a multi-Tesla co-axial solenoid the maximum rf gradient that can be achieved before breakdown is reduced. It is possible that, with further R&D, surface treatments can be found to mitigate this effect. Other solutions have also been proposed including:

(i) Using cavities filled with high pressure hydrogen gas[10]. An 805 MHz cell has been built and tested in a high field solenoid. No appreciable degradation of performance was observed with increasing magnetic field. In the coming months this technology will be tested in the presence of an intensely ionizing beam. It is possible that the ionization created in the cavity will limit its performance.
(ii) Using "magnetically insulated" cavities[11]. The magnetic field is designed so that it is parallel to surfaces where the rf gradients are maximum. This is expected to prevent energetic electrons from hitting these surfaces and causing problems.
(iii) Designing cooling channels in which the cavities are in regions of low magnetic field. This is not a preferred solution since it will mean longer less efficient channels.

Within the next couple of years the ongoing R&D is expected to determine which of these options are viable.

3.3. *Bunching, Phase Rotation and Cooling*

At the end of the decay channel the daughter muons have drifted some tens of meters, resulting in a time-energy correlation with the high-energy particles leading the low-energy particles. The decay channel is followed by a buncher section that uses rf cavities to form the beam into a bunch train, and a phase-energy rotating section that decelerates the early-rf-phase high energy bunches and accelerates the late-rf-phase low energy bunches, so that each bunch has the same mean energy. Present designs deliver a bunch train that is 50m long, captured within a 2T solenoid channel. The buncher parameters are determined by considering reference particles (1, 2) with velocities β_1 and β_2. The rf voltages are increased along the channel, with frequencies f_{rf} and phases set to place 1 and 2 at the center of bunches. This can be accomplished if the rf wavelength λ_{rf} increases along the buncher[12]:

$$N_B \lambda_{rf}(s) = N_B \frac{c}{f_{rf}(s)} = s\left(\frac{1}{\beta_2} - \frac{1}{\beta_1}\right) \tag{1}$$

where s is the distance from the target. N_B is an integer. In the present design, at the end of the channel all bunches have a mean momentum ~230 MeV/c, with μ^+ and μ^- bunches interleaved within the cycle. The reduction in overall energy spread effectively increases the number of useful muons by about a factor of 4.

The number of muons accepted by the downstream accelerators can be further increased by reducing the two-dimensional phase-space-area (emittance) in each transverse direction (ε_x and ε_y) by a factor of a few. This can be accomplished using "ionization cooling[13]" in which the muons lose energy by ionization as they pass through an absorber. This reduces their momenta in the longitudinal- and transverse-directions. An rf cavity then replaces the lost energy by reaccelerating in the longitudinal direction. After repeating the process many times, the transverse momenta (and transverse emittances) are reduced. The rate at which the normalized transverse emittance $\varepsilon_{xN} = \beta\gamma\, \varepsilon_x$ changes as muons with energy E_μ (GeV) lose energy by ionization loss dE_μ/ds within material with radiation length L_R is given by:

$$\frac{d\varepsilon_{Nx}}{ds} = -\frac{dE_\mu}{ds}\frac{\varepsilon_{Nx}}{E_\mu} + \frac{\beta_\perp (0.014)^2}{2E_\mu m_\mu L_R} \tag{2}$$

where $\beta_\perp$ is the "betatron function" which characterizes the focusing strength at the absorber. The second term describes heating due to scattering in the absorber which ultimately limits the cooling process. To minimize the impact of scattering it is desirable to use a low Z (high L_R) absorber (e.g. liquid hydrogen or LiH) and to focus the muons strongly (small $\beta_\perp$) so that the focusing angles are much larger than typical scattering angles. The present baseline cooling channel design consists of a sequence of LiH absorbers and 201 MHz rf cavities within a lattice of solenoids that provide the required focusing. Simulations show that the cooling channel increases the number of useful muons by about a factor of 2. To provide a proof-of-principle demonstration, the international Muon Ionization Cooling Experiment (MICE[14]) at RAL is preparing to test an ionization cooling channel cell in a muon beam. MICE will measure the response of individual muons to the cell as a function of the incident muon parameters (momentum, position, direction) and the various channel parameters (absorber type, magnetic fields, rf parameters). The initial phase of the experiment, which establishes the muon beam and measurement systems, has begun. It is anticipated that MICE will be completed by 2011-2012.

3.4. *Acceleration and Storage*

Since muons are short-lived, acceleration must occur at high average gradient. The accelerator must also accommodate the phase-space occupied by the beam. Typically the normalized transverse emittances are $\varepsilon_{xN} \cong \varepsilon_{yN} \sim 4$ mm-rad, longitudinal emittance $\varepsilon_L \sim 36$ mm, momentum spread $\sigma_{\Delta p/p} \sim 0.1$, and bunch length $\sigma_Z \sim 0.16$m. The need for large transverse and longitudinal acceptances favors using low frequency rf. Acceleration systems have been designed assuming Superconducting rf (SCRF) gradients at 201 MHz ranging from 11 MV/m (already demonstrated[15]) to 17 MV/m.

Various acceleration schemes have been studied. Typically they begin with a linear "pre-accelerator" that accelerates the beam to about 1 GeV. The muons are then sufficiently relativistic to use a Recirculating Linear Accelerator (RLA) in which arc-sections return the muons to the same linac several times. Higher energies can be obtained using further RLAs and/or FFAG (Fixed Field Alternating Gradient) accelerators. The ISS scheme uses a new accelerator type (a "non-scaling FFAG[16]") to accelerate to 25 GeV. The EMMA[17] experiment at Daresbury will study non-scaling FFAG beam dynamics, which are interesting because the particles are accelerated out of the rf bucket. EMMA results will enable the attractiveness of this particular scheme to be better assessed.

After acceleration the interleaved μ^+ and μ^- bunches are injected into the NF ring, where they circulate until they decay. Both racetrack (two straight sections) and triangular (three straight sections) geometries have been proposed. In principle both μ^+ and μ^- bunches can be stored in a single ring, injected in opposite directions. However, in the ISS baseline design there are two separate racetrack rings, one for each muon sign, with the beam-forming straight sections pointing to two distant detectors at different baselines.

4. Outlook

The next NF step, which has begun, is the so called International Design Study (IDS) which hopes to deliver a "Reference Design Report" by 2012. By this time it is anticipated that all of the proof-of-principle tests will be completed and a NF could then become part of the particle physics "road map". If the community wishes to proceed, after a few years of additional R&D, it is plausible that construction could start as early as the late 2010's.

References

1. S. Geer, Phys. Rev. D57 (1998) 6989.
2. S. Bilenky, Pontecorvo B. Phys. Rep. 41 (1978) 225.

3. A. Bross *et al.*, hep-ph arXiv:0911.3776v1
4. S. Geer, H. Schellman (Eds.) hep-ex/0008064; B. Autin, A. Blondel, J. Ellis, CERN-99-02; A. DeRujula, M. Gavela, P. Hernandez, Nucl. Phys. B547 (1999) 21; Y. Mori, J. Phys. G: Nucl. Part. Phys. 29 (2003) 1527; A. Blondel, Nucl. Instrum. Methods A451 (2000) 131.
5. S. King *et al.*, JINST 4:T05001 (2009).
6. N. Holtkamp, D. Finley (Eds.) Fermilab-Pub-00/108-E; [20] S. Osaki, R. Palmer, M. Zisman, J. Gallardo (Eds.) BNL-52623; S. Geer, M. Zisman (Eds), e-Print: physics/0411123; S. Berg *et al.* Phys. Rev. Special Topics-AB 9 (2006) 011001.
7. S. Berg *et al.*, JINST 4:P07001 (2009).
8. T. Abe *et al.* (ISS Detector Working Group) arXiv:0710.4947v1, 2007.
9. H. Kirk *et al.* (MERIT Collab.) Proc. 2007 Part. Accel. Conf., pp. 646:648.
10. K. Yonehara *et al.* Nucl. Phys. B (Proc. Suppl.) 149 (2005) 286.
11. R. Palmer *et al.* arXiv:0809.1633v1.
12. D. Neuffer, AIP CP 721, p. 407 (2004).
13. D. Neuffer, Advanced Accel. Concepts, AIP Conf. Proc. 156, 201 (1987).
14. R. Sandstrom (MICE Collab.), AIP Conf. Proc. 981, 107 (2008).
15. R. Geng *et al.*, Proc. of the 2003 Part. Accel. Conf. 1309:1311.
16. See for example C. Johnstone, S. Koscielniak, Proc. of EPAC 2004, 800.
17. S. Berg (EMMA Collab.) AIP Conf. Proc. 981, 330 (2008).

ADS AND ITS POTENTIAL

JEAN-PIERRE REVOL
Physics Department, CERN,
Geneva, 1211, Switzerland

Accelerator driven systems offer the possibility to make nuclear fission energy sustainable and acceptable to society. The Energy Amplifier proposed by Carlo Rubbia at CERN is used to illustrate the potential of ADS, and the resulting R&D activity wave it triggered in Europe and in the world is briefly discussed. ADS must be part of the strategy to provide clean, safe and abundant energy for a harmonious development of society.

1. A brief history of ADS

The basic process in accelerator driven systems (ADS) is nuclear transmutation, first observed by Ernest Rutherford in 1919 while bombarding nitrogen atoms with α particles, producing ^{17}O in the reaction: $^{14}N_7 + {}^4He_2 \rightarrow {}^{17}O_8 + {}^1p_1$. In this case, the particle accelerator was radioactive ^{210}Po.

In 1929, a major step forwards was achieved by Ernest O. Lawrence who designed the first cyclotron* opening the prospect for intense controlled particle beams.

In 1940, E. O. Lawrence in the USA and W. N. Semenov in the USSR independently proposed using a particle accelerator as neutron source. The first practical realization, setting an important milestone, was achieved in 1941, when Glenn T. Seaborg produced the first μg of ^{239}Pu using a cyclotron to bombard ^{238}U with 6 MeV deuterons.

Industrial scale applications started in 1950, when E. O. Lawrence boldly launched the Materials Testing Accelerator (MTA) project at the Lawrence Livermore Radiation Lab, to produce ^{239}Pu from depleted uranium. Two years later, in 1952, Wilfrid B. Lewis in Canada proposed using an accelerator to breed ^{233}U fuel from natural thorium for CANDU reactors. This can be considered as the birth of the electro-breeder concept.

* Credit for the invention also goes to Leo Szilard and Rolf Wideröe.

MTA was dropped and Lewis' project was slowed down when rich uranium deposits were discovered in the United States, and it was realized that it would require several hundred MW of beam power for these projects to be realistic. At that time, there was essentially no energy gain in the system. Nowadays, ADS considered either for nuclear waste transmutation or energy production require 10 times less beam power, therefore, a 10 MW proton beam, such as Fermilab Project X, is of interest.

Renewed interest in ADS arose again in the 1980s, when the US slowed down the development of fast critical reactors (Fast Flux Test Facility at Argonne National Lab finally stopped in 1993) and when it was realized that accelerator technology had made significant progress since Lawrence's time:

- Hiroshi Takahashi at Brookhaven National Lab submitted several proposals of ADS systems. One of these was the PHOENIX project whose purpose was to transmute nuclear waste, including minor actinides, with a fast neutron ADS with a high neutron multiplication factor $k \sim 0.99$;
- Charles D. Bowman at Los Alamos proposed a thermal neutron ADS for the Accelerator Transmutation of nuclear Waste (ATW) with thorium based fuel and chemistry on-line for extraction of fission products and ^{233}Pa;
- Japan launched Options for Making Extra Gains from Actinides (OMEGA) at JAERI. Today this program is continued as JPARC at JAEA, with the main purpose of burning minor actinides.

Major progress was made in the 1990s, when Carlo Rubbia became convinced that not only was accelerator technology then mature for a realistic exploitation of the ADS idea, but that tools existed for optimizing the parameters of such systems. He launched a vigorous research program at CERN based on three main elements:

1. Development of an innovative detailed simulation, inherited from particle physics, starting from the spallation process initiated by a proton hitting a lead or uranium target, producing neutrons transported throughout the system and describing fuel evolution as a function of burnup;
2. Specific experiments to test basic concepts (FEAT [1] and TARC [2] experiments at the CERN Proton Synchrotron (PS));
3. Design, construction and exploitation of an advanced neutron Time of Flight Facility (n_TOF) [3] at CERN, to acquire basic neutron cross-section data, crucial to simulate reliably any new reactor configuration with new materials. The remarkable performance of n_TOF is the subject of the presentation at this workshop by my colleague E. Gonzalez [4].

C. Rubbia named his system "Energy Amplifier" (EA), as it appears that the energy provided by the proton beam is "amplified" through fissions reactions in the system's core. His work at CERN triggered a major R&D wave on ADS in Europe [5] and in the entire world [6].

The Energy Amplifier is a fast neutron flux (typically 10^4 to 10^6 eV in the core) subcritical system driven by a proton accelerator, using fuel based on thorium rather than uranium. A thorium fuel cycle in a fast neutron ADS is quasi-sustainable, minimizes waste production, and is much less proliferating than the uranium cycle. Lead or lead-bismuth eutectic can be used as spallation target, moderator and coolant.

The properties of the EA are well described in several publications [7]; my objective here is to emphasize the choice and optimization of the strategic system parameters.

2. General guidelines for the choice of ADS parameters

Safety is the first concern with nuclear energy. By making ADS a subcritical system, criticality accidents are avoided. Parameters such as void and temperature coefficients or the fraction of delayed neutrons (β_{eff}) are no longer "critical", which greatly facilitates operation. The sensitivity to a sudden insertion of reactivity (corresponding for instance to the accidental extraction of control bars in a critical system) illustrates well the spectacular difference between critical and subcritical system behaviors (Fig.1).

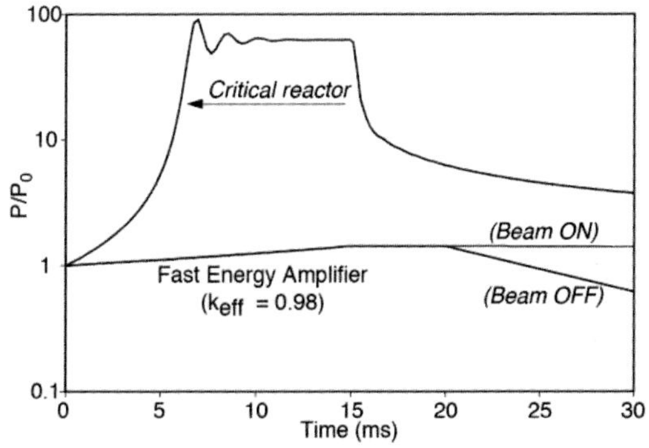

Figure 1. Comparison of power excursions between a lead-cooled critical reactor and a Fast EA for an accidental reactivity insertion corresponding to 170$[†]/s during 15 ms (Taken from Ref. [7]).

[†] The dollar is the reactivity expressed in units of β_{eff}, the fraction of delayed neutrons.

By equipping the system with passive safety elements borrowed from advanced US fast critical reactor designs, core meltdown accidents are highly improbable, and the large volume of lead would greatly limit their consequences. Lead as coolant is an additional safety asset compared to sodium.

The other main concern with nuclear energy is nuclear waste management. Therefore, the EA strategy consists of using (a) fast neutrons, (b) thorium fuel and (c) recycling of long-lived transuranic actinides which are reintroduced in the system as fuel. The result is that the necessary capacity of long-term storage of waste is considerably reduced.

The third important concern with nuclear energy is military proliferation. The fuel cycle used in the EA, based on thorium, is much less proliferating than the uranium fuel cycle, as the uranium mixture (^{232}U, ^{233}U, ^{234}U, ^{235}U and ^{236}U) coming out of the spent fuel is very difficult to use for manufacturing bombs. It has a hard gamma flux emitted by ^{232}U requiring remote handling to assemble a critical mass of 28 kg. The plutonium mixture produced in minute quantities contains mainly ^{238}Pu, making it of no interest for military usage [8]. ^{237}Np is produced in small quantities [7], less than 300 g in the first discharge after a burnup of 150 GW.day/ton (over a period of about 5 years), increasing to about 38 kg after 10 cycles (50 years), to be compared to a critical mass of 56.5 kg.

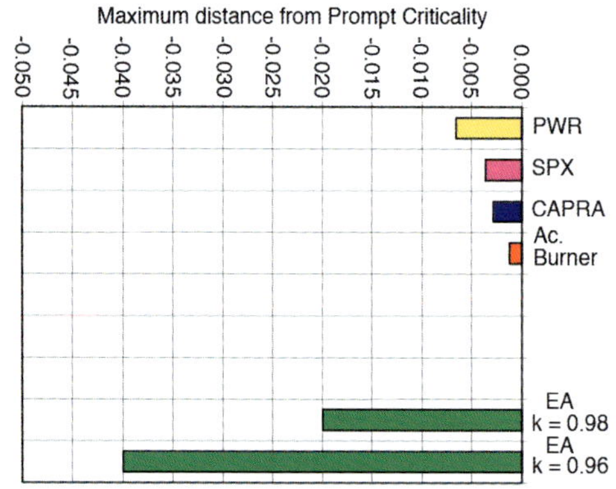

Figure 2. Comparison of prompt criticality margins for a PWR, Superphenix (SPX), the French CEA CAPRA system, the Japanese actinide burner at JAERI, and the Energy Amplifier.

In addition, the EA fuel cycle does not require separation of specific actinide components from the spent fuel (unlike the case of PUREX to extract plutonium from PWR spent fuel). In this context, the pyroelectric reprocessing developed at Argonne National Lab for the uranium cycle, which could be adapted to the thorium cycle, is an extremely precious American asset.

3. Physics of subcritical systems

The theory of subcritical systems is interesting in itself, as it provides invaluable insights into their behavior [9]. The general equation only differs from that of critical systems by the addition of a neutron source term C:

$$\frac{\partial n(\vec{r},t)}{\partial t} = \nu\Sigma_f \Phi(\vec{r},t) + C(\vec{r},t) - \Sigma_a \Phi(\vec{r},t) + D\nabla^2 \Phi(\vec{r},t) \qquad (1)$$

However, solutions of these equations are of different nature. In practice, Monte Carlo methods are needed to simulate all details of a real system. Nowadays, computing resources are no longer a limitation, as shown by CERN LHC detectors, which are described using tens of millions of geometry elements.

An important characteristic of ADS is the safety margin characterized by the distance to prompt criticality. For ADS, it is the result of a deliberate choice of system properties, while for critical systems it is imposed by Nature (Fig. 2). One can define a system energy gain (G), as the ratio of the energy produced by fissions in the core to the energy provided by the proton beam, usually parameterized as G = $G_0/(1\text{-k})$[‡], where k is the neutron multiplication factor with beam on, and G_0 is a constant that depends on the beam energy, the target material, and the geometry of the system. For a uranium target $G_0 \sim 3$ and for a lead target such as in the EA, $G_0 \sim 2.4$. Because of the 1/(1-k) behavior, there is an obvious trade-off between criticality margin and accelerator power. This is a design choice between making the system deeply subcritical, or using a k value closer to unity in order to demand less from the accelerator.

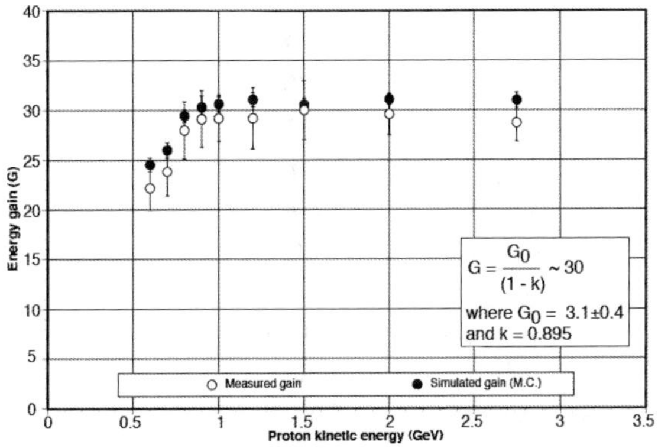

Figure 3. Energy gain in the FEAT experiment [1]: comparison of measurements with the EA simulation, as a function of the beam kinetic energy.

‡ This is an approximate formula valid for large values of k.

It also turns out that, by varying the beam intensity of such systems, one can in principle modulate the power output more easily than in critical systems. This may be an asset in case of massive deployment of fluctuating solar and wind energies, which would benefit from the complementarity with a modulated power source. In addition, neutronics with thorium is very favorable as the characteristic of the system's response to a change of beam power is driven by ^{233}Pa long half-life. This is twelve times longer than ^{239}Np in the corresponding uranium breeding chain. The concept of energy gain was tested at the CERN PS, in the First Energy Amplifier Test (FEAT) [1], which at the same time validated the innovative simulation developed, by C. Rubbia and his group. An assembly of 3.9 tons of natural uranium with k_{eff} ~ 0.9 was used to measure the energy produced in the system in two ways: by counting fissions and by measuring heat deposition, two methods that yielded consistent results shown in Fig. 3.

The main conclusion from FEAT is that optimum beam energy is reached at about 900 MeV. At lower energies, protons tend to loose their energy by ionization, and are therefore less efficient at producing neutrons, whilst above 900 MeV, an increase of proton kinetic energy is almost entirely transformed into a corresponding increase of the number of neutrons, and G_0 decreases slowly with the increasing production of new particles.

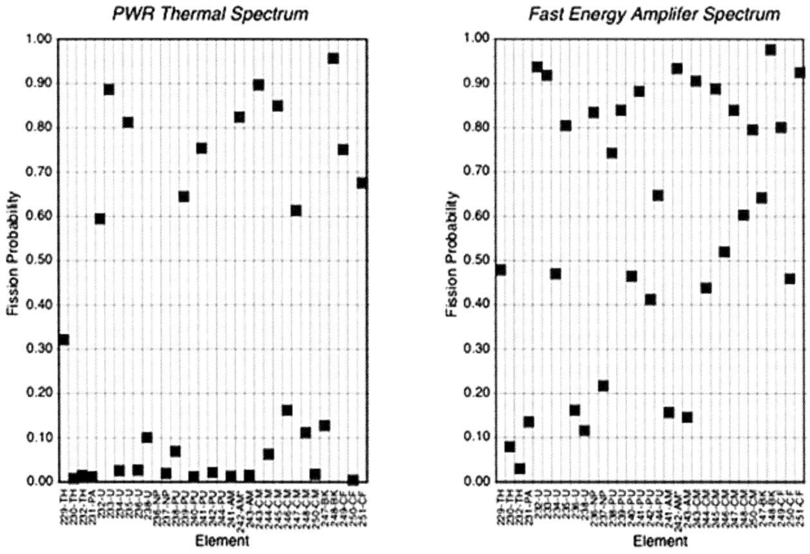

Figure 4. Fission probability of actinides, (ratio between the fission cross section and the sum of the fission and capture cross sections), for a PWR thermal neutron flux (left), and for the Energy Amplifier fast neutron flux (right), showing that with fast neutrons all TRU fission.

4. Elimination of nuclear waste

The two components of nuclear waste require different strategies for their elimination. Transuranic elements (TRU) such as Np, Pu, Am, Cm, Bk, etc., produced from neutron capture on the fuel and subsequent decays can only be eliminated by fission. This has the advantage of producing energy in return. Fission products can only be eliminated through neutron capture and subsequent decays into stable elements. In view of the 10000 times longer lifetime of TRU and of their danger as a potential source of fissionable elements, their destruction must have priority. This justifies the choice of a fast neutron system in order to enhance TRU fission probabilities (Fig. 4). Given that all TRU can fission, there is no need to separate Pu from minor actinides, a significant advantage to resist military proliferation. The use of fast neutrons comes with an added bonus of a much longer burnup due to the far lower neutron capture cross-section on fission products, hence a more efficient use of the fuel. Extended burnup up to 150 GW.day/ton were achieved at the Argonne electro-breeder, and of course confirmed by the EA simulation.

In order to obtain the fastest possible neutron flux, neutrons should be moderated as little as possible, hence the choice of lead. Nature was kind enough to also make lead an excellent spallation target as well as an excellent coolant and confinement medium. It is certainly less dangerous than sodium, not only for chemical reasons, but also because its boiling temperature (345°) is better separated from its fusion temperature (~ 1735°C). There is a drawback above 550°C due to material corrosion. However, we are witnessing rapid progress in the development of new corrosion and radiation resistant materials (Eurofer). Experience also exists from Russian Pb-Bi submarine reactors where corrosion is controlled by adjusting the oxygen contents. At the same time, progress in turbine efficiency using super critical CO_2, allow thermal efficiencies of 45% at only 550°C [10]. Neutron phenomenology was studied in great detail at CERN in the TARC experiment [2]. The PS proton beam hit a 350-ton assembly of lead, in which neutron fluxes were measured over 8 orders of magnitude in neutron energies, from thermal to few MeV neutrons (Fig. 5). TARC demonstrated the process of Transmutation by Adiabatic Resonance Crossing as a way of strongly enhancing neutron capture to transmute for instance ^{99}Tc into stable ^{100}Ru, or to industrially produce radioactive isotopes.

The next important element in the strategy to eliminate TRU is the use of thorium fuel. Natural thorium is isotopically pure and 4 to 5 times more abundant than uranium in the earth crust. The breeding chain produces fissionable ^{233}U starting from (n,γ) captures on ^{232}Th producing ^{233}Th with a 22.3

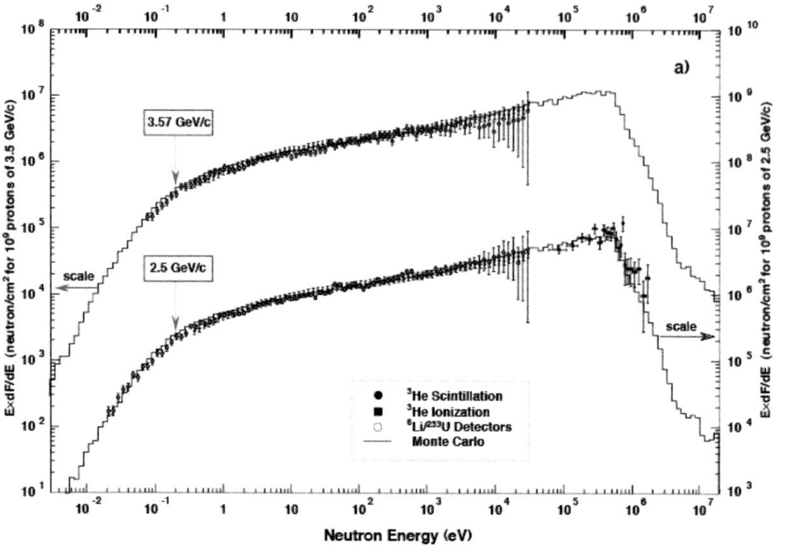

Figure 5. The measurement of the neutron flux in the TARC lead volume, as a function of neutron energy, for two proton beam energies.

mn half-life, which decays into ^{233}Pa with a 27d half-life, which in turn decays into ^{233}U. It takes 5 neutron captures to go from ^{233}U to ^{238}U, which explains why TRU production is much suppressed compared to uranium fuel, but also explains why the equilibrium concentration of TRU in thorium fuel is so low, that relatively modest concentrations of TRU places the system automatically in the incineration mode. For instance in the EA, the Pu equilibrium concentration is ~ 10^{-4} compared to 15% in uranium fuel.

The simulation validated by the FEAT and TARC experiments showed that an EA would destroy 36 kg of TRU per TW_{th}.h, to be compared to the production of 14kg of TRU by a PWR for the same energy output. In a mode in which the EA would recycle its own TRU, after 1000 years, the radiotoxicity of the spent fuel would be several orders of magnitude smaller than that of a PWR (Fig. 6), depending on the efficiency of reprocessing. This would similarly reduce the needs for long-term storage. Transmutation of some of the long-lived fission products using the TARC effect could reduce radiotoxicity further.

5. Energy production

The EA concept was developed in the 1990s, when the main concern was the destruction of existing nuclear waste. Today, the emphasis has shifted towards energy production, because of fears of global warming. The two options for an ADS are not incompatible, since in both cases one wants to minimize the production of waste. The difference might be in the size of the units, depending on local needs, and on the optimization of the thermal efficiency for electricity production, requiring the highest possible operating temperature.

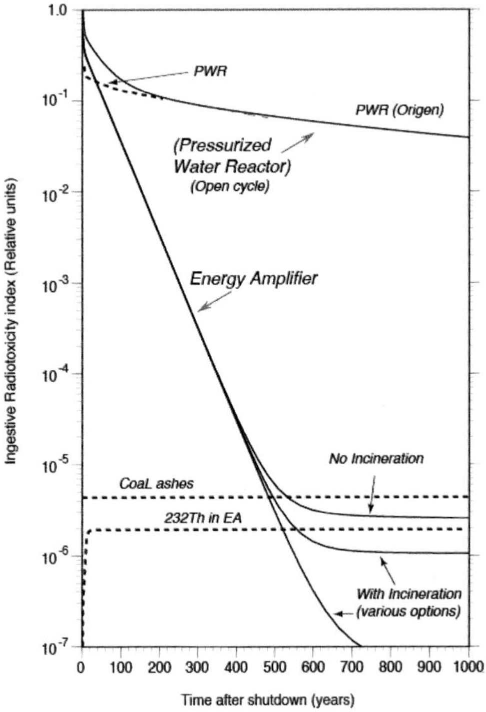

Figure 6. Comparison of ingestion radiotoxicities, in relative units, between PWR waste (open cycle) and Energy Amplifier waste, assuming a 99.99 reprocessing efficiency. Taken from Ref. [7].

6. The accelerator

From the point of view of ADS physics it does not matter how “external” neutrons are produced. Technological constraints will determine the properties of the beam. There is a simple relation between beam power (P_{beam}) and desired output power (P_{ADS}):

$$P_{beam} = \frac{(1-k)}{G_0} P_{ADS} \tag{2}$$

The highest beam power achieved at the PSI separate sector cyclotron [11] (approaching 3 mA and 1.8 MW, with 0.59 GeV protons) would already allow P_{ADS} = 243 MW with k = 0.98, or P_{ADS} = 486 MW with k = 0.99.

An ADS system, burning TRU enriched fuel or ^{233}U-Th fuel, would produce a substantial amount of energy. Taking the characteristics of the EA as defined by C. Rubbia et al. [7], assuming 43 % efficiency for a 10 MW beam accelerator, and 45% for heat to electricity conversion efficiency, an accelerator such as Project X would produce about half a GW of electric power (Fig. 7).

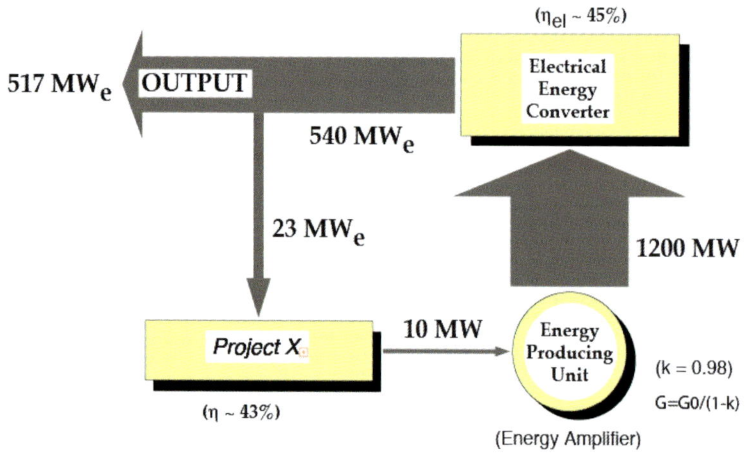

Figure 7. Sketch of the energy flow for a subcritical system with the characteristics of the Energy Amplifier driven by a 10 MW beam, such as Fermilab Project X.

The choice of accelerator parameters for an industrial system will depend on the desired power output and on the choice of k value. A beam power of 10–30 MW, a beam energy above 900 MeV, but probably below 2 GeV, depending on the interface between the accelerator and the core, and a broad beam spot seem to be the basic requirements. In principle, several technologies could produce the required beam: Linac, Cyclotron, Fixed-Field Alternating Gradient or hybrid solutions (Cyclotron injection into a c.w. linac). However, they will have to be evaluated against the following specifications:

1. Small beam losses (avoid irradiation of the accelerator and of environment);
2. Reliability (few beam trips per year; have several injectors?), beam stability and control, ease of maintenance;
3. Energy efficiency (maximize the fraction of electric power stored in the beam), physical size, and of course: cost.

The synergy with Project X would be in the R&D for the production of the low energy part of the beam achieving small beam losses and high reliability. This is a common issue for any accelerator application requiring high beam power.

7. R&D in Europe

Many projects were carried out within the European Union 5th and 6th Framework Programs, concerning the general field of partitioning and transmutation. All the aspects of ADS were addressed. A 1 MW spallation source (MEGAPIE [12]) was successfully tested at PSI (Switzerland). Within Eurotrans [5], thermodynamics and corrosion studies are carried out in as many as 9 research centers using molten lead or lead-bismuth loops (KALLA and COSTA at FZK, Germany, STELLA at French CEA, CIRCE and CHEOPE at ENEA, Italy, VICE at SCK-CEN, Belgium, CorrWett at PSI, Switzerland, TALL at KTH, Sweden, and CIRCO at CIEMAT, Spain).

The decision by the Belgium Government whether to build MYRRHA[§][13], a first ADS prototype of substantial power (100-200 GW_{th}), is expected next January.

8. Other developments concerning thorium

The example of India is instructive [6], a country with little uranium resources, but plenty of thorium, leading to a complicated three stage strategy:

1. PWR heavy water reactors (CANDU) and LWR to produce plutonium from their small uranium supply;
2. Sodium cooled U-Pu fast reactors with a Thorium blanket to breed ^{233}U;
3. Reprocessing and manufacturing of ^{233}U-Th fuel for heavy water reactors.

It is obvious in this case, that the use of ADS would considerably simplify the scheme, and make it sustainable. In addition, their mostly thermal neutron scheme is not addressing the waste issue.

As we heard in this workshop, many studies are taking place world-wide: in Europe, China, Russia, UK, Korea, Japan, etc., and we learned, in particular, that ADS R&D has the highest priority among China's scientific projects. Recently, thorium studies were carried out in thorium-rich Norway. In my view, it is only a matter of time before an ADS prototype of substantial power is built. Interesting questions are: who will take the lead in this strategic domain? What

---

[§] MYRRHA consists of a 4 mA, 600 MeV proton beam coupled to a liquid Lead-Bismuth Eutectic (LBE) spallation target located at the center of a subcritical fast-neutron reactor core cooled with liquid lead-bismuth.

is the best strategy for ADS: to use existing elements for a fast realization of a demonstrator or to go directly for a more ambitious industrial prototype?

9. Conclusion

In the past 15 years, a tremendous ADS R&D effort has been made worldwide. ADS is a challenging idea with no showstopper, with the potential for producing acceptable energy for society. The next step for ADS should be a "demonstrator" of significant power, preferably with international collaboration (we are facing a global problem that demands a global solution), to study the coupling between the accelerator and the core.

Nuclear fission energy is the only way we have now to produce quasi-unlimited amounts of clean energy to sustain harmonious growth on our planet:

- ADS can change the image of nuclear energy, making it "acceptable" to society; if nuclear energy is to be deployed on a large scale, it must be ADS;
- ADS would be complementary to renewable energies, allowing modulation of electric power, to follow fluctuations of wind and solar power;
- A proof of existence is needed for the accelerator. Project X could make an important contribution to the main accelerator issues: beam losses and reliability.

I also believe that it would be very good for particle physics to make a new important contribution to society in the crucial domain of the "energy problem".

References

1. S. Andriamonje et al., Phys. Lett. B348 (1995) 697-709.
2. A. Abánades et al., Nuclear Instrument and Methods in Physics Research A 478 (2002) 577-730.
3. C. Rubbia et al., "A High Resolution Spallation Driven Facility at the CERN-PS to Measure Neutron Cross Sections in the Interval from 1 eV to 250 MeV", CERN/LHC/98-02 (EET), May 30, 1998; "A High Resolution Spallation Driven Facility at the CERN-PS to Measure Neutron Cross Sections in the Interval from 1 eV to 250 MeV: A Relative Performance Assessment", CERN/LHC/98-02 (EET)-add. 1, June 15, 1998; S. Abramovich et al., "Proposal for a Neutron Time of Flight Facility", CERN/SPSC 99-8, SPSC/P 310, 17 March 1999.
4. E. Gonzalez, "CERN n-TOF system", WG3: SRF Linac driven subcritical systems (ADS), this Workshop.
5. Eurotrans, the Integrated Project on Transmutation (IP EUROTRANS), EURATOM 6th Framework Program.

6. Several world ADS efforts were presented at this workshop: Kenji Kikuchi, “ADS in Japan”; Shinian Fu, “ADS developments in China”; Srikumar Banerjee, “ADS efforts in India.
7. C. Rubbia et al., "Conceptual Design of a Fast Neutron Operated High Power Energy Amplifier", CERN/AT/95-44 (ET), Sept. 29, 1995; see also C. Rubbia, "A High Gain Energy Amplifier Operated with fast Neutrons", AIP Conference Proc. 346, Int. Conf. on ADT Technologies and Applications, Las Vegas, 1994.
8. Charles G. Bathke et al., “The Attractiveness of Materials in Advanced Nuclear Fuel Cycles for Various Proliferation and Theft Scenarios”, Proceedings of Global 2009 Paris, France, Sept. 6-11, 2009, Paper 9543.
9. C. Rubbia, “An analytical approach to the Energy Amplifier”, Internal Note CERN/AT/ET/Internal Note 94-036.
10. V. Dostal, M.J. Driscoll, P. Hejzlar, “A Supercritical Carbon Dioxide Cycle for Next Generation Nuclear Reactors”, MIT-ANP-TR-100, Mar. 10, 2004.
11. Paul Scherrer Institut (PSI), 5232 Villigen, Switzerland.
12. MEGAPIE, a liquid metal spallation target at a beam power level of 1 MW in the SINQ target station at the Paul Scherrer Institut.
13. MYRRHA, Multi-purpose hYbrid Research Reactor for High-tech Applications, a European XT-ADS at Mol, Belgium.

ADS HISTORY IN THE USA*

RICHARD L SHEFFIELD†
LANSCE-DO, MS H816, Los Alamos National Lab
Los Alamos, NM 87545 USA

ERIC J PITCHER
LANSCE-DO, MS H816, Los Alamos National Lab
Los Alamos, NM 87545 USA

Nearly all risks to future generations arising from long-term disposal of used nuclear fuel are attributable to the transuranic elements and long-lived fission products, about 2% of its content. The transuranic elements of concern are plutonium, neptunium, americium, and curium. Long-lived (>100,000-year half-life) isotopes of iodine and technetium are also created by nuclear fission of uranium. We can reduce the problem transuranics through accelerator-based transmutation. Accelerator Driven Systems (ADS) have been proposed for over two decades as one technique to transmute used nuclear fuel. This paper covers the history and some new possible applications of accelerator driven systems.

1. Introduction

Nearly all issues related to risks to future generations arising from long-term disposal of such spent nuclear fuel is primarily attributable to the transuranic elements and long-lived fission products, approximately 2% of its content. The transuranic elements of concern are plutonium, neptunium, americium, and curium, which are produced by neutron captures in the uranium-238 isotope that is the dominant heavy metal constituent of nuclear fuel burned in light water reactors. Long-lived (>100,000-year half-life) isotopes of iodine and technetium are created by nuclear fission of uranium. If we can reduce or otherwise securely handle this 2% of the spent fuel, the toxic nature of the remaining portion of spent fuel after a few centuries of cooling is below that of the natural uranium ore that was originally mined for nuclear fuel.

* This work was carried out under the auspices of the National Nuclear Security Administration of the U.S. Department of Energy under Contract No. DE-AC52-06NA25396.

Accelerator driven systems have four major technology elements: separations and waste forms, fuel fabrication, accelerators, and transmuters, see Figure 1. A well-designed accelerator-driven transmuter would operate in a sub-critical mode, and with limited excess reactivity such that the transmuter cannot reach criticality under any design basis accident [1]. For this type of transmuter, the fission rate is directly proportional to the source neutron production rate. The flexibility enabled by subcritical operation has several advantages:

1. Can drive systems with low fissile content (Th or minor actinides), or high burden of non-fissile materials
2. Unlike critical reactors, can safely operate with fuel having a relatively low delayed neutron fraction.
3. Can compensate for large uncertainties in initial reactivity or burnup reactivity swings by varying the source rate, which for an accelerator driven system is proportional to the beam current.

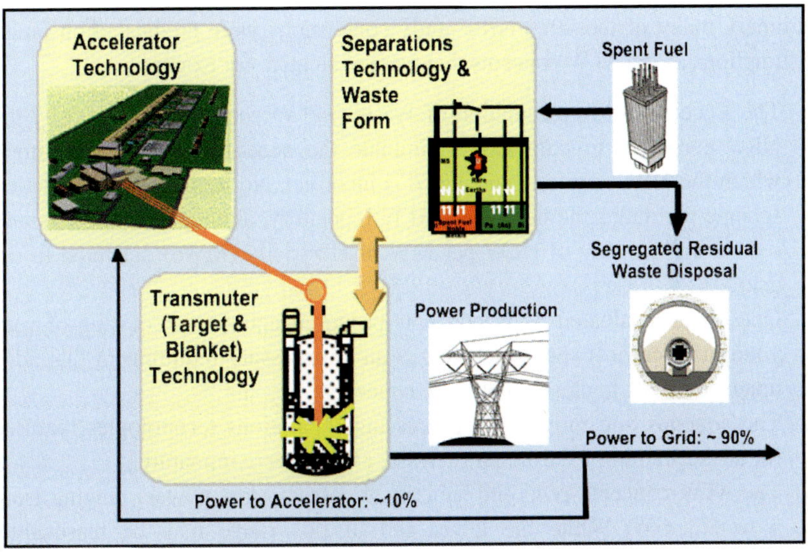

Figure 1. Accelerator driven systems includes four major technology elements: separations and waste forms, fuel fabrication, accelerators, and transmuters.

1.1. *Past Criticisms / Concerns about ADS*

In 2004, perceived issues with accelerator-based transmutation technology and budget limitations effectively eliminated accelerator-based transmuters as a component of U.S. fuel cycle research. The major points that have been raised in the past are identified in a 1996 National Research Council (NRC) Study [2] that was, from a system cost perspective, very negative on nuclear waste transmutation in general. For accelerator-based transmutation systems, the NRC concluded:

1. Accelerator-based waste tranmuters would not eliminate the need for a geologic repository.
2. Accelerator-based systems would be too expensive to be economically viable.
3. Poor economic performance due to the low thermal-to-electrical efficiency (19.3%) of one of the proposed ATW concepts (the pressure tube design).
4. The electrical power produced was too intermittent to be delivered to the electrical grid.
5. For linacs operating at up to 200 mA continuously, concern over beam losses precluding hands-on maintenance of the linac.

The points raised above are addressed in the rest of this section, but in summary many of the 1996 NRC study conclusions were predicated on faulty assumptions of the ATW concepts that were evaluated, for example:

1. The Accelerator Transmutation of Waste (ATW) concepts evaluated by the NRC purported to completely eliminate the need for a HLW repository, which the NRC correctly concluded is incorrect. No proposed transmutation scheme precludes the need for a HLW repository, although they can greatly reduce the quantity of HLW per MW of electricity (MWe) delivered to the grid.
2. One of the evaluated ATW concepts proposed aqueous processing of very young (hence hot) spent fuel; this is an unnecessary requirement that adds unneeded costs to closed fuel cycle concepts.
3. The aqueous and molten-salt processing separations technologies required for an Accelerator Transmuter of Waste (ATW) were immature.
4. The ATW concepts proposed subcritical core thermal powers ranging from 3 to 8.3 GW. While the lower end of this range may be reasonable, there have been no subsequent ATW concepts whose blanket powers approach 8 GW.
5. Superconducting accelerators were not widely used at the time of the study. Superconducting technology has a much lower capital and operating cost than the assumed room-temperature accelerator in the study. The study assumed a 1992 room-temperature accelerator design of 250 mA beam of protons at 1.6 GeV, an accelerator real-estate gradient of 1 MV/m, and a D_2O

moderated target generating 2×10^{15} n/cm^2-sec. Modern superconducting transmuter accelerator designs are 20 times the accelerator real-estate gradient and, with superconducting structures, much more electrically efficient, yielding significantly lower capital and operating costs.

1.2. *Response to criticisms*

The intermittent operation of accelerators was a large issue in previous studies based on the fault/trips experienced with existing research accelerators. Given the high accelerator trip rate and the GW of electrical power the transmuter could generate, the swings in grid power produced from an accelerator-based transmuter was deemed too great to be connected to the electrical grid.

Newer studies have been done to estimate the additional cost to electrical rates using an accelerator based transmuter. The NRC study concluded that conventional aqueous reprocessing alone, at $2,100/kg, would add about 8% to the wholesale cost of electricity. A study by JAEA [3] has estimated that an Accelerator Driven System (ADS) will result in a 2-3% increase in electricity cost. The cost included operating cost, partitioning, dedicated fuel fabrication facilities, reprocessing, and decommissioning (the retail price of electricity in Japan is about 17¢/kW-hr). A study from Sweden [4] gives another scenario that shows the cost of electricity for ADS + recycling is 30% more expensive than once through. The results they presented are not directly applicable to the fuel cycle presented in this paper, but provide a basis for doing more refined calculations on cost. However, even though the advanced fuel cycles studies show a higher cost of electricity, they are still low enough to be competitive in a future market with the anticipated higher electricity prices using other technologies, especially if carbon emissions taxes are implemented.

The capital cost of an accelerator-based transmuter is expected to be about 130% greater than an equivalently sized reactor, as given above. As will be seen later, we estimate that just 2 or 3 transmuters can support the entire existing fleet of US LWRs. So if the incremental cost to the present electrical rate is based on the adding of two to three more plants (that don't produce any power), the incremental cost to the electrical rate of the 104 LWRs should be 2 to 5%, not including reprocessing costs. Assuming reprocessing costs add 8% (as concluded in the NRC study), electricity costs would be 10 to 15% greater for a fully implemented ADS system as compared to the current once-through fuel cycle costs. Since operation of the accelerator-driven transmuters is expected to generate in excess of 7 GW of thermal heat, converting that power into a useable energy source is highly advantageous to help recover the facility capital and operating costs.

Electrical generation can be used to cover the transmuter facility's own operating needs but some storage capacity is needed to be able to run through the intermittent accelerator beam trips. We expect that improvements in energy storage devices allow for significant energy storage to cover fault interruptions: for example, superconducting coils: 100 MW for 100 s, flywheels ~ MWs, Vanadium Redux Battery ~ MWs, steam storage. A strong push will be made on high capacity storage technologies due to the expected increase in wind and solar power, and these technologies will have the potential for far greater grid interruptions than for an accelerator based system.

Consideration should still be given to converting the generated energy to another form useful for national consumption. One option is to sell the excess power to the grid. Based on recent experience with superconducting accelerator technology, the design of highly fault-tolerant accelerators is a reasonable expectation [5]. Storing energy with the use of the above mentioned energy storage devices could provide the electricity to run through faults if they can store enough energy to enable delivery of steady power to the grid through the intermittent interruptions of order 10 s in duration.

Other than intermittent operation affecting the quality of the power produced by the transmuter, transients will affect the lifetime of components in the transmuter assembly. Effect of transients on materials and fuels was evaluated at LANL for the proposed Material Test Station [6]. These studies show no significant deleterious effects for core clad or structural materials for the expected accelerator interruptions. Similar studies also show no significant concern for fuels. Effect of transients on reactor structures was evaluated at JAEA [3]. This study showed that the number of allowable trips were determined by the following major components: beam window: 10^5 trips per 2 yrs of <1 sec. reactor vessel: 10^4 trips per 40 yrs of 1 sec to 5 min. system availability: 1 trips per week of > 5 min (if used for electrical power generation).

1.3. *ADS Requirements*

The axial profile of the deposited power density in a spallation target is determined by the proton beam energy. A uniform profile reduces excessive local heating. The power density profile is characterized by a smooth curve ending with the Bragg peak where the proton stops in the tungsten. At low proton beam energies, the peak power density is defined by the maximum of the Bragg peak. At high beam energies, multiple scattering and nuclear processes lead to power peaking within a few centimeters of the front face of the tungsten target. Also, neutron production goes as beam energy less a 240-MeV offset, which favors higher beam energy. A beam energy of 1500 MeV distributes the neutron source over an appropriate length (about 90 cm) along the direction of proton beam propagation, and yields high neutron production per unit beam power.

The beam current of the linac is determined by the required neutron yield. The number of spallation neutrons produced per proton is shown in Figure 2 as a function of proton beam energy. Beyond 500 MeV, neutron yield per proton increases linearly with proton beam energy. This linear dependence, extrapolated towards low proton beam energy, has a 240-MeV offset because of nuclear reaction thresholds of neutron production. Simulations show that spallation from heavy sub-actinide metals (Ta, W, Hg, Pb, Bi) all produce about the same number of neutrons. Neutron production from uranium is about 60% greater than heavy metals due to a ~1-barn fission cross-section above 1 MeV. Using the neutron yields from Figure 2 and assuming a subcritical blanket reactivity of k_{eff} = 0.94, a 1500-MeV, 77-mA beam will produce a transmuter fission power of 3 GW, as shown in Figure 3. Note that lead-bismuth eutectic (LBE) target material exhibits low parasitic neutron capture, so that production within and leakage from the spallation target are nearly equal; neutron leakage for spallation targets made of other materials is less than the production yields shown in Figure 2.

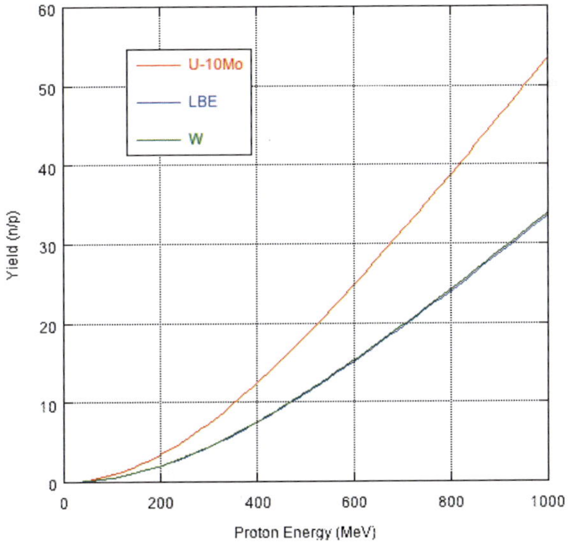

Figure 2. Number of neutrons produced per proton incident on target versus proton beam energy.

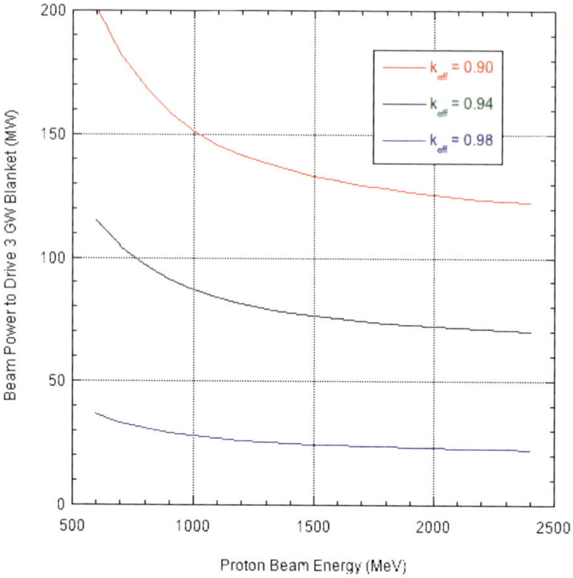

Figure 3. Beam power required to drive a 3-GW blanket versus beam energy.

The beam current i_b required to drive a subcritical core with multiplication factor k_{eff} at a thermal power level P_{th} is given by

$$i_b=P_{th}/Q * (1-k_{eff})/k_{eff} * v/S , \tag{1}$$

where Q is the energy released per fission (200 to 210 MeV), v is the number of neutrons released per fission (2.43 to 3.2 depending on the nuclides undergoing fission), and S is the number of source neutrons leaking into the blanket per proton delivered to the target.

For a 3-GW blanket,

$$i_b = 14.4 * (1-k_{eff})/k_{eff} * \nu/\mathrm{S}. \tag{2}$$

The accelerator interface requirements must consider material stress limits and safety considerations. Thermal stresses in structural components arise due to beam trips that cause coolant exit temperature to quickly drop to the coolant inlet temperature. Conversely, thermal stresses in nuclear fuel and fuel cladding generally *decrease* when the beam trips off. A ramped restart of the beam current will likely be required to keep transient thermal stresses below steady-state stresses. The major concern is that repetitive thermal cycling will induce cyclic fatigue and potential failure.

Safety considerations that must be addressed in any ADS include beam overpower (excessive current or energy) could potentially cause a power/flow mismatch in the target or multiplier, an unplanned beam shutdown (i.e. for loss of flow or heat sink), and failure of the beam expansion system (rastering or higher-order magnets).

No significant deleterious effects for fuel or structural components are expected when structures are properly designed and ramped beam turn-on is employed.

The present status for ADS accelerator requirements is:

1. Beam current: up to 60 mA (CW for operations, pulsed for tune up and diagnostics)
2. Beam energy: 1-2 GeV nominal energy
3. Current stability: ≤ 1 % *(5% bad, 2% a concern, 1% OK)*
4. Safety-class beam shutdown
5. Hard limit on maximum beam current and limit on rate of change
6. Low incidence of beam trips (based on analysis to date, including margin)
7. Beam trips with duration 0.3 s to 100 s: < 1000 trips/yr
8. Beam trips with duration > 100 s: < 30 trips/yr
9. Active current control: automatic, independent-feedback, dynamic current-control mechanism driven by blanket power feedback
10. Must have ability to smoothly ramp up (or down) beam current over seconds to minutes
11. Availability during scheduled operation: >95%.

2. Summary

If the ADS mission is limited to transmuting the component of high-level nuclear waste that drives intermediate-term (300 to 100,000 years) geologic repository performance (that is, transmuting only the Am in used nuclear fuel), the number of accelerator-driven systems needed to address the waste stream from the current U.S. nuclear fleet is two or three 3-GW systems. Estimates on the impact of reprocessing used LWR fuel and operating these three systems is in the range of 10 to 15% greater cost in the wholesale price of electricity.

The whole national strategy towards addressing used nuclear fuel is changing. The Obama Administration plans to convene a "Blue Ribbon Panel" to study approaches to addressing the nation's high level waste.

In summary, the accelerator needed to drive an ADS is based on demonstrated technologies. Spallation targets have been demonstrated at 1 MW and are scalable to ~100 MW. The major challenges to deploying a closed fuel cycle are related to fuel forms, separations, and waste forms for ultimate disposal. Accelerator driven systems offer potentially cost competitive options for dealing with certain components of a closed nuclear fuel cycle.

Reference

1. G. Rimpault, *EUROTRANS DM1 Safety Meeting*, FZK, RFA, 27-28 November (2008).
2. *"Nuclear Wastes: Technologies for Separations and Transmutation,"* National Academy Press Washington, D.C. (1996).
3. H. Oigawa, *OECD/NEA 9th Information Exchange Meeting on Partitioning & Transmutation*, September 26-28, (2006).
4. D. Westlén, *M.Sc. Thesis*, KTH Royal Institute of Tech., Stockholm (2001).
5. I. E. Campisi, F. Casagrande, M. Crofford, M. Howell, Y. Kang, S. H. Kim, Z. Kursun, P. Ladd, D. Stout, W. Strong, *AIP conference proceedings [0094-243X]* vol:985 iss:1 pg:1586 -1593 (2008).
6. E. Pitcher, et al., *PHYSOR08 (Proc. of the Int. Conf. on the Physics of Reactors, Interlaken, Switzerland, 2008),* log 574.

ACCELERATOR DRIVEN TRANSMUTATION OF WASTE: HIGH POWER ACCELERATOR FOR THE EUROPEAN ADS DEMONSTRATOR

J. L. BIARROTTE and T. JUNQUERA*
Institut de Physique Nucleaire, CNRS-IN2P3, 91400 ORSAY, France

An Accelerator Driven System (ADS) for transmutation of nuclear waste typically requires a 600 MeV - 1 GeV accelerator delivering a proton flux of a few mA for demonstrators, and a few tens of mA for large industrial systems. This high power accelerator requires an exceptional reliability: because of the induced thermal stress to the subcritical core, the number of unwanted beam-trips should not exceed a few per year, far above usual linac performances. This paper describes the reference solution adopted for such a machine, based on a so-called "fault-tolerant" concept for linear accelerators using superconducting cavities.

1. The European ADS Demonstrator

The basic purpose of Accelerator Driven Systems (ADS) is to reduce – by orders of magnitude – the nuclear wastes' radio-toxicity, volume and heat load before their underground storage in deep geological depositories [1]. The EUROpean research programme for the TRANSmutation of high-level nuclear waste in accelerator driven systems (EUROTRANS) is funded by the European Commission within the 6th FrameWork Programme, and involves more than 40 partners (research agencies, universities and nuclear industries). It is a 5-year programme (2005-2010), extending previous R&D (e.g. the PDS-XADS project), and which activities are split into several main domains, devoted to: the advanced design of a transmuter demonstrator including all its sub components, the generic conceptual design of an industrial transmutation facility (see Table 1), experimental work dealing with the coupling of an accelerator, a spallation target and a sub-critical core, like the zero-power GUINEVERE experiment [2], and finally, studies on advanced fuels for transmuters, investigations on suited structural materials and collection of nuclear data for transmutation.

* Work supported by EU FP6 through the EUROTRANS program

Table 1: European Transmuter Main Specifications.

Transmuter demo (XT-ADS / MYRRHA project)	Industrial transmuter (EFIT)
50 – 100 MWth power	Several 100 MWth power
keff value ~ 0.95	keff value ~ 0.97
Highly-enriched MOX fuel	Minor Actinide fuel
Pb-Bi Eutectic coolant & target	Pb coolant & target
600 MeV – 2.5mA CW proton beam	800 MeV – 20mA CW proton beam
Allowed beam trips number (> 1sec): < 5 per 3-month operation cycle	Allowed beam trips number (> 1sec): < 3 per year

The main objective of the EUROTRANS programme is actually to pave the road towards the construction of an eXperimental facility (MYRRHA) willing to demonstrate the technical feasibility of Transmutation in an Accelerator Driven System (XT-ADS concept). SCK.CEN (Mol, Belgium) has initiated the MYRRHA project in 1998 [3]. Its purpose is to design and later to build a flexible irradiation facility as suitable replacement for the existing Material Testing Reactor BR2 that is operating since 1962. The new facility would serve both as a test-bed for transmutation and as a fast-spectrum facility for material and fuel developments. It would operate first as a sub-critical (accelerator driven) system and later as a critical reactor. The Central Design Team project is presently being settled to further develop the engineering design of such a facility, which should be fully operational in 2020. (CDT project is supported by the EU FP7)

2. The Reference ADS Accelerator

The European ADS concepts requires a high-power proton accelerator operating in CW mode, ranging from 2.4 MW (XT-ADS operation) up to 16 MW for the industrial EFIT. The main beam specifications are shown in Table 1. At first glance, the extremely high reliability requirement (beam trip number) can immediately be identified as the main technological challenge to achieve.

The conceptual design of the accelerator has been developed during the PDS-XADS project [4]. It is a superconducting linac-based solution, leading to a very modular and upgradeable machine (same concept for demonstrator and industrial scale), an excellent potential for reliability (see dedicated section here after), and a high RF-to-beam efficiency thanks to superconductivity (optimized operation cost). A more advanced reference design is presently under final definition – to be frozen by mid 2010 – in the MYRRHA context. The main characteristics, still subjected to slight changes, are detailed below. The investment cost for such a machine has been assessed around 200 M€ (manpower included, buildings and general utilities excluded).

2.1. *The Linac Front-End*

The linac injector is composed of a 50 kV ECR proton source, a short magnetic Low Energy Beam Transport line and a 3 MeV 4-vane copper RFQ operating at 352 MHz, hanling up to 30 mA CW beams with close to 100% transmission.

This "classical" injection section is then followed by a more "exotic" but promising energy booster [5], that is a combination of normal conducting and superconducting CH (Crossbar H-mode) DTL structures as shown in Figure 1, bringing the beam up to 17 MeV. Focusing is ensured by quadrupole triplets and superconducting solenoids inside the cryomodule containing the 4 superconducting CH cavities, and a couple of re-bunchers is used to perform the longitudinal beam adaptation. The design of the DTL structures is based on the KONUS beam dynamics concept [6], which allows to exhibit excellent accelerating efficiency at these low energies with a net energy gain of 14 MeV in 7.5 metres, while having very low power consumption in CW operation. In this front-end line, the beam beta-profile is frozen by design, so that any accelerating section failure will inevitably lead to a beam interruption. For this reason and in order to enhance the machine reliability, it is proposed to duplicate the injector (at least the ion source, at most the whole 17 MeV front-end) to provide a hot stand-by injection line able to relieve the main one in case of failure

2.2. *The Independently-Phased SC Linac*

From 17 MeV, a fully modular superconducting linac then accelerates the proton beam up to the final energy (600 MeV), with a total length of ~220 meters. It is composed (see Table 2) of an array of independently powered spoke and elliptical cavities with high energy acceptance and moderate energy gain per cavity – low number of cells and conservative accelerating gradients

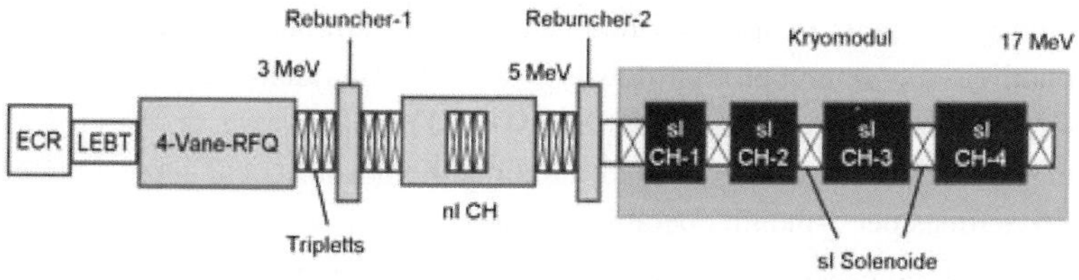

Figure 1: The reference linac front-end.

Table 2: Independently-Phased Linac Overview.

SC cavity type	#Cavities #Cryomodules	Energy range	Section length
352 MHz 2-gap β 0.35 Spoke	63 cav 21 cryo	17 ~ 90 MeV	~60 m
704 MHz 5-cell β 0.50 Elliptical	30 cav 15 cryo	90 ~ 190 MeV	~50 m
704 MHz 5-cell β 0.65 Elliptical	64 cav 16 cryo	190 ~ 600 MeV	~100 m

(around 50mT and 25 MV/m peak fields nominal operation point) – in order to increase as much as possible the tuning flexibility and provide sufficient margins to allow the implementation of fault-recovery scenarios.

The linac design is based on the use of regular focusing lattices, with short cryostats – easy maintenance and fast replacement – and room-temperature quadrupole doublets. The beam tuning has been performed keeping a constant longitudinal acceptance, and limiting phase advances below 90° per lattice [7]. This "conservative" optical design leads to very safe beam behaviours, with low sensitivity to mismatched conditions or current fluctuations, and producing very low emittance growths (below 5%).

The objective of the final vertical transport line is to safely inject the proton beam with the specified footprint – donut-shaped – onto the spallation target located inside the reactor (~30 metres deep). It is composed of two non-dispersive 2×45° achromats, and the footprint is obtained by raster scanning, using a redundant set of fast steering magnets operated at frequencies of 50 to a few hundreds Hz, and acting in the two transverse directions.

3. The Accelerator Reliability Issue

The ADS accelerator is expected – especially in the industrial scenario – to have a very limited number of unforeseen beam interruptions per year. This requirement is motivated by the fact that frequently-repeated long enough beam interruptions induce high thermal stresses and fatigue on the reactor structures, the target and the fuel elements, with possible significant damages especially on the fuel claddings; moreover these beam interruptions decrease the plant availability, implying plant shut-downs of tens of hours in most of the cases. The present rough requirement in the XT-ADS/MYRRHA case is therefore that beam trips in excess of one second duration should not occur more frequently than five times per 3-month operation period.

Many thermo mechanical calculations of the sub-critical reactor have been recently performed, in particular of the fuel elements cladding. Assessment of this critical design is essential for defining the working conditions in terms of stress and fatigue of the various structures. In the Fig. 2, a simulated behavior of a fuel element induced by a beam trip of different time duration is presented. Given the calculated temperature transients, it appears that frequently repeated beam trips can significantly damage these structures, reducing the ADS plant availability.

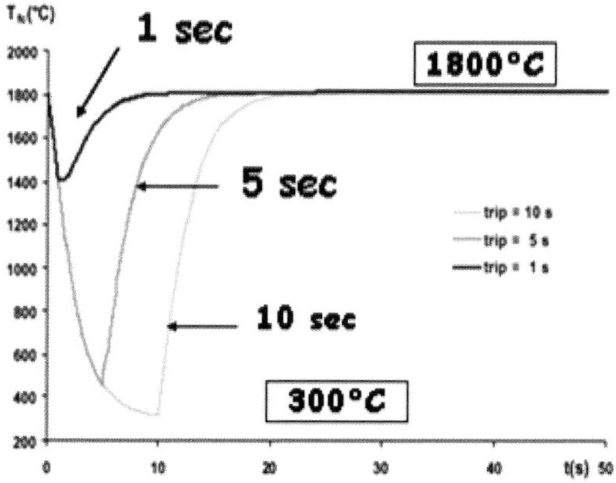

Figure 2: Thermal response to a beam interruption.

If the transient duration is small, the energy stored in the fuel will allow a restart without noticeable change of the fuel temperature. But if the transient lasts a few seconds, the fuel temperature will strongly decrease. This kind of transient need to be examined in detail, to determine the allowable transient duration and the allowable power increase rates after transient. The order of magnitude of the allowable duration of the beam trips is about 1 second. Regarding the damage on the ADS structures, the allowable number of such long beam interruptions (> 1 s) depends on the technology of these equipments (window concept, materials, primary coolant). This number can be quite large, as shown in several preliminary studies [8]: the order of magnitude is hundreds per year, depending on ADS type and design, and assuming the plant is designed for a life-time of 40-60 years. The evaluation of these numbers still needs to be revisited, adopting more realistic operating conditions (stress, lead environment, irradiation damage…).

From the point of view of availability (and associated cost), in the perspective of an industrial application, the tolerable number of long-term beam interruptions is much lower. The number of unexpected fast shutdown for the present nuclear plants dedicated to the electricity production is a few per year

(1 or 2 in PWR plants, about 15 for the experimental PHOENIX fast reactor). From this comes the XT-ADS allowed beam trip values of 5 beam trips per cycle (3 months).

To address in a more quantitative manner this issue, new sets of experiments are planned to study the integrity of the oxide layer over the clad, and the thermal shocks under different stress conditions and within lead environment.

3.1. *Reliability-Oriented Design of the Accelerator*

To reach such an extremely ambitious goal, it is clear that reliability-oriented design practices needed to be followed from the early design stage. In particular strong component design ("over-design" and de-rating). A high degree of redundancy needs in critical areas is needed, i.e. the linac injector area and the RF power systems, where solid-state amplifiers should be used as much as possible. The accelerator should be able to pursue operation despite some major faults in basic components ("fault-tolerance" capability); it has been shown in [9] that a solution based on a modular independently-phased SC linac is indeed capable to easily adapt its nominal tuning in the case of a loss of any RF cavity or powering unit (see here after), and even of a quadrupole doublet, while keeping beam dynamics and beam properties on target into specifications.

3.2. *Tolerance to RF Faults in the SC Linac*

Because we deal with a non-relativistic proton beam, any RF cavity fault implying beam energy loss will also lead to a phase slip along the linac, pushing the beam out of the stability region and leading to a complete loss of the beam. To recover such RF faults conditions, the philosophy is to re-adjust the accelerating fields and phases of some non-faulty RF cavities to recover the nominal beam characteristics (phase and energy) along the linac,. A simple way to perform it is to react on the accelerating cavities neighbouring the failing one. This so called "local compensation method" has the advantage of involving a small number of elements, and therefore of being able to compensate multiple RF faults in different sections of the machine at the same time. Beam dynamics simulations show that nominal beam parameters at the target can always be restored using such a retuning method, given the condition that a 20 to 30% rise in accelerating field and RF power can be sustained in the few (4 to 8) retuned elements [9]. This method is of course rather demanding in terms of linac length and installed RF power budget, but is on the other hand totally in-line with the ADS over-design criterion, and in any case required to try to reach the required reliability level.

Transient beam dynamics have been performed to better analyse what happens to the beam during such retuning procedures, keeping in mind that they have to be performed in less than 1 second ideally. A new simulation tool has been developed, based on the TraceWin code [10], allowing analysing the effect of time-dependent perturbations on the beam through RF control loop modelling. From this work [11], a reference "fast failure recovery scenario" has been defined, that consists in stopping the beam for 1 sec maximum while achieving the retuning. The following sequence (~100ms duration) is proposed: (1) the RF fault is detected (or anticipated) via suited dedicated diagnostics and interlocks, and a fast beam shut-down is triggered. (2) the new correcting field and phase set-points (previously stored in the low level RF cards' memory during the commissioning phase) are updated. (3) the failed cavity is quickly detuned (using piezoactuators) to avoid the beam loading effect, and the associated failed RF loop is cut off. (4) once steady-state is reached, beam re-injection is triggered.

3.3. *Accelerator Reliability Analysis and Discussion*

Two independent integrated reliability analyses have been performed so far to try to estimate the number of malfunctions of the XT-ADS accelerator that could cause beam/plant shutdowns per 3-month operation cycle, and to analyze the influence of MTBFs (Mean Time Between Failures), MTTRs (Mean Time to Repair), and of the whole system architecture on the results.

These studies have been respectively performed by means of a reliability block diagram analysis using the Relex© software [12] and by home-made Monte-Carlo simulations [13] with slight differences in the hypotheses. In both cases, the results show that such linacs have a high potential for reliability improvement if the system is properly designed with this particular objective: from about 100 unexpected beam shut-downs per 3-month operation period for a classical "all-in-series" SC linac, this figure falls around 5 beam interruptions – which is actually the XT-ADS goal – in the case where a second redundant injector stage with fast switching capabilities is used, and when fault-tolerance is included in the independently-phased linac via fast fault-recovery scenarios. Nevertheless, the obtained absolute figures remain highly questionable, because of the somewhat crude modelling used for such a complex system, and because of the lack of a well-established component reliability figures database. Having a look at present high-power hadron facilities – like the SNS, not specifically designed for reaching a high reliability – it appears that the experienced number of beam trips longer than 1 second is much higher, by at least one order of magnitude [14], showing that there is still a long way to go on this topic. But on the other hand, facilities like ESRF, which is very much concerned by the reliability issue, prove that overall MTBF of several days can already be obtained routinely [15], leading to an equivalent of about 20 beam interruptions per 3-month operation. As a conclusion, it seems at least not completely unrealistic to approach and ultimately reach the ADS accelerator reliability goal. It will imply, as underlined before, to include in the linac design de-rating, redundancy and fault-tolerance, and to have a few years of commissioning and training to identify and fix the weak elements. Approaching the goal "from the other side", i.e. relaxed specifications on beam trip numbers and their durations by appropriate design measures in the,target/reactor system, would also help.

References

1. European Technical Working Group, "The European Roadmap for Developing ADS for Nuclear Waste Incineration", ISBN 88-8286-008-6, ENEA 2001.
2. A. Billebaud et al, "The GUINEVERE project for ADS physics", GLOBAL 2009, Paris, France.
3. H. A. Abderrahim et al, "MYRRHA: A multipurpose accelerator driven system...", Nucl. Instr. and Meth. in Phys. Res. A 463 (2001), pp. 487-494.
4. J-L. Biarrotte et al, "A reference accelerator scheme for ADS applications", Nucl. Instr. and Meth. In Phys. Res. A 562 (2006), pp. 565-661.
5. C. Zhang et al, "Conceptual studies for the...", Proc. PAC2007, Albuquerque, USA.
6. U. Ratzinger and R. Tiede, "Status of the HIIF RF linac study ...", Nucl. Instr and Meth. in Phys. Res. A 415 (1998), pp. 229-235.
7. R. Duperrier et al, "Impact of a RF frequency change...", Proc. LINAC 2006, Knoxville, USA.
8. H. Takei et al, "Comparison of beam trip frequencies between ..." HPPA5 workshop proceedings mol Belgium, May 2007.
9. J-L. Biarrotte & al., "Beam dynamics studies for the fault...", Proc. HPPA 2004, Daejeon, Korea.
10. http://irfu.cea.fr/Sacm/logiciels/index.php
11. J-L. Biarrotte, D. Uriot, "Dynamic compensation of an rf cavity failure ...", Phys. Rev. ST – Accel. & Beams, Vol. 11, 072803 (2008).
12. L. Burgazzi, P. Pierini, "Reliability studies of a high power proton accelerator...", Reliability Engineering & Systems Safety, Vol. 92, n°4 (2007), pp. 449-463.
13. R. Brucker et al, "Integrated reliability analysis of the ...", Eurotrans DEL n°1.69 (Empres. Agrup.), 2009.
14. J. Galambos et al., "Commissioning strategies...", Proc. ICFA HB 2008, Nashville, USA.
15. L. Hardy et al., "Operation and recent developments at the ESRF", Proc. EPAC 2008, Genoa, Italy.

MYRRHA, TECHNOLOGY DEVELOPMENT FOR THE REALISATION OF ADS IN EU: CURRENT STATUS & PROSPECTS FOR REALISATION

R. FERNANDEZ, H. AÏT ABDERRAHIM, P. BAETEN, D. DE BRUYN, D. MAES, A. AL MAZOUZI, B. ARIËN, E. MALAMBU, P. SCHUURMANS, M. SCHYNS, V. SOBOLEV, G. VAN DEN EYNDE, D. VANDEPLASSCHE

Belgian Nuclear Research Centre (SCK•CEN)
Boeretang 200, BE-2400, Belgium

The coupling between an accelerator, a spallation target and a subcritical core has been studied for the first time at SCK•CEN in collaboration with Ion Beam Applications (IBA, Louvain-la-Neuve) in the frame of the ADONIS project (1995-1997). ADONIS was a small irradiation facility, based on the ADS concept, having a dedicated objective to produce radioisotopes for medical purposes and more particularly 99Mo as a fission product from highly enriched 235U (HEU) fissile targets. The ad-hoc scientific advisory committee recommended extending the purpose of the ADONIS machine to become a Material Testing Reactor (MTR) for material and fuel research, to study the feasibility of transmutation of the minor actinides and to demonstrate at a reasonable power scale the principle of the ADS. The project, since 1998 named MYRRHA, has then evolved to a larger installation.

MYRRHA is now conceived as a flexible irradiation facility, able to work as an Accelerator Driven (subcritical mode) and in critical mode. In this way, MYRRHA will allow fuel developments for innovative reactor systems, material developments for GEN IV systems, material developments for fusion reactors, radioisotope production for medical and industrial applications and industrial applications, such as Si-doping.

MYRRHA will also demonstrate the ADS full concept by coupling the three components (accelerator, spallation target and subcritical reactor) at reasonable power level to allow operation feedback, scalable to an industrial demonstrator and allow the study of efficient transmutation of high-level nuclear waste.

Since MYRRHA is based on the heavy liquid metal technology, the eutectic lead-bismuth, it will be able to significantly contribute to the development of Lead Fast Reactor Technology. Since MYRRHA will also be operated in critical mode, MYRRHA can even better play the role of European Technology Pilot Plant in the roadmap for LFR.

1 Introduction

Since its creation in 1952, the Belgian Nuclear Research Centre (SCK•CEN) at Mol has always been heavily involved in the conception, the design, the realisation and the operation of large nuclear infrastructures. The Centre has even played a pioneering role in such type of infrastructures in Europe and worldwide. SCK•CEN has successfully operated these facilities at all times thanks to the high degree of qualification and competence of its personnel and by inserting these facilities in European and international research networks, contributing to the development of crucial aspects of nuclear energy at international level.

One of the flagships of the nuclear infrastructure of SCK•CEN is the BR2 reactor [1], a flexible irradiation facility known as a multipurpose materials testing reactor (MTR). This reactor is in operation since 1962 and has proven to be an excellent research tool, which has produced remarkable results for the international nuclear energy community in various fields such as material research for fission and fusion reactors, fuel research, reactor safety, reactor technology and for the production of radioisotopes for medical and industrial applications. BR2 has been refurbished twice, in the beginning of the eighties and in the mid nineties. The BR2 reactor is now licensed for operating until 2016 but around that period, it will have to be decided whether another refurbishment around 2020 will have to be done or whether BR2 will have to be replaced by another facility. Therefore, the Belgian Nuclear Research Centre at Mol is working since several years at the pre- and conceptual design of MYRRHA (Multi-purpose hYbrid Research Reactor for High-tech Applications), a multi-purpose flexible irradiation facility that can replace BR2 MTR and that is innovative to support future oriented research projects needed to sustain the future of the research centre.

2 Scope of MYRRHA

At international level, there is a clear need to obtain a sustainable solution for the high level long-lived radioactive waste (HLLW) consisting of minor actinides (MA) and long-lived fission products (LLFPs). These MA and LLFP stocks need to be managed in an appropriate way. Reprocessing of used fuel followed by geological disposal or direct geological disposal are today the envisaged solutions depending on national fuel cycle options and waste management policies. The required time scale for geological disposal exceeds the time span of profound historical knowledge and this creates problems of public acceptance. The Partitioning and Transmutation (P&T) concept has been pointed out in numerous studies in the past [2, 3 and 4] and more recently in the frame of the GEN IV initiative as the strategy that can relax the constraints on the geological disposal and that can reduce the monitoring period to technological and manageable time scales. The reduction of the volume and the half-life of HLLW and LLFP's can be achieved conceptually by using a park of fast critical reactors that will simultaneously produce electricity and transmute the actinides, or by a 'double strata' reactor park with a first stratum of reactors dedicated to electricity production using 'clean fuel' containing only U and Pu and systems devoted to transmutation of HLLW and LLFPs. These systems of the second stratum will be based on special fast critical reactors or more probably subcritical fast systems (Accelerator Driven Systems (ADS)) loaded with homogeneous fuels with high MA content.

Even when considering the phase out of nuclear energy, the combination of P&T and dedicated burner technologies such as ADS is needed to relax the constraints on the geological disposal and reduce the monitoring period to technological and manageable time scales for existing waste. Hence, since ADS represent a possible major component in the P&T framework, the demonstration of the subcritical dedicated burner concept is needed and this was indicated in the EU vision document [5] and in the strategic research agenda (SRA)[6].

Since 2000, the Generation International Forum (GIF) [7] has selected 6 Generation IV (GEN IV) concepts of which 3 are based on the fast spectrum technologies namely; the sodium fast reactor (SFR), the lead cooled fast reactor (LFR) and the gas cooled fast reactor (GFR). The SNETP community has at present given a higher priority to the SFR technology but indicated also the need for the development of an alternative coolant technology being lead or gas. The technological development of the fuel and materials of these concepts request the availability of a flexible fast spectrum irradiation facility. The vision document and the strategic research agenda (SRA) of the SNETP has also stated that Europe should be in a front-runner position for GEN IV reactor development.

Since MYRRHA is based on the heavy liquid metal technology, the eutectic lead-bismuth, it will be able to significantly contribute to the development of Lead Fast Reactor Technology. Since MYRRHA will also be operated in critical mode, MYRRHA can even better play the role of European Technology Pilot Plant in the roadmap for LFR.

Taking into account these national, European and even worldwide needs in terms of demonstration and irradiation capabilities, SCK•CEN is proposing MYRRHA as a flexible fast spectrum irradiation facility able to operate in sub-critical and critical mode and is positioning MYRRHA within the European Research Area of Experimental Reactors (ERAER). The flexible irradiation capacity within the ERAER is then based on three pillars:

- **JULES HOROWITZ reactor (JHR), France**: a flexible thermal spectrum irradiation facility that will be answering the needs for industrial applications for GEN II & III in terms of structural material and fuel performance improvement as well as some generic GEN IV research. It will also be acting as backup irradiation facility for radioisotopes production.
- **MYRRHA, Belgium:** a flexible fast spectrum irradiation facility, operating as a sub-critical (accelerator driven) system, and as a critical reactor for material and fuel developments for GEN IV and fusion reactors and in a back-up role for radioisotopes production.
- **PALLAS, the Netherlands:** a dedicated irradiation facility for securing the radioisotopes production for medical applications in Europe and as a complementary facility in support of the industrial needs for technological development for present and future reactors.

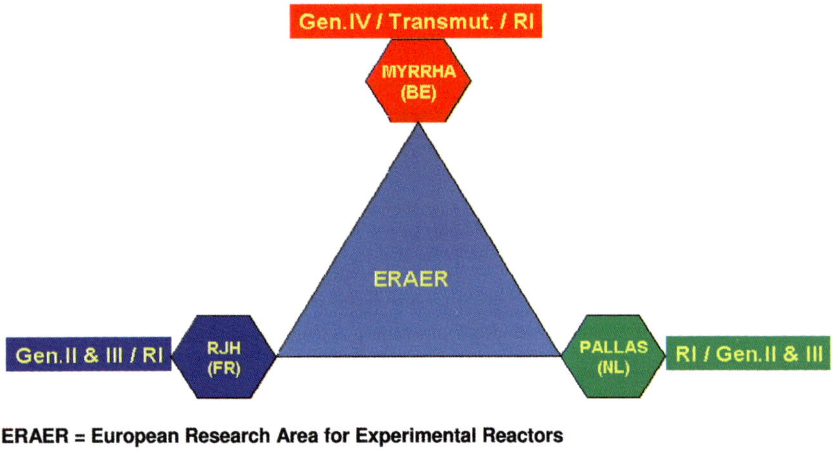

Figure 1: The position of MYRRHA in the ERAER.

Since Europe does no longer have a fast spectrum irradiation facility, there is a clear need for a flexible fast spectrum irradiation facility in support of the development of the different fast reactor systems: SFR, LFR and GFR. MYRRHA will play this role.

3 MYRRHA/FASTEF objectives

MYRRHA with the objectives described in the previous paragraph should therefore target the following applications catalogue:

- To demonstrate the ADS full concept by coupling the three components (accelerator, spallation target and subcritical reactor) at reasonable power level to allow operation feedback, scalable to an industrial demonstrator.
- To allow the study of efficient transmutation of high-level nuclear waste, in particular minor actinides that would request high fast flux intensity (Φ>0.75MeV = 1015n/cm2s),
- To be operated as a flexible fast spectrum irradiation facility allowing for:
  - fuel developments for innovative reactor systems, which need irradiation rigs with a representative flux spectrum, a representative irradiation temperature and high total flux levels (Φtot = 5x1014 to 1015n/cm2s); the main target will be fast spectrum GEN IV systems which require fast spectrum conditions;
  - material developments for GEN IV systems, which need large irradiation volumes (3000 cm^3) with high uniform fast flux level (Φ>1MeV = 1 ~ 5.1014n/cm2s) in various irradiation positions, representative irradiation temperature and representative neutron spectrum conditions; the main target will be fast spectrum GEN IV systems;
  - material developments for fusion reactors which need also large irradiation volumes (3000 cm^3) with high constant fast flux level (Φ>1MeV = 1 ~ 5.1014n/cm2s), a representative irradiation temperature and a representative ratio appm He/dpa(Fe) = 10;
  - radioisotope production for medical and industrial applications by:
    - holding a backup role for classical medical radioisotopes,
    - focusing on R&D and production of radioisotopes requesting very high thermal flux levels (Φth = 2 to 3.1015n/cm2s) due to double capture reactions;
  - industrial applications, such as Si-doping need a thermal flux level depending on the desired irradiation time: for a flux level Φthermal = 1013n/cm2s an irradiation time in the order of days is needed and for a flux level of Φthermal = 1014n/cm2s an irradiation time in the order of hours is needed to obtain the required specifications;
  - the demonstration of LBE technology;
- To contribute to the demonstration of LFR technology and the critical mode operation of heavy liquid metal cooled reactor as an alternative technology to SFR.

4 Genesis of MYRRHA and Its Evolution

4.1 The ADONIS-project (1995-1997)

The coupling between an accelerator, a spallation target and a subcritical core has been studied for the first time at SCK•CEN in collaboration with Ion Beam Applications (IBA, Louvain-la-Neuve) in the frame of the ADONIS project (1995-1997). ADONIS was a small irradiation facility, based on the ADS concept, having a dedicated objective to produce radioisotopes for medical purposes and more particularly 99Mo as a fission product from highly enriched 235U (HEU) fissile targets. The proposed design was of limited size with an accelerator of 150 MeV and a core with a power of around 1.5MWth. The subcritical core was made of the 235U targets for production of the 99Mo without other driver fuel. The system was a thermal spectrum machine and therefore water was used as coolant and moderator.

4.2 From ADONIS to MYRRHA (1998-2005)

The ad-hoc scientific advisory committee recommended extending the purpose of the ADONIS machine to become a Material Testing Reactor (MTR) for material and fuel research, to study the feasibility of transmutation of the minor actinides and to demonstrate at a reasonable power scale the principle of the ADS.

This decision was taken around the same time of the last BR2 refurbishment. It was then clear that a next BR2 refurbishment would be compulsory around 2020. Even if such refurbishment remains technically possible, a new machine would be better adapted to the new needs, in particular for GEN IV and fusion research. The project, since 1998 named MYRRHA, has then evolved to a larger installation.

At mid-2002, a first pre-design file of MYRRHA, namely the 'MYRRHA Draft – 1' file [8] with a core nominal power of 30 MWth, was submitted to an International Technical Guidance Committee (ITGC) for reviewing. This international panel consisted of experts from research reactor designers, ADS development, reactor safety authorities and spallation target specialists. No show stopper was identified in the project but some recommendations were made. The design was upgraded and the MYRRHA project team has favoured as much as possible mature or less demanding technologies in terms of research & development. In its 2005 version, MYRRHA consisted of a proton accelerator delivering a 350MeV * 5mA beam to a windowless liquid PbBi spallation target that in turn couples to a PbBi cooled, subcritical fast core of 50MW thermal power. The so-called 'Draft-2' design, published early 2005 [9], is summarized in reference [10], for which also a Business plan [11] has been written.

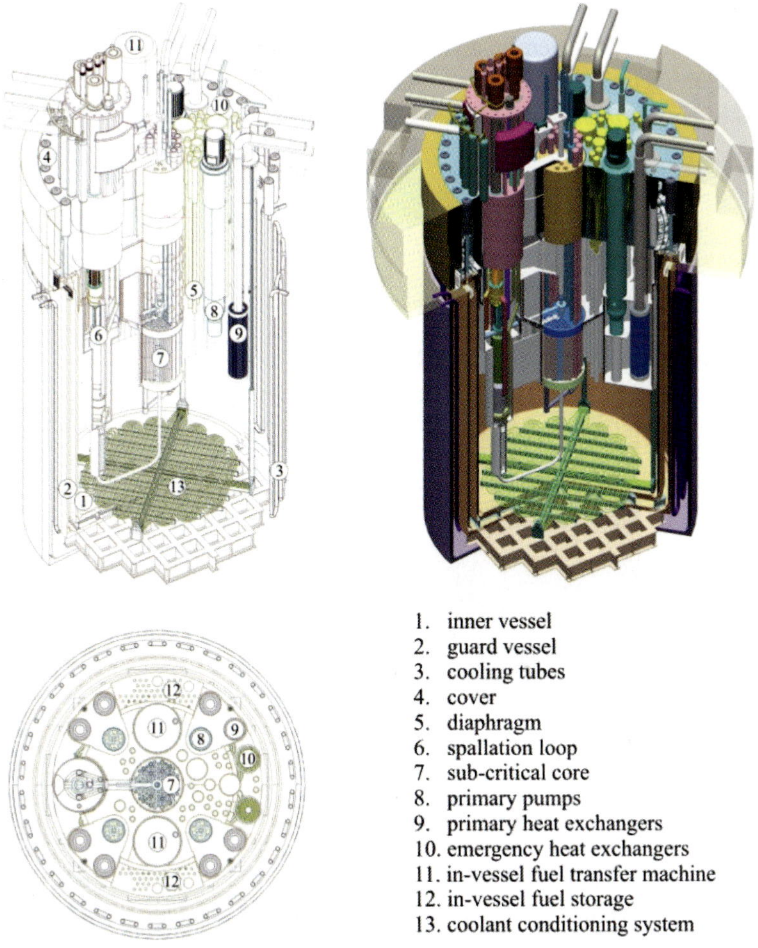

Figure 2: MYRRHA draft 2.

4.3 From MYRRHA to XT-ADS (2005-2009)

SCK•CEN offered to use the MYRRHA 2005 design file as a starting basis for the XT-ADS (eXperimental facility demonstrating the technical feasibility of Transmutation in an Accelerator Driven System) design in the FP6 EUROTRANS integrated project. This allowed optimizing an existing design towards the needs of XT-ADS and within the limits of the safety requirements instead of starting from a blank page. MYRRHA and XT-ADS differ however on several points such as the accelerator (600MeV beam instead of 350MeV), the temperature level, the plutonium vector in the fuel, the fuel pitch and the simplification of several major components resulting in a lower power density core, an increased core power of 57MW and a neutron flux of 7x1014n/cm^{3}s instead of 1x1015 n/cm^{3}s (>0.75MeV) [12].

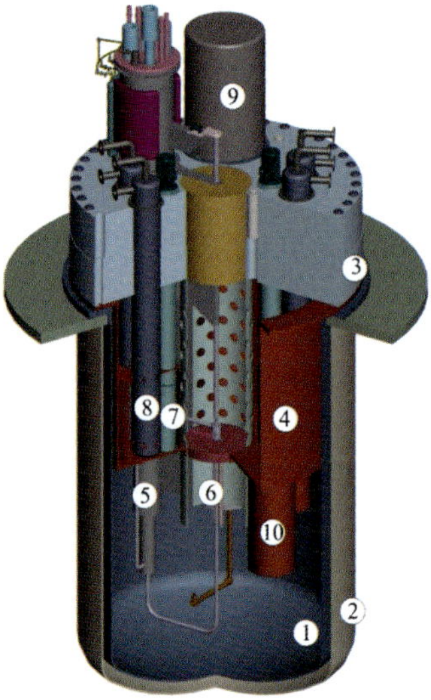

1. inner vessel
2. guard vessel
3. cover
4. diaphragm
5. spallation loop
6. sub-critical core
7. primary pumps
8. primary heat exchangers
9. in-vessel fuel transfer machine
10. in-vessel fuel storage

Figure 3: MYRRHA/XT-ADS.

4.4 From XT-ADS to FASTEF (2009-2011)

The design of MYRRHA presented in this document is the MYRRHA/XT-ADS concept as designed within IP-EUROTRANS, which is based on the MYRRHA Draft-2 concept [9].

The objectives as described in the previous paragraph do not fully correspond to the objectives for MYRRHA/XT-ADS. Therefore, it is clear that future design work is needed to bring the design of MYRRHA in line with the objectives, applications catalogue as described in previous paragraph.

First of all, since the objective of MYRRHA is to operate both in a sub-critical mode and a critical mode, an analysis will be performed to establish to which extent the design of MYRRHA/XT-ADS (which only considers sub-critical mode operation) needs to be modified to respond to the objective to operate also in critical mode. In this respect, it is clear that reactor control and scram systems have to be included in the design. A primary reactor shutdown system based on shutdown rods and a secondary shutdown system based on another type of system need to be incorporated to guarantee not only redundancy but also diversification. Control rods need to be foreseen as well. Also, an analysis will have to be performed to determine which further design changes might be induced by removing the spallation loop, especially with regard to the operation of MYRRHA in critical mode as a flexible fast spectrum irradiation facility. Working in critical mode will significantly modify the safety characteristics of the facility and the impact of the safety feedback on the design needs to be implemented.

A second major topic for the update of the MYRRHA design is to obtain the targets set in the applications catalogue. The MYRRHA/XT-ADS concept is able to reach the different targets in terms of irradiation conditions listed in the applications catalogue except for the targeted fast flux level above E = 0,75MeV of 1015n/cm2s. In MYRRHA/XT-ADS, this flux level reaches Φ>0.75MeV = 0,7x1015n/cm2s. To increase this flux level, it is obvious that the total power and power density will have to be increased. Therefore, already a preliminary study [13] was performed to investigate what would be the most optimal design changes to reach this target of Φ>0.75MeV = 1015n/cm2s given the different design constraints. Preliminary conclusions of this study resulted in preserving the existing design choices of MYRRHA/XT-ADS except for:

- An increase of the power from 57MWth to 85MWth;
- an increase of the power density by decreasing the number of fuel assemblies from 72 to 64 for the equilibrium core;
- an increase of the allowable temperature increment over the core from 100°C to 130°C, resulting in a decrease of the inlet temperature from 300°C to 270°C (the outlet temperature remains at 400°C);
- an increase of the beam current from 2.2mA to 3.0mA for ADS operation (during normal operation);
- an increase of the Pu content from 33% in a depleted uranium matrix to 35% in a natural uranium matrix.

The study pointed out that the implications of these modifications can be accommodated by the design, but a detailed confirmation is still needed.

These aspects for the upgrade of the MYRRHA design will together with a progress towards an advanced engineering design level for all components make up the work for the Central Design Team (CDT) within the accepted FP7 project CDT [14].

Table 1: MYRRHA design evolution.

Draft 2 2005	XT-ADS 2006-2009	XT-ADS-HF end 2009	FASTEF 2012
LINAC 350MeV	LINAC 600MeV	LINAC 600MeV	Subcritical design optimisation and critical operation
High power density core → Safety 'ULOF'	Low power density core	Low power density core	
Low core power 52MW	Low core power 57MW →	High core power 85MW	
Neutron flux 10^{15}n/cm^{2}s (>0.75MeV)	Neutron flux 7.10^{14}n/cm^{2}s (>0.75MeV)	Neutron flux 10^{15}n/cm^{2}s (>0.75MeV)	

5 Project schedule and budget

At the end of the EUROTRANS project, in which MYRRHA is serving as the basis, in March 2010, the conceptual design of the machine and of its different components will be available.

During the period 2009-2013, the following tasks will be accomplished in parallel:

- the detailed engineering design, including 2009-2011, part of this design work will be performed through the FP7-project, Central Design Team (CDT) for MYRRHA;
- drafting of the technical specifications of the different lots, the publication of the call for tenders, the comparison of the technical and financial proposals from the contractors and finally the awarding of the manufacturing contracts (2012-2013);
- development and testing of key innovative components (for the accelerator and for the reactor);
- licensing and permitting activities, namely the writing of a Preliminary Safety Assessment Report (PSAR), the Environment Impact Assessment (EIA) report and the Preliminary Dismantling Plan (PDP). The objective being to obtain the permit for construction at the end of 2013.

The construction period of the components and the civil engineering work is to be accomplished in a three-year period (2014-2016) followed by a one year assembling together of the different components in 2017. The commissioning at progressive levels of power will accomplished in two year period (2018-2019) with the final objective to be in full power operation in 2020.

Figure 4: Project schedule.

MYRRHA as a flexible fast spectrum irradiation facility able to operate in subcritical and critical modes is intended to be started around 2020 at full power. In the design and licensing of MYRRHA both modes of operation are foreseen from the beginning. A minimum of two years preparatory period is considered for the progressive commissioning of the facility which will include both operation modes.

To obtain an answer on the viability of the ADS concept in an international context as soon as possible and thereby to allow evaluating the realism of subcritical burners in a double strata fuel cycle approach, it was opted to start in MYRRHA with the demonstration of ADS and hence the full power coupling with an accelerator. The MYRRHA facility is thus projected to be initially operated as an ADS in sub-critical mode for the demonstration of the full power ADS operation and the efficient transmutation of minor actinides (with available fuels). During this period, MYRRHA will also be used as a flexible fast spectrum irradiation facility with the applications catalogue as described in previous paragraph. It is also envisaged to deflect a small part (about 5 % of the nominal beam) for nuclear physics applications at the proton accelerator.

Up till now, all flexible irradiation facilities have been operated very efficiently in critical mode. Working efficiently and reliable in sub-critical mode introduces heavy constraints on the accelerator. Since working in critical mode is not conditioned by this fact, it might be advantageous to work in critical mode. The facility will then be converted into a critical fast spectrum irradiation facility based on heavy liquid metal coolant technology by separating the accelerator from the reactor, by removing the spallation source and adding fuel assemblies to reach criticality. The entire accelerator facility can then be used for nuclear physics applications by the user-club of Belgian and other universities.

In the central positions, which hosted the spallation source, an in-pile section with a maximum high flux can be inserted. MYRRHA will from then on be operated as a flexible fast spectrum irradiation facility with the applications catalogue as described in the previous paragraph, except for the demonstration of the ADS concept. Further demonstration of efficient demonstration of minor actinides can be pursued in the critical MYRRHA and new minor actinide fuels can be tested and qualified.

Since MYRRHA is based on the heavy liquid metal technology, the eutectic lead-bismuth, it will be able to significantly contribute to the development of Lead Fast Reactor Technology. Since MYRRHA will also be operated in critical mode, MYRRHA can even better play the role of European Technology Pilot Plant in the roadmap for LFR.

Presently nearly all major research facilities built worldwide are from the very beginning conceived as open users' facilities. Even the existing facilities that were built as national infrastructures are moving towards open-users organisations. This means that the research facility proposed is of interest for the international research community and is unique or having characteristics making it very attractive to a large number of researchers and users.

This leads to a formation of an owner consortium group that develops and later on operates the facility.

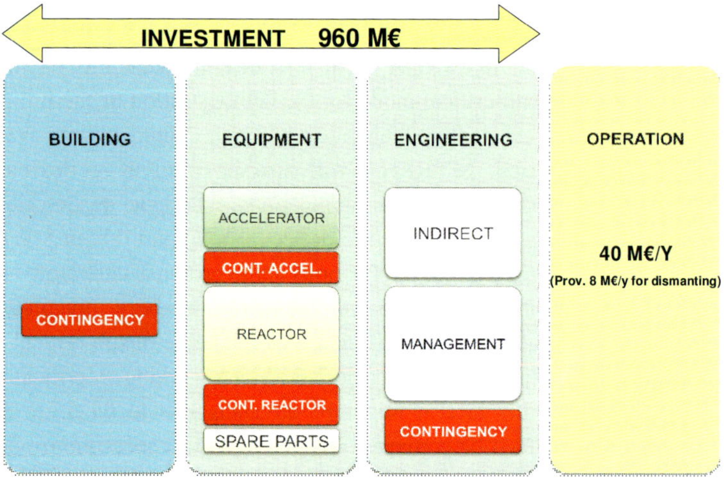

Figure 5: Investment budget (in €2009).

Apart from the traditional sources of funding, e.g. the contribution of the owner consortium group members during the investment phase of the project and the endowment for the operation phase, SCK•CEN management, as the promoter of the project, intends to call on the recent financing facility jointly developed by the European Commission (EC) and the European Investment Bank (EIB) in support of higher risk financing of Research Infrastructures such as the MYRRHA infrastructure.

6 Conclusion

Europe, aims to become the most competitive and dynamic knowledge-driven economy, as decided in Lisbon in March 2000, by encouraging and promoting research and innovation. In addition, Europe is also facing the challenges of its long term sustainable development by use of non-CO2 emitting energy technologies that can play a significant role in guaranteeing the energy independence of Europe.

SCK•CEN is proposing to replace its ageing flagship, the Material Testing Reactor, BR2 by a new flexible irradiation facility MYRRHA. In view of its innovative character and its field of application MYRRHA contributes to the realisation of EU's goal towards a sustainable future.

MYRRHA is conceived as a flexible fast spectrum irradiation facility able to work in sub-critical and critical mode. MYRRHA is foreseen to be in full operation by 2020 and it will be operated in the first years as an Accelerator Driven Systems to demonstrate the ADS technology and the efficient demonstration of Minor Actinides in subcritical mode. Afterwards it is intended to decouple the accelerator from the reactor and run MYRRHA as a critical flexible fast spectrum irradiation critical facility by removing the spallation loop. As a fast spectrum irradiation facility, it will address fuel research for innovative reactor systems, material research for GEN IV systems and for fusion reactors, radioisotope production for medical and industrial applications, such as Si-doping. Since MYRRHA is based on the heavy liquid metal technology, the eutectic lead-bismuth, it will be able to significantly contribute to the development of Lead Fast Reactor Technology. Since MYRRHA will also be operated in critical mode, MYRRHA can even better play the role of European Technology Pilot Plant in the roadmap for LFR.

This project is intended to be organised as an international open user facility to be constructed by an owner consortium made of European and even larger, research centres and European member states representative agencies or international bodies such as the European Commission.

References

1. BR2 reactor, (http://www.sckcen.be/en/Our-Research/Research-facilities/BR2-Belgian-Reactor-2)
2. OECD/NEA 2002, Accelerator-driven Systems (ADS) and Fast Reactors (FR) in Advanced Nuclear Fuel Cycles, A Comparative Study, ISBN 92-64-18482-1.
3. OECD/NEA 2006, Physics and Safety of Transmutation Systems, A Status Report, ISBN 92-64-01082-3.

4. PATEROS, Partitioning and Transmutation European Roadmap for Sustainable nuclear Energy, FI6W – Contract Number 036418.
5. European Commission, The Sustainable Nuclear Energy Technology Platform, A vision report, EUR 22842.
6. European Commission, Strategic Research Agenda of the Sustainable Nuclear Energy Technology Platform under publication.
7. US DOE Nuclear Energy Research Advisory Committee, A Technology Roadmap for Generation IV Nuclear Energy Systems, December 2002, GIF-002-00.
8. H. Aït Abderrahim & P. Kupschus (2002) "MYRRHA, A Multipurpose Accelerator Driven System (ADS) for Research & Development, March 2002 Pre-Design Report", Report prepared for reviewing of the project by the "International Technical Guidance Committee", SCK•CEN Report R-3595, HAA-PK/svg.32.B043000 85/02-64, March 2002.
9. H. Aït Abderrahim et al (2005), "MYRRHA Pre-Design File – Draft-2", SCK•CEN Report R-4234.
10. H. Aït Abderrahim, D. De Bruyn, et al., MYRRHA Project – Technical description, SCK•CEN Report reference ANS/HAA/DDB/3900.B043000/ 85/07-17bis, April 2007, 57 p.
11. H. Aït Abderrahim, D. De Bruyn, M. Giot, P. Van Doorslaer, F. Hardeman, MYRRHA Project, Multi-purpose hYbrid Research Reactor for High-tech Applications at Mol (Belgium), Business Plan 2007, SCK•CEN report ANS/HAA/DDB/31.B043000/85/07-17.
12. D. De Bruyn, D. Maes, L. Mansani, B. Giraud, From MYRRHA to XT-ADS: the design evolution of an experimental ADS system, Eighth International Topical Meeting on Nuclear Applications and Utilization of Accelerators (AccApp'07), Pocatello (Idaho), United States, 30 July - 2 August 2007, La Grange Park, Illinois, United States, American Nuclear Society, 2007, p. 848-854
13. K. Nishihara et al., MYRRHA update for high performance, SCK•CEN internal report, ANS/KN/3910.B043000 85/08-19.
14. P. Baeten et al., The Central Design Team FP7 proposal, FP7-232527.

HIGH INTENSITY PROTON BEAM PRODUCTION WITH CYCLOTRONS

JOACHIM GRILLENBERGER and MIKE SEIDEL

Paul Scherrer Institut, PSI
5232 Villigen, Switzerland

The cyclotron was invented back in the 1930th and is one of the first resonant accelerator concepts. Cost effective cyclotrons are used today in a broad variety of industrial applications with low and medium beam-powers. However, the resonant acceleration principle and the continuous wave operation mode also permit the production of very intense beams with cyclotrons. In particular the TRIUMF cyclotron with 200 kW beam power and the PSI Ring Cyclotron that generates 1.3 MW are examples of high intensity cyclotron accelerators. In this article we describe basic aspects of cyclotrons optimized for this purpose as well as practical experience gained in the PSI facility.

1. Basic Concepts of Cyclotrons

The cyclotron is a classical example for the principle of resonant acceleration of charged particles. The classical cyclotron consists of two D-shaped hollow electrodes and a magnet producing a vertically oriented bending field. Ions generated in a central ion source are repetitively accelerated by an RF voltage that is applied to the Dee's. First operating cyclotrons were built in the 1930's by Lawrence and Livingston [1]. Disadvantages of the original concept are the monolithic layout and the lack of sufficient vertical focusing.

To overcome these difficulties today's high intensity cyclotrons are realized as modular accelerators that consist of sector magnets for bending and RF resonators to accelerate the beam. Sufficient vertical focusing is achieved by utilizing two effects, the periodic variation of the bending field strength in sector magnets and intermediate gaps, and by edge focusing at magnet boundaries. Edge focusing is enhanced by introducing a spiral magnet shape. The principle of the azimuthally varying field (AVF) cyclotron was proposed by L.H. Thomas in 1938, and it was implemented already in compact cyclotrons by appropriate shaping of the single magnets pole. In a sector cyclotron with spiral magnets the squared vertical betatron frequency is given approximately by the relation:

$$\nu_z^2 \approx -\beta^2\gamma^2 + F(1+2\tan^2\xi);\quad F = \frac{\overline{B^2} - \overline{B}^2}{\overline{B}^2}. \tag{1}$$

Here, F is the relative mean squared variation of the bending field along the circumference, the so called flutter factor and ξ the mean spiral angle of the magnet boundaries. In order to allow continuous acceleration at constant RF frequency (revolution time) the average magnetic field varies with radius as $\frac{r}{B}\frac{\partial B}{\partial r} = \beta^2\gamma^2$. This condition of isochronicity gives rise for the negative first term in (1). Without Thomas- and edge-focusing stable vertical orbits would not be possible. Besides such beam dynamics arguments the sector cyclotron exhibits several technical advantages.

2. The PSI High Intensity Proton Accelerator Facility

2.1. *The accelerators*

The PSI High Intensity Proton accelerator facility consists of a Cockroft-Walton pre-accelerator and a chain of two cyclotrons. A continuous proton beam is pre-accelerated to 870 keV. The beam is then bunched to a 50.63 MHz continuous-wave (CW) structure using a buncher cavity superimposed by a 3rd harmonic to linearize the bunching voltage [3]. The 2.2 mA of protons left after bunching and collimation are further accelerated to 72 MeV in the Injector 2 Cyclotron and then transferred to the Ring Cyclotron. The final beam energy is 590 MeV at a beam current of 2.2 mA in standard operation in 2009. The beam power amounts to 1.3 MW which is at present the highest average beam power generated by any proton accelerator. In both isochronous cyclotrons the revolution time of the particles is kept constant throughout the acceleration process, although their velocity is changed significantly. In consequence, the orbit radii vary strongly, e.g. from 2.1 m to 4.5 m in the PSI Ring Cyclotron. Vacuum chamber and dipole magnets have to accommodate this large lateral variation of the beam position. This fact is often considered as a disadvantage of the cyclotron concept in comparison to a synchrotron. For high power beams it actually presents also an advantage because it allows the individual turns to be separated and to continuously extract the beam from the cyclotron. For the purpose of extraction the electrode of an electrostatic deflector is placed in-between the orbits of the last two turns. Scattering of protons from beam tails in this electrode is the most severe loss mechanism in a high power cyclotron.

2.2. *Targets and experimental facilities*

At PSI the 590 MeV beam is used to produce pions and muons by interaction with two graphite targets. After the second target, with a thickness of 40 mm, a fraction of 30% of beam current is lost because of nuclear reactions, Coulomb scattering, and subsequent collimation. The remaining beam is transferred to the neutron Spallation source SINQ to produce neutrons in a target that contains lead filled Zircaloy tubes. The research based on the PSI proton accelerator covers a broad range of applications involving neutron scattering experiments, muon spin resonance spectroscopy and particle physics experiments. For the purposes of neutron production a high beam power is particularly important to maximize the neutron flux. The total electrical power consumed in the PSI facility is 10 MW when the full beam power of 1.3 MW is produced. From that the overall efficiency of the PSI facility is approximately 13%. At zero beam current but with magnets and RF systems in operation there are still 8 MW drawn from the grid. Many of these magnets and other auxiliary systems contributing to the power balance are not needed for the basic accelerator operation.

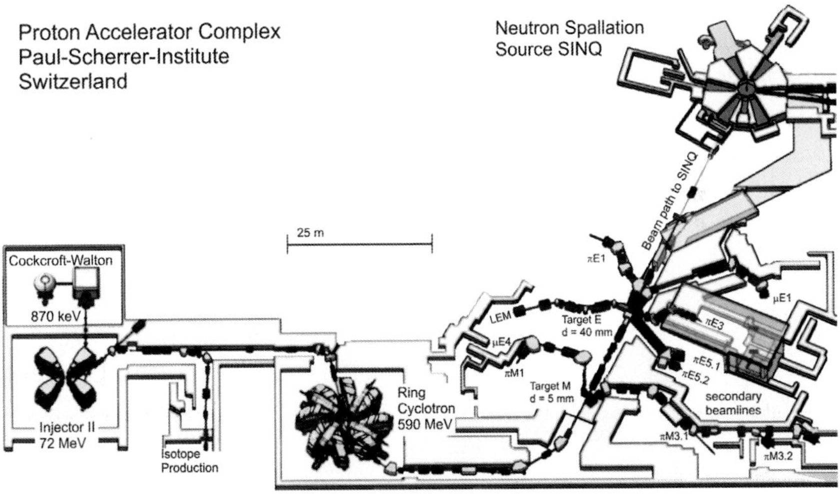

Figure 1: View of the PSI high intensity proton facility including the experimental facilities, i.e. the secondary beamlines for meson production and the neutron Spallation source SINQ.

2.3. *Performance and limitations of the PSI-proton accelerator*

The intensity limitation of the PSI facility is given by the beam losses that cause activation and potentially limit the serviceability of the accelerators.

Intensity upgrades were realized only if the relative beam losses could be lowered in proportion, thus keeping the absolute losses constant. Over the long term history of the accelerator the typical collective radiation dose received by the personnel during maintenance and modification of the accelerator was continuously decreased. With respect to activation the critical components in a high power cyclotron are the extraction elements and the magnets in the subsequent beamline. The electrostatic deflector channel of the PSI Ring Cyclotron employs an inner electrode that consists of 50 μm thick tungsten foils, which are placed between the last and the second last turn. Protons in the beam tails may hit the tungsten foils and as a result of scattering and energy loss these particles are separated from the beam core and hit components in the extraction beamline. The activation levels in this region amount to several mSv/h with a local peak value of 9 mSv/h. Under optimized conditions the losses in the extraction beamline are at the level of $2{\cdot}10^{-4}$ relative to the total beam current. In order to keep those losses at a minimum the generation of beam tails originating from space charge effects is minimized. Joho has argued convincingly that the extraction losses are a strong function of the number of turns in the cyclotron, and in fact they scale with the third power of the number of turns [4]. Under the constraint to keep the absolute losses constant, the beam current was raised in inverse proportion to the third power of the number of turns in the Ring Cyclotron, as shown in Fig. 3b.

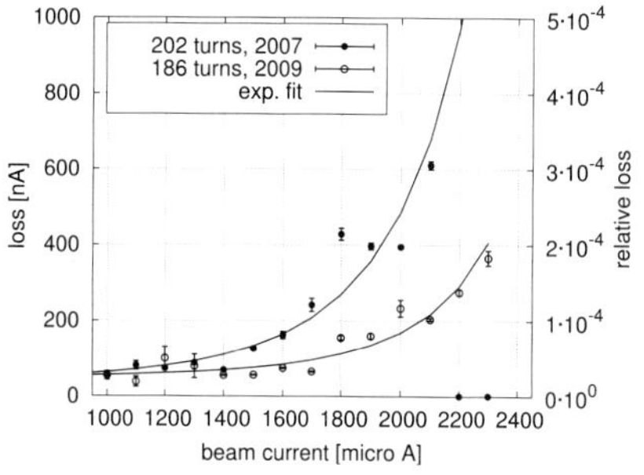

(a)

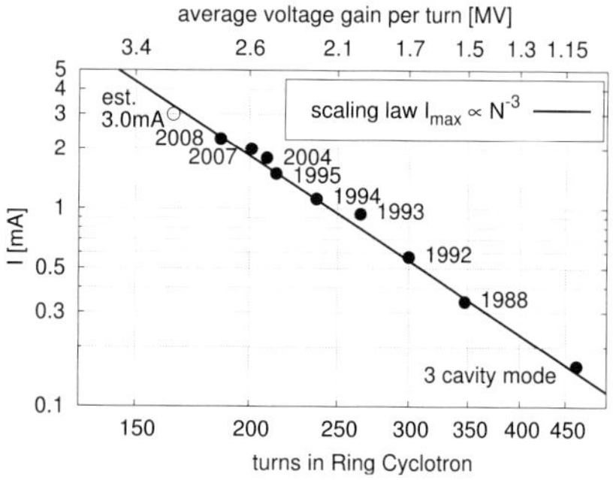

(b)

Figure 3: a) Scaling of the beam losses as a function of current. b) Maximum beam current extracted from the PSI-Ring Cyclotron as a function of the number of turns. The linear fit within the double logarithmic diagram nicely confirms the scaling law $I_{max} \sim n^{-3}$.

Evidently, the number of turns can be decreased by raising the gap voltage of the accelerating structures, i.e. the RF-cavities. Hence, the most important upgrade of the PSI-proton accelerator in order to reach higher beam currents is the installation of more powerful resonators. During the shutdowns between 2004 and 2008 the 4 original aluminium resonators were successively replaced by 4 new structures made of copper. The new resonators allow higher gap voltages and thus a reduction of the turns from 202 to only 186 at present. This has led to a significant reduction in the extraction losses by a factor of 2 (Fig. 3a).

2.4. *Operation statistics and reliability*

The operation of the PSI proton accelerator is characterized by sudden interruptions, with durations ranging from 30 seconds to failures which require repair before restart. Most of the short term interruptions in the PSI-facility are caused by high voltage breakdowns of the electrostatic elements which deflect the injected and extracted beam. Other triggers of the interlock system are intermittent spikes in the loss rates or trips of the RF system. In most cases the system that triggered the interruption is automatically reset and the beam current is ramped up again within 30 seconds. Therefore, the minimum time required for recovery is determined by the ramping procedure even though triggers may only last for a few milliseconds. In order to quantify the trip statistics we have analyzed the run periods in 2007 and 2008 [5, 6].

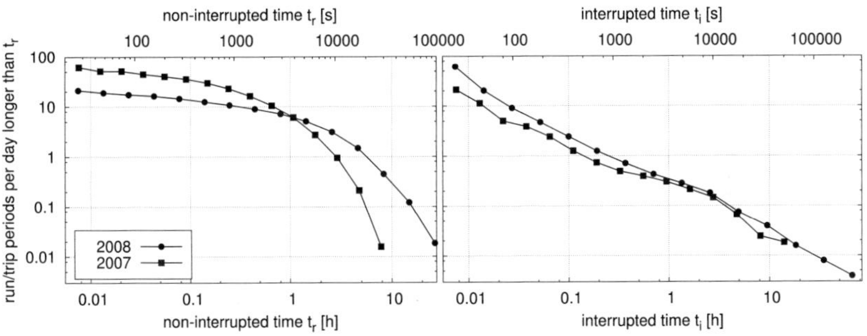

(a) (b)

Figure 4: a) The number of interruptions per day lasting longer than *t* scale inversely in contrast to b) the number of run periods longer than *t* [5].

Fig. 4 shows integrated histograms for the statistical occurrence of run and interruption periods. At the very left end of each graph the total number of interruptions a) respectively runs b) per day can be extracted. The overall availability of the PSI accelerator is defined as the ratio of delivered and scheduled run time. Over the past two years the average availability was 90% which presents a relatively good performance in comparison with other high power accelerators. The total trip rate in 2008 was ~20/d versus ~60/d in 2007.

2.5. *Upgrade to 1.8 MW*

The PSI facility follows an upgrade path to further increase the beam power to an ultimate level of 1.8 MW at a beam current of 3 mA. Since the Ring Cyclotron has now been equipped with more powerful resonators (Fig. 5) the Injector Cyclotron will be upgraded with two additional accelerating resonators [7]. Those resonators will replace the two existing 3rd harmonic flattops. Because a short bunch length was achieved using a recently added 3rd harmonic buncher in the 870 keV injection beam line, the flattop resonators are redundant. Due to strong space charge effects the bunches will stay compact in circular shape [3]. This fact and further reduction of the turn number will enable us, also for this cyclotron, to reduce the losses and to enhance the beam quality. Analog, the installation of a 10th harmonic buncher between the Injector and the Ring Cyclotron will improve the quality of the beam injected into the Ring Cyclotron.

Figure 5: One of the 4 new copper resonators before installation into the Ring Cyclotron. The resonators operate at a frequency of 50.63 MHz and provide a maximum gap voltage of 1.2 MV.

3. Proposal for a 10 MW cyclotron

With 1.3 MW power at a beam energy of 590 MeV the presently operated facility has reached a performance less than one order of magnitude below the useful range for accelerator driven systems (ADS). For a facility with 10 MW beam power the relative beam losses have to be reduced further. This can be achieved by even faster acceleration in the main cyclotron in combination with a larger extraction radius.

In Ref. [2] a proposal has already been worked out for a 1 GeV/10 MW cyclotron which is in principle an up-scaled version of the presently operating

PSI machine. The proposed cyclotron employs 12 sector magnets with 8 embedded accelerating resonators to be operated at peak gap voltages of 1 MV. The projected number of turns would then be around 140. Because of relativistic effects the energy of 1 GeV seems to be the ultimate limit for an isochronous cyclotron. On the other hand, there exist no definite limits for the beam power, although at some point technical challenges will arise from limitations in the power couplers of the accelerating resonators. The limited maximum energy would not pose a problem to an AD system based on a cyclotron because the neutron production rate in a spallation source is not a strong function of energy but scales almost linearly with beam power. In Fig. 6 the concept for a 10 MW/1GeV cyclotron is shown with only slight modifications compared to the original proposal by Stammbach et al. In comparison to the original proposal it is suggested here to use a 10th harmonic buncher in the transfer line between the Injector cyclotron and the booster cyclotron to provide a short bunch for injection into the main cyclotron. This will relax the requirements for 3rd harmonic resonators or even make them obsolete. In the case of nine accelerating cavities the average energy gain per revolution would be ~7 MeV.

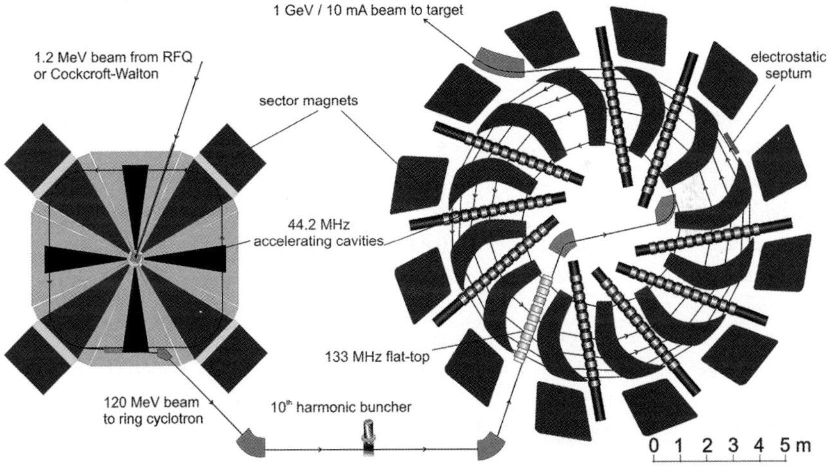

Figure 6: Schematic view of a proton driver based on two isochronous sector cyclotrons. An RFQ would act as a pre-accelerator. A similar proposal for the Ring Cyclotron on the right hand side of the figure was given by Stammbach et al. [2]

The higher energy gain will lead to a large turn separation at extraction in an isochronous field of the 1 GeV cyclotron. In addition, the beam should be accelerated into the fringing field to further widen the turn separation. Large turn separation at the electrostatic elements will improve also their trip rate.

Fault tolerance against RF trips can be achieved by an automated redistribution of the RF power, thereby guaranteeing a constant integral gap voltage even when a single resonator fails. Because of geometrical and technical restrictions the energy gain factor in a sector cyclotron is limited. For the 10 MW facility as shown in Fig. 6 we propose a three stage design, where an injector cyclotron produces a 120 MeV beam for injection into the main cyclotron. In this injector cyclotron four resonators operated at 44.2 MHz and a peak voltage of approximately 600 kV would be sufficient to push the maximum beam current up to 10 mA. Pre-acceleration of the protons will be accomplished by a radio-frequency quadrupole rather than a Cockcroft-Walton as currently used at PSI. Since wall losses scale quadratically with the gap voltage it is more efficient to distribute the total circumference voltage over many resonators. Assuming about 5 MW power for the magnets and auxiliary systems, the power conversion efficiency of the 1 GeV / 10 MW cyclotron would thus be dominated by the RF-system. If the AC to RF conversion is optimized to 75%, the overall efficiency would be around 35%.

Table 1: Parameters for the proposed 1 GeV / 10MW proton driver in comparison to the existing PSI 590 MeV / 1.3 MW Ring Cyclotron.

Parameter	*1 GeV Ring*	*PSI Ring*	*120 MeV Inj.*	*PSI Inj.*
kin. energy	**1000 MeV**	**590 MeV**	**120 MeV**	**72 MeV**
max. current	**10 mA**	**2.2 mA**	**10 mA**	**2.7 mA**
magnets	12	8	4	48
max. field	2.1 T	1.1 T	1.3 T	1.2 T
Cavities	8 (9)	4	4	2
Peak voltage	1 MV	0.8 MV	0.6 MV	0.3 MV
Frequency	44.2 MHz	50.6 MHz	44.2 MHz	50.6 MHz
3rd harm.res.	2(1)	1	--	2
peak voltage	650 kV	460 kV	--	40 kV
harm. number	6	6	10	10
inj. energy	120 MeV	72 MeV	1.2 MeV	0.87 MeV
inj. radius	2.8 m	2.1 m	0.6 m	0.4 m
extr. radius	5.7 m	4.5 m	4.0 m	3.5 m
turns	140	186	68	86
ΔE at extraction	6.3 MeV	2.4 MeV	2.3 MeV	1.4 MeV
Beam power	**10 MW**	1.3 MW	1.2 MW	0.2 MW

4. Summary

The resonant acceleration principle and the continuous operation mode with constant bending fields and RF frequency make the cyclotron concept a good candidate for high intensity ion beam production. With 1.3 MW power at a beam energy of 590 MeV the PSI-facility has reached a respectable performance which can still be improved by further upgrades to 1.8 MW. This is already within the useful range for experimental AD systems. For the cyclotron based facility as well as with other accelerator concepts it will be difficult to achieve the very low trip rates of 0.01 d^{-1} that are desirable for AD systems. Measures like redundancy of vital systems and the operation of components at conservative parameters would improve the situation. As a next step in cyclotron development the design of a facility that provides a 1 GeV/10 MW beam with improved reliability seems feasible.

References

1. E.O. Lawrence, N.E. Edlefsen, On the production of high speed protons, Science 72, 376 (1930)
2. Th. Stammbach, S. Adam, H.R. Fitze, W. Joho, M. Msrki, M. Olivo, L. Rezzonico, P. Sigg, U. Schryber, “the feasibility of high power cyclotrons”, Nuclear Instruments and Methods in Physics Research B 113 (1996) 1-7
3. J. Grillenberger, M. Humbel, J.Y. Raguin, P.A. Schmelzbach, Commissioning and Tuning of the new Buncher System in the 870 keV Injection Beamline, 18th Int. Conf. on Cyclotrons and their Application, Catania (2007) 464
4. W. Joho, Proc. 9th Int. Conf. on Cyclotrons and their Application, Caen (1981)
5. M. Seidel, A.C. Mezger, Performance of the PSI High Power Proton Accelerator, International Topical Meeting on Nuclear Research Applications and Utilization of Accelerators, Vienna (2009)
6. A.C. Mezger, private communication on failure statistics (2009)
7. L. Stingelin, M. Bopp, H. Fitze, Development of the New 50 MHz Resonators for the PSI Injector II Cyclotron, Cyclotrons and Their Applications, Catania (2007) 467

FFAG FOR HIGH INTENSITY PROTON ACCELERATOR

Y. MORI

Kyoto University, Research Reactor Institute,
Kumatori, Osaka 590-0494, Japan
E-mail: mori@rri.kyoto-u.ac.jp
www.rri.kyoto-u.ac.jp/

Fixed Field Alternating Gradient accelerators(FFAGs) have been developed for various applications recently. Contrary to the ordinary synchrotron, FFAGs have a large capability of accelerating high current beams because of strong beam focusing to have a large beam acceptance and static magnetic field which allows to have very fast beam acceleration and large repetition rate in operation. In this paper I will present the expected characteristics and possible performance of FFAG as the proton drivers for ADSR and muon source, and also show the recent preliminary experiment on ADSR with FFAG proton accelerator complex at Kyoto University Research Reactor Institute (KURRI).

Keywords: FFAG;ADSR

1. Introduction

The idea of FFAG accelerator was originally invented by Ohkawa in 1953[1] and also by Symon and Kolomenskii, independently. Electron models of FFAG accelerator have been built by Kerst, Cole and Symon at MURA in the early 1960s.[2] However, since then, no FFAG proton accelerator has been built until recently. A proton FFAG accelerator, in particular, has some severe technical difficulties such as its complicated configuration of magnetic field and difficulty of rf acceleration. In 2000, the world's first proton FFAG accelerator (POP-FFAG) was demonstrated using a novel broad-band RF cavity with high permeability MA(magnetic alloy) cores at KEK.[3] Following this success, it has been recognized that FFAGs have large advantages in rapid acceleration with large momentum acceptance, which are useful both for muon accelerators and for high power proton drivers. Since then, intensive studies and discussions have been made, and various works for the development of FFAGs have also started for many applications at many institutes. The FFAG accelerator has distinctive features compared with other types of accelerators in beam focusing and acceleration. In transverse beam focusing, two types of beam focusing can be categorized: scaling and non-scaling. The scaling type of FFAG accelerator has non-chromatic beam optics in which a large amount of non-linear field is essential. In the non-scaling type of FFAG accelerator, fields are linear. However, a number of betatron resonances have to be crossed during acceleration. As for beam acceleration, various schemes with new ideas such as broad-band MA rf cavity, stationary bucket acceleration, serpentine acceleration and harmonic number jump acceleration have been studied intensively and some of them have been realized.

2. Beam focusing and acceleration in FFAG accelerator

2.1. *Beam focusing*

2.1.1. *Scaling*

In the scaling type of FFAG accelerator, each beam orbit for different beam momentum has similarity in shape (curvature), and zero chromaticity in the beam optics is realized. Thus, the betatron tunes for both horizontal and vertical directions keep constant during beam acceleration without any problems caused by resonance crossing. The transverse linearized betatron equation of motion can be expressed by the following equation.

$$\frac{d^2X}{ds^2} + K(s)X = 0, \tag{1}$$

where s is the distance along the reference trajectory and K(s) can be given for horizontal (x) and vertical (z) directions by,

$$K_x = \frac{1 - n(s)}{\rho^2(s)} \tag{2}$$

$$K_z = -\frac{n(s)}{\rho^2(s)}, \tag{3}$$

respectively. Here n is a field index,

$$n = -\frac{\rho}{B}. \tag{4}$$

The scaling condition requires that the equation of motion of eq.(1) is independent of momentum both for horizontal and vertical directions, respectively. In cylindrical coordinates,

$$r = \rho$$
$$\frac{r}{B}(\frac{B_z}{z})_{z=0} = k. \quad (5)$$

It can be shown analytically that the configuration of the magnetic field at the median plane can be expressed in the following equation including azimuthal direction.[4]

$$B(r,\theta) = B_0(\frac{r}{r_0})^k f(\theta - \zeta ln\frac{r}{r_0}) \quad (6)$$

where ζ=tanξ, and ξ is a spiral angle of the magnet in the azimuth plane. Accordingly, two schemes for beam focusing are invoked by this magnetic field configuration: one is radial sector focusing and the other spiral sector focusing. Radial sector focusing uses a combination of positive and negative bending magnets to make strong beam focusing with a FODO lattice configuration. In spiral sector focusing, edge focusing is used efficiently.

2.1.2. *Non-scaling*

The non-scaling type of FFAG accelerator has a basically simple structure composed of linear optical elements such as dipole and quadrupole magnets. The transverse betatron tune, however, is not constant during acceleration and fast acceleration is essential in the non-scaling type of FFAG accelerator to avoid beam losses caused by resonance crossing. To ensure this, the RF cavity phases are fixed for beam acceleration and the range of times-of-flight over the energy range of the accelerator is minimized when the path length difference has a parabolic dependence on beam momentum. In Fig. 1, a typical cell structure of a non-scaling FFAG accelerator are presented.[5] Based on the concept of linear non-scaling FFAG accelerators, some variations such as isochronous or semi-achromatic lattice [6,7] including small non-linear elements(semi-scaling) have also been proposed.

2.1.3. *Advancements of scaling FFAG optics*

Scaling law of FFAG accelerator to satisfy the zero chromaticity is used to be applied to circular accelerator as described above and each cell provides a total bending angle of a ring. If a scaling FFAG unit cell with straight line where a overall bending angle is zero is created, insertion/matching optics with straight lines can be cooperated to the ordinary FFAG ring. The

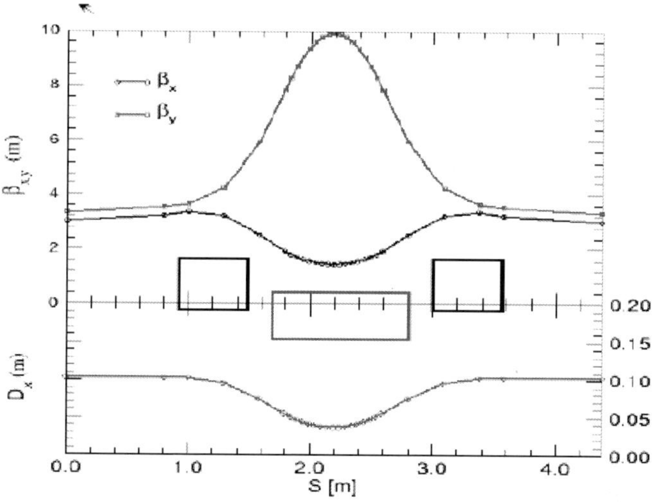

Fig. 1. Typical cell structure of a non-scaling FFAG accelerator are presented.

scaling condition for the straight line unit cell, i.e. a same phase advance per cell at every energy, leads to a different field law as shown below.

$$\frac{d(\frac{1}{\rho^2})}{dp} = 0 \tag{7}$$

$$\frac{d(K\rho^2)}{dp} = 0 \tag{8}$$

Geometrical similarity is given by $\rho = const.$ and it leads to a solution for magnetic field configuration.[8]

$$B_z = B_0 \exp^{\frac{n}{\rho}(X - X_0)} \tag{9}$$

With a scaling FFAG straight line, insertion/matching optics to the ordinary scaling FFAG rings have been examined for various applications.[9,10] Figure 2 shows one of the examples for the dispersion suppressing insertion of the muon FFAG ring which is designed to accelerate muons up to 10GeV.[10]

2.2. *Acceleration*

One of the most distinctive features in FFAG accelerators compared with ordinary strong focusing synchrotrons is that very fast acceleration can be allowed because the magnetic field is static. Accordingly, various new RF

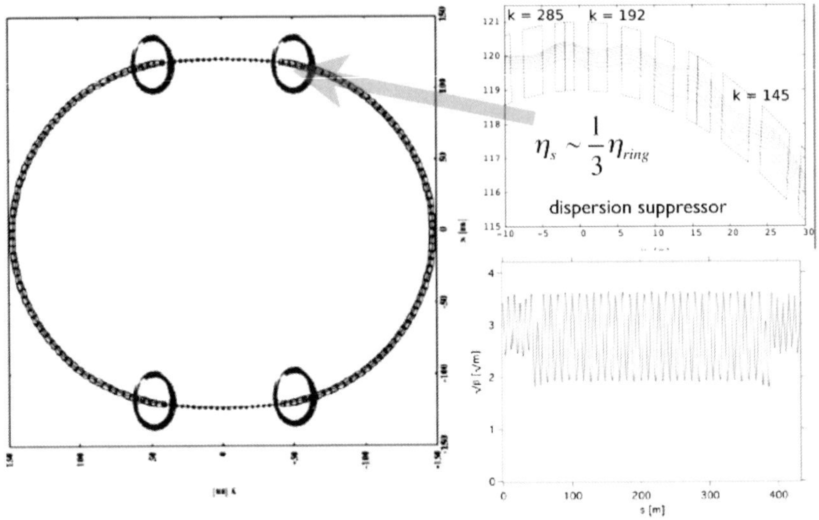

Fig. 2. Scaling FFAG for muon acceleration with dispersion suppressing insertions.

acceleration schemes, which are adequate either or both for scaling and non-scaling FFAG accelerators, have been proposed for fast acceleration, and some of them have already been realized.

2.2.1. *Broad-band RF cavity with MA cores*

Broad-band RF cavities with high permeability magnetic alloy (MA) cores have been used mostly in scaling type of proton FFAG accelerators such as the POP-FFAG developed at KEK. No frequency tuning synchronized with beam revolution is needed and a very rapid acceleration cycle of several 100Hz to 1kHz becomes possible, however, above than that, it seems rather difficult with the practical rf voltage obtained by this type of rf cavity.

2.2.2. *Stationary RF bucket acceleration*

In longitudinal phase space, RF bucket of the scaling type of FFAG accelerator for relativistic particles such as high energy muons is not distorted for large momentum range because the momentum compaction keeps constant and has no momentum dependence. If the RF voltage is large enough, the particles can be accelerated in a stationary RF bucket after a half period of synchrotron oscillation.[11]

2.2.3. *Serpentine acceleration*

In the non-scaling FFAG accelerator, a fixed frequency RF system is installed and acceleration out of longitudinal buckets can be possible for relativistic particles when the path length difference has a parabolic dependence on momentum as described above. The revolution time (time-of-flight per turn) changes and has a minimum during beam acceleration. Figure 3 shows an example of serpentine acceleration.[12]

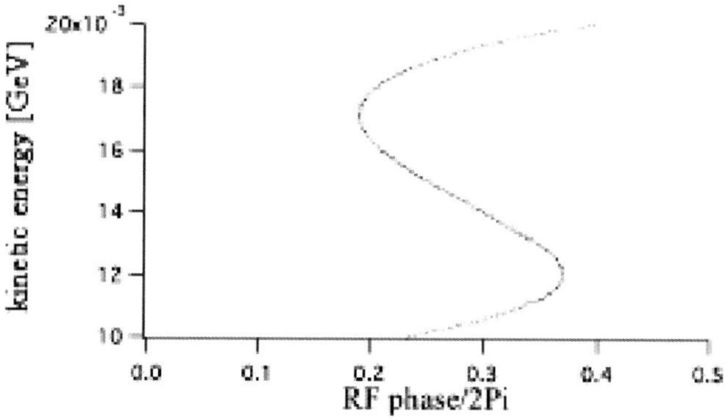

Fig. 3. Serpentine acceleration in non-scaling FFAG

In the scaling FFAG accelerator, the similar acceleration scheme with a fixed frequency RF is possible where the beam is accelerated passing through the transition energy where the time-of-flight per turn can be minimum. The longitudinal Hamiltonian in the case of scaling FFAG accelerator can be analytically obtained as,

$$\frac{H}{m_0c^2 f_s} = 2\pi\left[-\frac{A}{2(1-\lambda)}(\gamma^2-1)^{1-\lambda}+\gamma\right]+\frac{eVh}{m_0c^2}cos\phi. \tag{10}$$

Here,

$$\lambda = \frac{k}{2(k+1)} \tag{11}$$

$$A = \frac{(\gamma_s^2-1)^{\lambda}}{\gamma_s} \tag{12}$$

Figure 4 shows an example of the phase space contour for the longitudinal motion of the scaling FFAG accelerator. As can be seen from this figure, the serpentine path through the transition energy exists and beam acceleration for non-relativistic particles can be possible with this scheme in scaling FFAG accelerator.

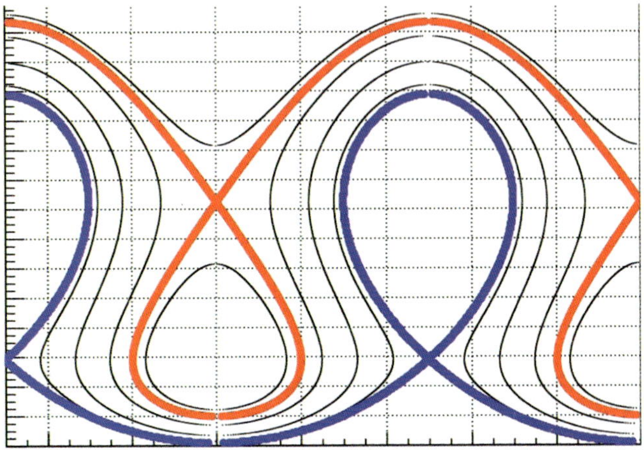

Fig. 4. Longitudinal Hamiltonian contour for serpentine acceleration in scaling FFAG accelerator.

2.2.4. *Harmonic number jump*

In FFAG accelerators, the transverse beam positions change during acceleration. If the energy gain varies properly with transverse position to achieve an integer change in the harmonic number with each turn, the beam can be accelerated with a fixed frequency RF system.[13] The difficulty of this scheme is how to vary the energy gain properly and ideas such as a position dependent RF voltage configuration have been proposed to overcome this difficulty.

3. FFAG for high intensity proton accelerator

Next generation of high intensity proton accelerator (HIPA) is, needless to say, a proton driver for producing an order of magnitude more intense secondary particles such as neutron, muon, meson etc. compared with the present HIPA such as J-PARC, SNS and ISIS. Obviously, the beam energy of next generation HIPS should be more than 1GeV because the nuclear interaction generating hadron shower can be very efficient there and the beam power of more than 10MW is now under discussion. Requirements for the time structure of the beam from next generation HIPS, however, have so many varieties which depend on each application. Neutrino factory/muon collider requires a very short beam bunch (nsec) and low repetition rate operation (10Hz), on the other hand, ADSR requires very large repetition rate of more than kH or even cw operation. It must be favorable that next generation HIPS should satisfy these various requirements as much as possible.

The types of accelerator for next generation of HIPA, from this point of view, are limited. Ordinary synchrotron and cyclotron could be excluded as next generation of HIPS. The former is difficult for high repetition operation and the latter for short bunched and high peak intensity pulse operation. The candidates of the next generation HIPS could be a super conducting H- linac + proton storage ring (LINAC+PSR), and a FFAG.proton accelerator. The FFAG accelerator could be operated in various repetition pulsed mode and even in cw mode as described in the previous section.

Obviously, when the beam repetition is less than 1kHz, the braod-band RF system as described in the previous section can be used for either scaling or non-scaling FFAGs. However, in order to realize cw operation, fixed-frequency RF acceleration scheme is essential and, in particular, for the proton driver where the particle velocity is still non-relativistic, a scaling FFAG accelerator with serpentine acceleration seems to be a good candidate. Table 1 presents an example of parameter list of proton driver based on scaling FFAG accelerator using serpentine beam acceleration shown in Fig. 5.

Table 1. Parameters of an example of proton driver based on scaling FFAGs with serpentine acceleration

Item	Value
average radius(m)	15
field index	3
injection energy	300MeV
extraction energy	2.2GeV
RF voltage per turn (h=1)	38MV

4. First ADSR experiment with FFAG accelerator at Kyoto University Research Reactor Institute (KURRI)

The Research Reactor Institute of Kyoto University started the KART (Kumatori Accelerator-driven Reactor Test facility) project in fiscal year

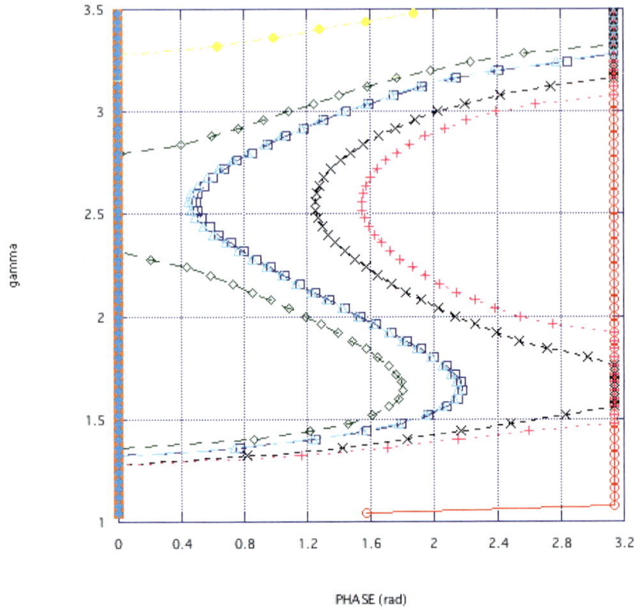

Fig. 5. Serpentine acceleration for HIPA(0.3-2.2GeV) with scaling FFAG accelerator.

2002 under the Contract with the Ministry of Education, Culture, Sports, Science and Technology of Japan and in March of 2009,[14] the first experiment of Accelerator-Driven Sub-critical Reactor system (ADSR) with the Kyoto University Critical Assembly (KUCA) has been successfully carried out. The purpose of this research project is to demonstrate the basic feasibility of accelerator-driven system (ADS), studying the effect of incident neutron energy on the effective multiplication factor in a subcritical nuclear fuel system. For this purpose, a variable-energy FFAG (Fixed Field Alternating Gradient) accelerator complex was developed and constructed, and coupled with the Kyoto University Critical Assembly (KUCA).[15] On March 4th, 2009, the world's first injection of spallation neutrons generated by the high-energy proton beams into a reactor core was successfully accomplished. By combining the Fixed Field Alternating Gradient (FFAG) accelerator (Fig. 6) with the A-core (Fig. 7) of KUCA, a series of ADSR experiments was carried out under the condition that the spallation neutrons were supplied to a subcritical core through the injection of 100 MeV protons onto a tungsten target, whose size was 80 mm diameter and 10 mm thickness.

Fig. 6. FFAG accelerator complex at KURRI.

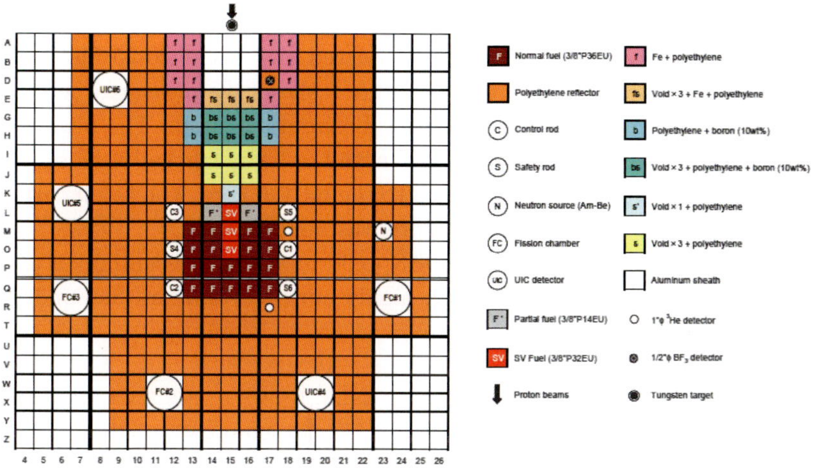

Fig. 7. A-core of KUCA at KURRI.

The accelerator complex is composed of three FFAG rings; injector, booster, and main ring.[16] The specifications of the each FFAG are summarized in Table1. As a first stage, the accelerator complex has been planned to output 100 MeV-0.1 nA proton beams. The injector is a spiral type of scaling FFAG accelerator which is composed of eight spiral sector magnets, two acceleration gaps, and electric septa for injection and extraction, respectively. The field distribution in the radial direction can be controlled by trim-coils, and which makes variable energy acceleration possible. Maximum beam energy is 2.5 MeV in design. The booster FFAG is a radial sector type of scaling FFAG accelerator, which adopts multi-turn beam injection using horizontal space. The beam repetition rate is 60 Hz. For such a rapid-cycling accelerator, adiabatic capture cannot be used because relatively long beam capture time compared to the period of the synchrotron oscillation is required. Thus, fast longitudinal matching with bunch rotation method was tried. Saw-tooth rf was employed in order to minimize the filamentation. The main ring is also a radial-sector type of scaling FFAG and its design is essentially same with the one that was developed at KEK in 2004. Proton beam of 11.6 MeV was injected in the main ring. The injected beam is kicked into the closed orbit by an electrostatic septum and a magnetic kicker. Figure 5 shows the circulating beam picked up by an electrostatic bunch monitor in the main ring. The betatron-phase advance between the kickers is 550 deg, so that the kickers work on phase. This means that the strengths of them is complementary. The beam was accelerated with the rf gap-voltage of 2.5 kV and the synchronous phase of 30 degree. A rapid beam-loss was observed at 4 ms at 20-25 MeV. The beamloss is related to ν_x - 2 ν_y=1. Figure 8 shows the measured betatron tune trajectory in the main ring during beam acceleration. The beam was successfully accelerated up to 100MeV with the beam current of more than 0.1nA and extracted and transported to the KUCA core. The beam current at the target placed in the KUCA core was about 10pA.

The A-core employed in the ADSR experiments was essentially a thermal neutron system composed of a highly enriched uranium fuel and the

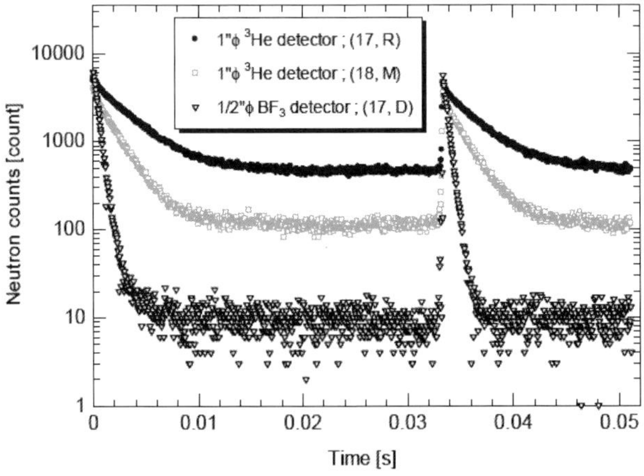

Fig. 8. Measured betatron tune trajectory during beam acceleration in the main ring.

polyethylene moderator/reflector. In the fuel region, a unit cell is composed of a 93 % enriched uranium fuel plate 1/16" thick and polyethylene plates 1/4" and 1/8" thick. In these ADSR experiments, three types of fuel rods designated as the normal, partial and special fuel were employed. From the reason of the safety regulation for KUCA, the tungsten target was located not at the center of the core but outside the critical assembly, and an outside-location was similar to the previous experiments using 14 MeV neutrons. To obtain the information on the detector position dependence of the prompt neutron decay measurement, the neutron detectors were set at three positions shown in Fig. 6: near the tungsten target (Position of (17, D)experimentally confirmed by observing the time evolution of neutron density in ADSR: an exponential decay behavior and a slowly decreasing one, respectively. These behaviors clearly indicated the fact that the neutron multiplication was caused by an external-source: the sustainable nuclear chain reactions were induced in the subcritical core by the spallation neutrons through the interaction of the tungsten target and the proton beams from the FFAG accelerator. In these kinetic experiments, the subcriticality was deduced from the prompt neutron decay constant by the extrapolated area ratio method. The difference of measured results of 0.74 % Δ k/k and 0.61 % Δ k/k at the positions of (17, R) and (18, M) in Fig. 2, respectively, from the experimental evaluation of 0.77 % Δ k/k, which was deduced from the combination of both the control rod worth by the rod drop method and its calibration curve by the positive period method, was within about 20 % . Note that the subcritical state was attained by a full insertion of C1, C2 and C3 control rods into the core. The generation of the spallation neutrons was included in the MCNPX calculation bombarding the tungsten target with 100 MeV proton beams. Since the reactivity effect of the In wire is considered to be not negligible, the In wire was taken into account in the simulated calculation: the reaction rates were deduced from tallies taken in the In wire setting region. The result of the source calculation was obtained after 2,000 active cycles of 100,000 histories, which led the statistical error in the reaction rates of less than 10 % . The measured and the calculated reaction rate distributions were compared to validate the calculation method. The calculated reaction rate distribution (Fig. 3) agreed approximately with the experimental results within the statistical errors in the experiments, although these experimental errors were rather larger than those of the calculations. These larger errors in the experiments were attributed to the current status of the proton beams, including the weak beam intensity and the poor beam shaping at the target.

5. Summary

Beam focusing and acceleration in Fixed Field Alterenating Gradient Accelerator (FFAG) have been briefly described and a possible concept of FFAG accelerator for next generation High Intensity Proton Accelerator (HIPA) is discussed. The first experiment of ADSR with FFAGs at KURRI is also presented.

References

1. T.Ohkawa, Proc. annual meeting of JPS(1953).
2. K.R.Symon et al., Phys. Rev 103,,1837(1956).
3. Y.Mori, Proc. EPAC1998,,289(2000).
4. A.A.Kolomenski, "Theory of Circular Accelerators", 340(1966).
5. D.Trbojevic, Proc. of the International Workshop on FFAG Accelerators (FFAG05), Osaka, Japan(2005).
6. G.Rees, Proc. of the International Workshop on FFAG Accelerators (FFAG04), Tsukuba, Japan, 2004,page 77.
7. A.G. Ruggeiero, Proc. of the International Workshop on FFAG Accelerators (FFAG05), Osaka, Japan, 2005, page 47.
8. JB.Lagrange, Proc. of PAC09, FR5PFP002, Vancouver, 2009.
9. Y.Mori, talk at the International Workshop on FFAG Accelerators (FFAG09) at FNAL, Chicago, USA, 2009.
10. T.Planche, Proc. of PAC09, FR5PFP003, Vancouver, 2009.
11. Y.Mori, talk at the International Workshop on FFAG Accelerators (FFAG2006) at FNAL, Chicago, USA, 2006.
12. S.Machida, Proc. of the International Workshop on FFAG Accelerators (FFAG05), Osaka, Japan, 2005, page 68.
13. S.Berg, Proc. of the International Workshop on FFAG Accelerators (FFAG05), Osaka, Japan, 2005, page 93.
14. K.Mishima et al, "Research Project on Accelerator-driven Subcritical System Using FFAG Accelerator and Kyoto University Critical Assembly" , J. Nucl. Scie. Technol., 44, 499(2007).
15. C.H.Pyon et al., "First Injection of Spallation Neutrons Generated by High-Energy Protons into the Kyoto University Critical Assembly", J. Nucl. Scie. Technol. rapid communication, (2009).
16. Y. Mori, "Development of FFAG accelerators and their applications for intense secondary particle production", Nucl. Instrum. Methods, 562, 591, (2006).

Working Group 1

SRF Linac for Neutrino, Muon and Kaon Physics

Conveners

David Jaffe (Brookhaven)

Robert Tschirhart (Fermilab)

KAON YIELDS FOR 2 TO 8 GEV PROTON BEAMS*

K. K. GUDIMA‡, N. V. MOKHOV†, S. I. STRIGANOV

Fermilab, MS 220, P.O. Box 500
Batavia, IL 60510, U.S.A.

Production rates of kaons and accompanying particles from nuclear targets are modeled with LAQGSM and MARS15 for low-energy proton beams.

1. Introduction

The next generation of rare-decay experiments – to operate at the intensity frontier – requires kaon beams of extraordinary quality. The proton kinetic beam energy (T_p) threshold for producing kaons is 1.7 GeV (on protons) and the kaon yield fraction grows with the increasing number of exclusive production channels that open and saturate around T_p of 6 GeV. The research potential of recently proposed accelerator architecture at Fermilab, which is based on a 2-GeV CW proton linac, turns to be very high [1]. It is strongly linked to energy dependence of kaon production rates in this near-threshold region. The rates of kaons and accompanying particles from light nuclear targets at $T_p < 10$ GeV are studied in this paper with the improved LAQGSM/MARS15 codes.

2. LAQGSM Event Generator

The Los Alamos Quark-Gluon String Model code (LAQGSM) [2] is an improved version of the time-dependent Dubna intranuclear Cascade Model, combined with isobar model at $E < 4.5$ GeV and Quark Gluon String Model (QGSM) for elementary interactions inside nucleus at $E > 4.5$ GeV, coupled with the Fermi breakup, coalescence and generalized evaporation-fission models. It has recently been enhanced in the intermediate 2 to 8 GeV region of interest for the Project X experiments. It is used in the MARS15 code [3] for photon, hadron

---

* Work supported by Fermi Research Alliance, LLC under contract No. DE-AC02-07CH11359 with the U.S. Department of Energy.

‡ Visiting from Institute of Applied Physics, Academy of Sciences of Moldova.

† E-mail: mokhov@fnal.gov

and heavy-ion projectiles at a few MeV/A to about a few TeV/A, providing a power of full theoretically consistent modeling of exclusive and inclusive distributions of secondary particles, spallation, fission, and fragmentation products in realistic experimental setups. In the LAQGSM code, K, Λ, and Σ are produced via all meson-baryon (MB) and baryon-baryon (BB) interactions. With a focus on low-Z targets in this study, special efforts were put on model improvements in case of a deuterium target. The momentum distribution of nucleons inside deuteron is taken in the form $N(q)dq = Cq^2dq/[\alpha^2 + q^2]^{\beta}$, fitting experimental data.

3. Benchmarking

The code predictions on kaon production in the near-threshold region on deuterium and carbon targets have been benchmarked against experimental data from the ANKE spectrometer at COSY-Julich [4]. The results are shown in Fig. 1 with absolute predictions by LAQGSM. The agreement is amazingly good.

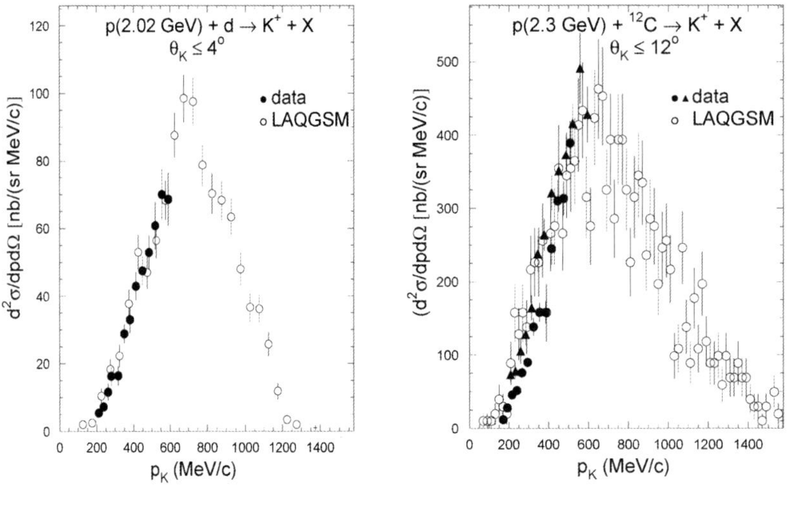

Figure 1. Double differential cross sections of produced K^+ mesons in interactions of protons with deuterium (left) and carbon (right) nuclei.

4. Kaon Production Optimization

LAQGSM calculations reveal that in pp and pA interactions, meson and hyperon yields Y grow rapidly with proton energy T_p starting from the threshold energies to a transition energy of about 2 GeV (pions) and 6 GeV (kaons and hyperons), slowing down at higher energies and reaching a logarithmic growing mode at about 12-15 GeV. As a result, the yield over proton energy, Y/T_p, has a broad maximum at the above transition energies. The kaon yield fraction grows with T_p and saturates at T_p ~ 6 GeV. Fig. 2 shows energy dependence of Y/T_p for a carbon target and for various angular and momentum regions of interest for the kaon program [1]. It was also found that at the considered T_p, the yield of kaons with momentum p>0.3-0.4 GeV/c is higher from low-Z targets. At the same time, the Fermi-motion in nuclei makes the kaon production threshold energy lower compared to the hydrogen's one of 1.7 GeV. As found in MARS15(LAQGSM) calculations for the case considered, the appropriate target materials range from deuterium to carbon and the effect of secondary interactions in a thick target is rather minor.

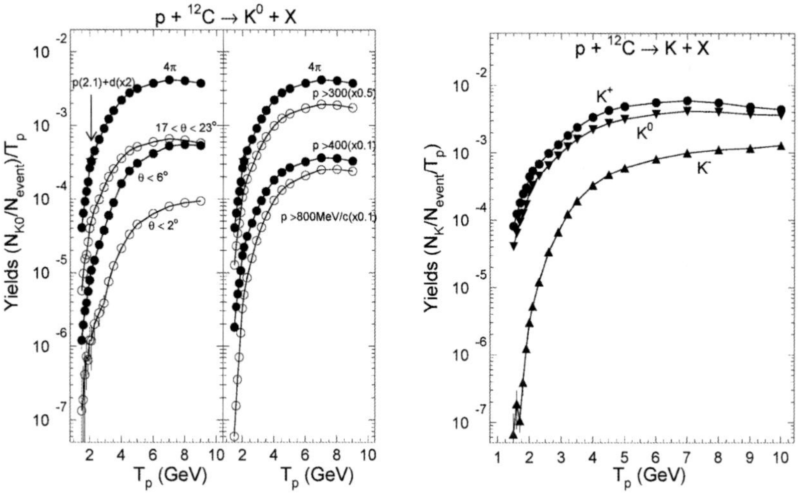

Figure 2. Energy dependence of kaon yields per incident proton kinetic energy T_p for different angle and momentum windows (left) and total yields of kaons (right) produced by proton in carbon target.

The K^+ and K^0 yields become quite interesting at beam energy T_p ~ 2 GeV, while a modest increase of proton energy to T_p ~ 2.5 GeV makes the entire kaon program – from a particle production standpoint – definitely feasible. Some loss in the Y/T_p ratio from not going to a higher energy of 5-6 GeV can easily be compensated by a beam power of the CW linac [1].

Momentum and angular distributions of kaons, pions and neutrons generated on a deuterium target by a 3-GeV proton beam are shown in Fig. 3. Neutral kaon distributions coincide with those for K^+ (not the case for hydrogen target), while anti-K^0 yield is 20-30 times lower. There are practically no angular dependence of the kaon momentum distributions at $\theta < 15$deg at $T_p < 3$ GeV, while it is quite substantial at higher proton energies. At $T_p = 8$ GeV, the K^0 yields on a deuterium target are slightly lower than the K^+ ones and the anti-K^0 yield is at a 30% level of those for K^+ and K^0. These results, generated with MARS15(LAQGSM), can be used in designing the experimental setup and planning the experiment.

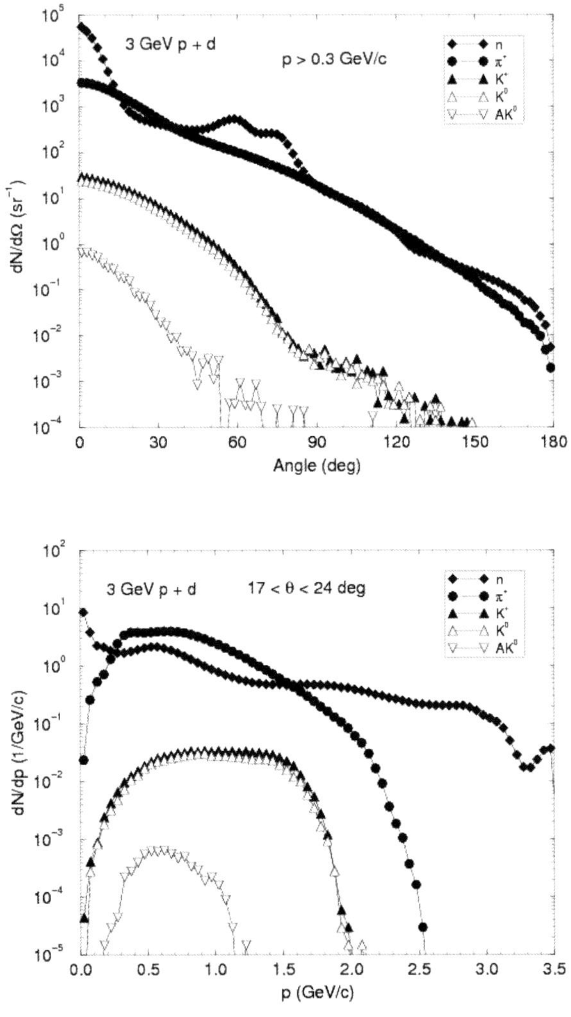

Figure 3. Neutron, pion and kaon angular (left) and momentum (right) distributions calculated with MARS15(LAQGSM) for 3-GeV proton-deuteron inelastic interactions.

References

1. Report from ICD-2 Research Program Task Force, Oct. 2009, http://www.fnal.gov/pub/projectx/pdfs/.
2. S.G. Mashnik, K.K. Gudima et al., LANL Report LA-UR-08-2931 (2008); arXiv:0805.0751v1 [nucl th] 6 May 2008.
3. N.V. Mokhov, Fermilab-FN-628 (1995); N.V. Mokhov, S.I. Striganov, "MARS15 Overview", in Proc. of Hadronic Shower Simulation Workshop, Fermilab, September 2006, AIP Conf. Proc. **896**, pp. 50-60 (2007); http://www-ap.fnal.gov/MARS/.
4. M. Buscher, V. Koptev et al., arXiv:nucl-ex/0401031v2 (2004).

PION YIELD STUDIES FOR PROTON DRIVER BEAMS OF 2-8 GeV KINETIC ENERGY FOR STOPPED MUON AND LOW-ENERGY MUON DECAY EXPERIMENTS

S. I. STRIGANOV*

Fermilab, Batavia, IL 60510, USA
E-mail: striganov@fnal.gov
www.fnal.gov

Energy dependence of slow pion yield in proton-nucleus interactions is analyzed. Experimental data of the HARP collaboration is fitted using two-fireball model with $\chi^2/n_d \sim 1$ from 3 to 8 GeV/c. It is shown that low momentum negative pion yield rises almost linearly with proton kinetic energy. Normalized low momentum positive pion yield is larger at lower proton energy.

Keywords: pion production; proton-nucleus interaction.

1. Introduction

There are a variety of experiments using intense muon sources and offering the possibility of probing New Physics in ways complementary to the collider program. Most of such experiments require a source of stopped muons, and thus require an intense source of low energy (momentum $\leq$ 100 MeV/c) muons, which in turn originate from low energy (momentum $\leq$ 200 MeV/c) pions. The two accelerator designs (with beam energies at 2 and 8 GeV) are under consideration at Fermilab. Estimate of the slow pion yields at these energies is important to explore the opportunities presented by above designs.

2. Experimental data

Very few pion production measurements in the challenging region of 2-8 GeV and wide angle range are reported in literature. HARP collaboration [1] presented comprehensive data set of double differential $\pi^{\pm}$ cross sections in the range of momenta from 100 to 800 MeV/c and angles from 0.35 to 2.15 radian. Recently low angle (from 0.025 to 0.25 radian) measurements [2] in the range of pion momenta from .5 to 8 GeV/c were published.

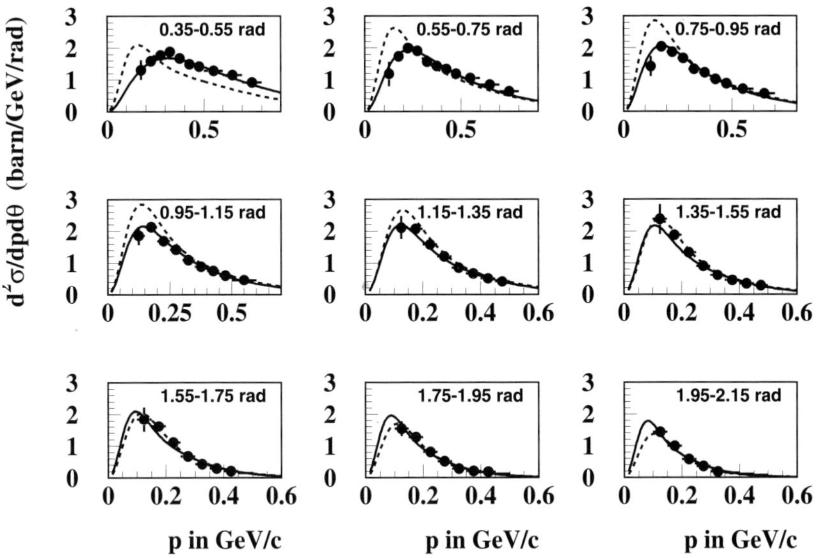

Fig. 1. Cross section of π^- production in proton-tantalum interaction at 8 GeV/c. Symbols – HARP data, full lines – two-fireball fit, dashed lines – FANCY fit.

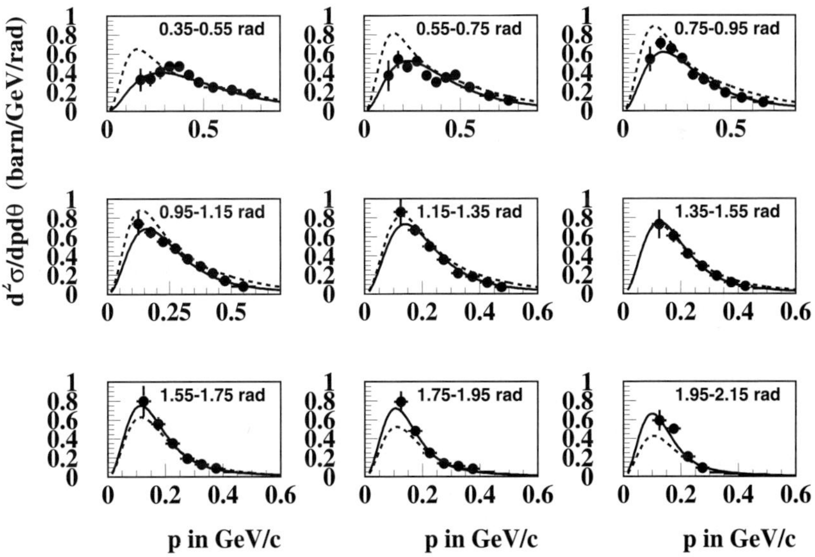

Fig. 2. Cross section of π^- production in proton-tantalum interaction at 3 GeV/c. Symbols – HARP data, full lines – two-fireball fit, dashed lines – FANCY fit.

In both cases the results were obtained mainly at four incident proton beam momenta (3,5,8 and 12 GeV/c). The HARP collaboration performed comparisons with publicly available Monte Carlo simulations: GEANT4 [3] and MARS15 [4]. None of the considered models describe fully HARP data.

Negative pion production has been measured at KEK using a FANCY spectrometer [5]. The spectrometer covers angular and momentum ranges of 0.64-1.57 radian and 0.2-0.9 GeV/c, respectively. The measurements were performed for 3 and 4 GeV/c proton beam and aluminum/lead targets. In order to take into account the different angular binnings in HARP and FANCY results, which prevent direct comparison, a two-moving-source fit is using to present the FANCY measurements. To take into account different beam momenta, FANCY fit was multiplied by a ratio of the beam kinetic energies. As shown in Figs. 1–2 FANCY fit is in a reasonable agreement with HARP data at large angles where FANCY measurements were performed.

3. Two-fireball parameterization

Experimental data of the HARP collaboration were fitted using a two-component model

$$E\frac{d^3\sigma}{dp^3} = \sum_{i=1}^{2} a_i(1+\alpha_i z)exp\{-\frac{T(1-\beta_i z)}{T_i}\}, \tag{1}$$

where T is pion kinetic energy, z is cosine of pion azimuthal angle, $a_i, \beta_i, T_i, \alpha_i$ are model parameters. HARP large angle data [1] are fitted with $\chi^2/n_d \sim 1$ for charged pion production in proton-tantalum and proton-carbon interactions at 3 and 8 GeV/c. Accuracy of the fit becomes worse when low angle data [2] at 8 GeV/c were included ($\chi^2/n_d \sim 3$). Statistical errors of the data are large at low angles and 3 GeV/c proton momentum, so, including this data does not change the fit quality. Comparison of the fit with HARP data is presented in Figs. 1–2.

4. Thick target effects

A pion production target in low energy muon experiments usually is long enough (about 1.5 interaction length) with a small radius. Secondary/tertiary interactions and ionization energy losses could modify pion spectra in comparison with proton-nucleus interaction. MARS simulation shows that track length distribution of low energy pions has maximum near target radius. In the gold target with 3 mm radius, pions lose only about 10 MeV before they reach the target surface.

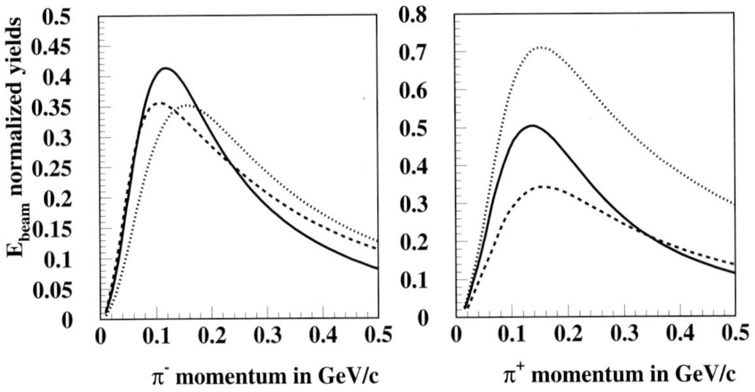

Fig. 3. Normalized pion yield in proton-nucleus interactions. Full lines – pTa at 3 GeV/c, dashed lines – pTa at 8 GeV/c, dotted lines – pC at 3 GeV/c.

In the interactions of 8 GeV/c protons with a 16 cm gold target, 20-30 % of low energy negative pions are produced in secondary/tertiary interactions. But a large part (about 50%) of these interactions is quasi-elastic negative pion scattering. So, low energy pion yield from the "thick" target is within $\sim$ 10 % proportional to low energy pion yield in proton-nucleus interactions.

5. Conclusion

Charged pion momentum distributions (divided by primary proton kinetic energy) at 3 and 8 GeV/c are presented in Fig. 3. They were obtained by angular integration of the two-fireball parameterization. It is seen that energy dependence of slow negative pion production from heavy (tantalum) target is almost linear. Normalized positive pion yield from heavy target is larger at 3 GeV/c than at 8 GeV/c. More positive pions are produced from light (carbon) target at low proton momentum.

References

1. M. G. Catanesi et al., *Phys. Rev. C* **77**, 055207 (2008)
2. M. G. Catanesi et al., *Phys. Rev. C* **80**, 035208 (2009)
3. S. Agostinelli et al., *Nucl. Instrum. Meth.* **A506**, 250 (2003)
4. N. V. Mokhov, *Fermilab-FN-628* (1995)
   N. V. Mokhov et al, *Rad. Prot. Dos.* **116**, 99 (2005)
   http://www-ap.fnal.gov/MARS/
5. H. En'yo et al., *Phys. Lett.* **B159**, 1 (1985)

J-PARC ACCELERATOR STATUS AND FUTURE PLANS

HITOSHI KOBAYASHI*

KEK, High Energy Accelerator Research Organization, 1-1 Oho, Tsukuba, Ibaraki 305-0801, Japan

The phase-1 construction of J-PARC was complete by the end of March 2009. Since then all accelerators and experimental facilities have started operations. This paper will summarize the present status of the J-PARC accelerator. Upgrade plans for the Main Ring are discussed.

1. Introduction

The J-PARC accelerator [1] is composed of a 400 MeV H^-linac (but initially only 181 MeV), a 3-GeV RCS (Rapid Cycling Synchrotron), a 50-GeV MR (Main Ring) and related experimental facilities. The RCS provides a proton beam to the MLF (Materials and Life science experimental Facility) to generate intense pulsed neutron beams. A muon target is installed at a position before the neutron target to generate intense muon beams for use in various experiments.

The MR has two experimental facilities: One is a neutrino facility, which uses a fast extraction system and the other a hadron beam facility, which uses a slow extraction system.

2. Status of the LINAC

The first beam in the linac was successfully accelerated January 2007. The linac has been in routine operation injecting beams into the RCS, which itself is being used for both MLF user runs and MR beam studies. For a high power test of the RCS, single shot beams of 2.93×10^{13}protons/pulse from the linac were accelerated, which corresponded to an RCS output power of 352 kW.

Unfortunately in September 2008 RF breakdown discharges in the RFQ started happening frequently. We decided to temporarily fix the pulse width at 100 μs and beam current at 5 mA. Since then, we have done many tasks such as

* http://www.kek.jp

increasing vacuum system diagnostics and installing three cryo-pumps were carried out. These measures were very effective. The operational condition of the RFQ has been drastically improved and wider microwave pulses can be stably applied since October 2009.

In parallel with routine operation of the linac, construction of the ACS (annular ring coupled structure) accelerator for raising the linac energy from 180 MeV to 400 MeV has been advancing. According to the current schedule, installation of all the ACS structures will be carried out during the summer of 2012.

3. Status of the RCS

The operation of the RCS in support of the MLF was initiated in December 2008, and the RCS has been providing beams for the commissioning of the MR as well. At the same time, high intensity operation of the RCS was tested in detail using a single pulse mode. In the case of optimum painting and optimal parameters for the rf 2nd harmonic, beam loss as measured by a beam current transformer was about 1% for a beam intensity of 2.6 x 10^{13} protons/pulse. This corresponded to a RCS output power of about 310 kW for continuous operation at 3 GeV and a repetition rate of 25 Hz.

Continuous operation of the RCS has been limited by the front end to about 20 kW, as described previously. Since November of 2009, operation of the RCS at over 100 kW has restarted reflecting the recovery of the RFQ.

4. MR

4.1. *First Beam Commissioning Results*

Beam commissioning of the MR started May 2008. In this run, injection of beam and circulation without acceleration were carried out. At that time we noticed a very large common mode current riding on the output of the power supplies for the dipoles and quadrupoles, resulting in both orbit and tune fluctuations. Thus the first beam commissioning results showed that some measures would be needed because of the common mode ripple issue.

After the first stage tests, we changed the power cabling configuration for both dipoles and quadrupoles to mitigate the magnetic field fluctuations induced by the common mode current ripple. They all are re-configured to be symmetrical [2], [3], [4].

On December 23, 2008, the first proton beam was accelerated from 3 GeV to 30 GeV, the design energy of the first phase of the project. Tune fluctuation after the reconfiguration was improved by an order of magnitude. The J-PARC

MR is the first large proton-accelerator designed with an imaginary gamma. Therefore it is of interest that the acceleration was carried out very smoothly and the beam was very stable through the acceleration period. Acceleration of the beam in the MR is shown in Fig. 1.

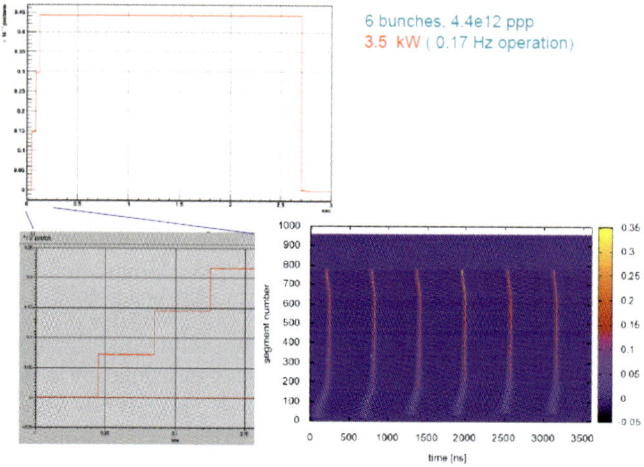

Figure 1. 6 bunches beam acceleration of the MR.

On January 27th, 2009, a 30-GeV proton beam was successfully extracted to the Hadron Experimental Hall by a slow extraction system using a third-order resonance.

April 2009, beams were guided to the neutrino target successfully. We are continuing to guide higher power beams to the neutrino target and in the machine study of November 2009, neutrinos were detected in the pre-detector system.

4.2. *Toward High Power*

In the first stage of the J-PARC project, the maximum output of the MR was set to be 750 kW. An upgrade scenario is shown by a dotted line in Fig. 2. Time of the upgrade has not yet been decided. A doubling of the output power is now under discussion. Our basic idea is to increase repetition rate of the MR to double the MR output power. Our discussions are still at a very preliminary stage but the following has already become our common consensus: a higher RF voltage is crucial for high repetition rate operation, and power supplies for the main magnets such as dipoles, quadrupoles and sextupoles should work at more than double of present repetition rate.

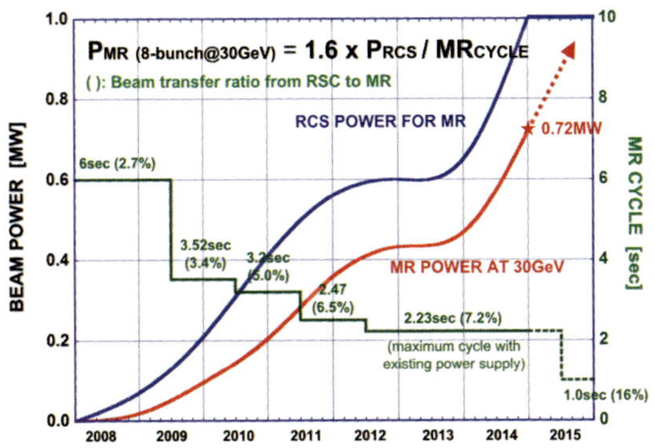

Figure 2. Power upgrade scenario under KEK road map.

Magnetic alloys are used to generate higher acceleration fields in the RF sources of both the RCS and the MR. New nano-technology cores could make an important contribution to generating 22.5 kV/m [5]. We are presently using cores of FT3M material, FT3L, which has a little bit different crystallizing process from the present FT3M, is a promising candidate for higher acceleration voltage. Its shunt impedance is expected to be twice that of the present cores. If we can double the shunt impedance of the core, then we can double accelerating voltage by re-designing the acceleration cavity and increasing the number of cavities [6].

We have set 1 Hz to be our target repetition rate. Of necessity the output voltages required from the power supplies become very high and would exceed the voltage rating of the presently used cables. For the future, our idea is to divide each family into several groups, each group would have almost same total inductance. Then, a standardized unit power supply could drive each of the magnet groups. For such a system to work successfully, precise tracking of the power supplies would be required absolutely. R&D on candidate power supplies has just begun.

References

1. Accelerator Technical Design Report for J-PARC, KEK-Report2002-13 and JAERI-Tec 2003-044H.
2. K.Sato and H. Toki, Nucl. Instrum. Meth. **A565**(2006)351-357.
3. H. Kobayashi, Proceedings of PAC09, Vancouver Canada.
4. H. Toki and K. Sato, J. Phys. Soc. Japan **78** (2009) 094201.
5. M. Yoshii et al., Proc. of EPAC2002.
6. C. Ohmori, private communication.

SIMULATION AND VERIFICATION OF DPA IN MATERIALS*

N. V. MOKHOV†, I. L. RAKHNO, S. I. STRIGANOV
Fermilab, MS 220, P.O. Box 500
Batavia, IL 60510, U.S.A.
E-mail: mokhov@fnal.gov

A recent implementation and verification of consistent modeling of displacements per atom (DPA) in the MARS15 code are described for high-energy particles and heavy ions.

1. Introduction

Radiation damage is displacement of atoms from their equilibrium position in a crystalline lattice due to irradiation with formation of interstitial atoms and vacancies in the lattice. Resulting deterioration of material (critical) properties is measured – in the most universal way – as a function of displacements per target atom (DPA). DPA is a strong function of projectile type, energy and charge as well as material properties including its temperature. The phenomenon becomes very serious for high-intensity beams especially for high-charge heavy ions ($\sim z^2$), being identified, for example at FRIB and FAIR, as one of the critical issues, limiting lifetime of targets to as low as a few weeks. A recent implementation of consistent DPA modeling into the MARS15 code [1] and its verification are described in this paper.

2. DPA Modeling in MARS15 Code

A model used in the MARS15 code for DPA calculations in electromagnetic processes [2] has been extended to an arbitrary projectile of energy ranging from 1 keV to 10 TeV. A primary knock-on atom (PKA) created in nuclear collisions can generate a cascade of atomic displacements. This is taken into account via damage function ν(T). DPA is expressed in terms of damage cross section σ_d :

---

* Work supported by Fermi Research Alliance, LLC under contract No. DE-AC02-07CH11359 with the U.S. Department of Energy.

$$\sigma_d(E) = \int_{T_d}^{T_{max}} \frac{d\sigma(E,T)}{dT} \nu(T) dT,$$

where E is kinetic energy of the projectile, T is kinetic energy transferred to the recoil atom, T_d is the displacement energy, and T_{max} is the highest recoil energy according to kinematics. In a modified Kinchin-Pease model [3], ν(T) is zero at $T<T_d$, unity at $T_d<T<2.5T_d$, and $k(T)E_d/2T_d$ at $2.5T_d<T$, where E_d is "damage" energy available to generate atomic displacements by elastic collisions (Fig. 1). T_d is an irregular function of atomic number (~40 eV). The displacement efficiency, k(T), introduced as a result of simulation studies on evolution of atomic displacement cascades [4], drops from 1.4 to 0.3 once the PKA energy is increased from 0.1 to 100 keV, and exhibits a weak dependence on target material and temperature.

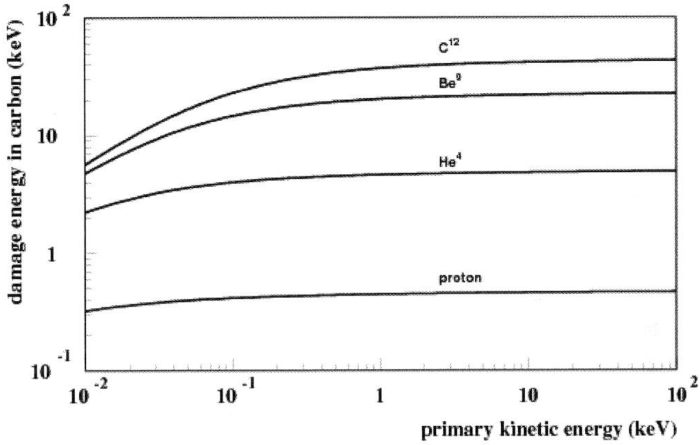

Figure 1. Damage energy E_d versus PKA energy.

For electromagnetic elastic (Coulomb) scattering, the Rutherford cross-section with Mott corrections and nuclear form-factors (a factor of two effect) are used in our model. Resulting displacement cross-sections due to Coulomb scattering are shown in Fig. 2 for various projectiles on silicon and carbon targets. For elementary particles, energy dependence of σ_d disappears above 2-3 GeV, while it continues to higher energies for heavy ions. For projectiles heavier than a proton, σ_d grows with a projectile charge z as z^2/β^2 at $\gamma\beta > 0.01$, where β is a projectile velocity. All products of elastic and inelastic nuclear interactions as well as Coulomb elastic scattering of transported charged particles (hadrons, electrons, muons and heavy ions) from 1 keV to 10 TeV contribute to DPA in the MARS15 model.

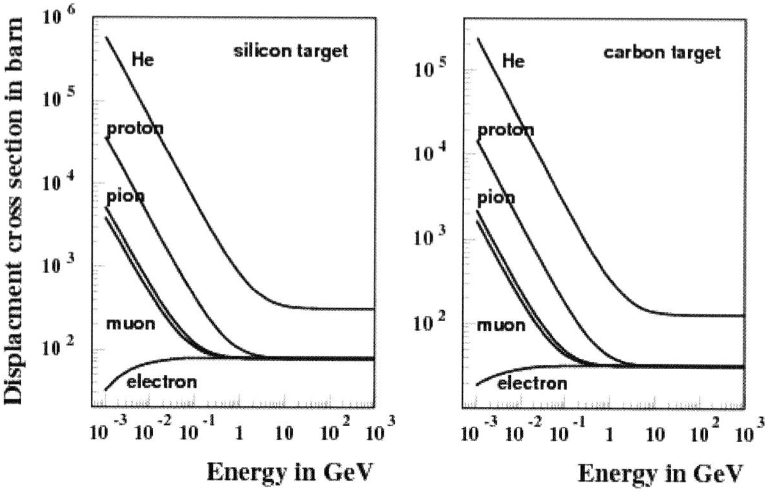

Figure 2. Displacement cross-section in silicon and carbon for various projectiles.

3. DPA Calculation Comparison

In this section, results on DPA calculated with the new MARS15 model for three cases are compared with those obtained with other DPA-capable codes, SRIM/TRIM, PHITS and MCNPX. The first case is a 1-GeV proton beam of 1-cm^2 area on a 3-mm thick iron target. SRIM, PHITS and MCNPX results are courtesy of Susana Reyes. As one can see in Table 1, there is a quite substantial difference between the predictions, with SRIM giving a very small value and the MARS15 result being a factor of 2.6 to 2.9 above those by PHITS and MCNPX. Calculated with MARS15 contributions to DPA of physics processes are as follows: 75.5% nuclear inelastic, 16% nuclear elastic, 2.75% electromagnetic elastic, 5.5% low-energy neutrons, and 0.25% electrons. The dominance of nuclear interactions in this case explains the above differences.

Table 1. DPA for 1-GeV protons on 3-mm iron.

Code	SRIM	PHITS	MCNPX	MARS15
DPA/pot	1.18e-22	2.96e-21	3.35e-21	8.73e-21

The second case is a 0.32-GeV/u Uranium beam of 9-cm^2 area on a 1-mm thick beryllium target. SRIM and PHITS results are again a courtesy of Susana Reyes. Table 2 shows that SRIM and MARS15 results are now very close to each other, while those calculated with PHITS are a factor of 70 lower. Calculated with MARS15 contributions to DPA of physics processes are as follows: 0.3% nuclear inelastic, 99.06% electromagnetic elastic, 0.02% low-energy neutrons, and 0.62% electrons. The dominant role of Coulomb scattering in this case explains the similarity of the SRIM and MARS15 predictions.

Table 2. DPA for 0.32-GeVu uranium on 1-mm beryllium target.

Code	SRIM	PHITS	MARS15
DPA/pot	2.97e-20	5.02e-22	2.13e-20

The third case is a 0.13-GeV/u Germanium beam of 0.004-cm^2 area on a 1.2-mm thick tungsten target. TRIM and PHITS results are a courtesy of Yosuke Iwamoto. Table 3 gives calculated DPA values in the first hundred microns of the target. The difference between TRIM and MARS15 needs to be understood.

Table 3. Entrance DPA for 0.13-GeVu germanium on 1.2-mm tungsten target.

Code	TRIM	PHITS	MARS15
DPA/pot	8.04e-16	1.25e-17	1.43e-16

4. BLIP Beam Tests for 0.7-MW NuMI/LBNE Target

A majority of data on radiation damage is available for reactor neutrons. Studies with hundred MeV protons [5] have revealed that a threshold of about 0.2 DPA exists for carbon composites and graphite. MARS15 studies helped realize that the BLIP beam tests with 0.165-GeV protons can emulate the NuMI neutrino target situation for a 120-GeV proton beam (Table 4). It turns out that despite a substantial difference in the beam energies in these cases, nuclear interactions and Coulomb scattering contribute about the same way (45-50% each) to the peak DPA in thick graphite targets irradiated at these two facilities.

Table 4. Peak DPA in POCO graphite targets at BLIP and NuMI.

Target	E_p (GeV)	Beam σ (mm)	N_p (1/yr)	DPA (1/yr)
NuMI	120	1.1	4.0e20	0.45
BLIP	0.165	4.23	1.12e22	1.5

References

1. N.V. Mokhov, Fermilab-FN-628 (1995); http://www-ap.fnal.gov/MARS/.
2. I.L. Rakhno, N.V. Mokhov, S.I. Striganov, Fermilab-FN-0817-APC (2008).
3. G.H. Kinchin, R.S. Pease, Rep. Prog. Phys., 18, 1 (1955).
4. R. E. Stoller, *J. Nucl. Mat.*, **276**, 22 (2000).
5. N. Simos et al., MOPC093, in *Proc. of EPAC08*, Genoa, Italy, June 2008.

PERFORMANCE AND OPERATIONAL EXPERIENCE OF THE CNGS FACILITY

EDDA GSCHWENDTNER, KAREL CORNELIS, ILIAS EFTHYMIOPOULOS, ALFREDO FERRARI, ANS PARDONS, HEINZ VINCKE, JOERG WENNINGER
CERN, 1211 Geneva 23, Switzerland

DARIO AUTIERO
IPNL Lyon, Lyon, France

PAOLA SALA
INFN Sez. Milano, Milano, Italy

ALBERTO GUGLIELMI
INFN Sez. Padova, Padova, Italy

The CNGS facility (CERN Neutrinos to Gran Sasso) aims at directly detecting muon to tau neutrino oscillations. An intense muon-neutrino beam (1E17 muon neutrinos/day) is generated at CERN and directed over 732 km towards the Gran Sasso National Laboratory, LNGS, in Italy, where two large and complex detectors, OPERA and ICARUS, are located. CNGS is the first long-baseline neutrino facility in which the measurement of the oscillation parameters is performed by observation of tau-neutrino appearance. In this paper, an overview of the CNGS facility is presented. Highlights on CNGS beam performance since during physics run in 2008 and 2009 are given.

1. Introduction

The CNGS facility was first operational in July 2006 for an approved physics program of five years with a total of 22.5E19 protons on target (4.5E19 protons/year). The 400GeV/c CNGS beam is fast extracted from the CERN SPS accelerator. The nominal intensity is 2.4E13 protons on target per 10.5μs extraction. During the 6s cycle, there are two extractions separated by 50ms. The beam is sent down an 840m long proton beam line with a slope of 5.6% onto a carbon target producing kaons and pions, corresponding to an average power at the target of 510kW. The positively charged pions and kaons are energy-selected and guided with two focusing lenses, the so-called horn and reflector, in the direction towards Gran Sasso. These particles decay in a 1000m long, 2.5m diameter decay vacuum tube into muon-neutrinos and muons. All the hadrons, i.e. protons that have not interacted in the target, pions and kaons that have not decayed in flight, are absorbed in a hadron stopper. Only neutrinos and muons can traverse this 18m long block of graphite and iron. The muons, which are ultimately absorbed downstream in around 500m of rock, are measured in two muon detector stations. They are arranged in a cross-shaped array, measure the muon intensity and the vertical and horizontal muon profiles which allows concluding on the quality and intensity of the neutrino beam produced and on the beam profile. A schematic overview of the CNGS neutrino beam facility at CERN is shown in Fig. 1.

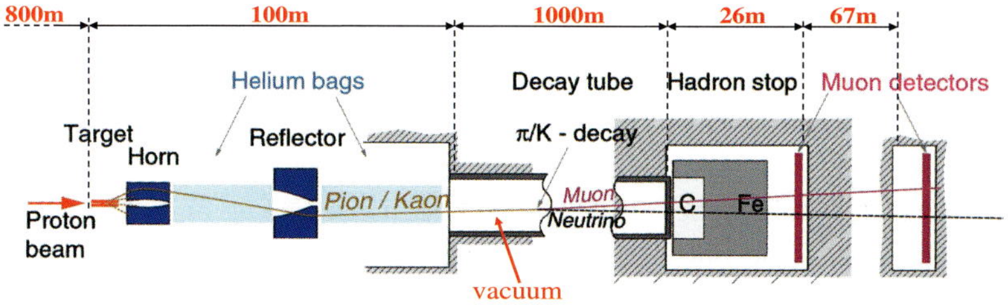

Figure 1: Main components of the CNGS facility.

2. CNGS Operation

CNGS was commissioned successfully in 2006 [1]. During 2007 CNGS was running for 6 weeks. After the completion of the OPERA detector [2] and finishing successfully some initial issues that occurred in the facility, CNGS had its first complete year of physics in 2008 with 1.78E19 protons on target. In 2009 in total 3.52E19 have been cumulated and OPERA has collected more than 21400 on-time events and more than 3700 candidate interactions in the bricks.

Table 1: Cumulated protons on target for CNGS to date.

	protons on target
2006	7.56 E17
2007	6.13 E17
2008	1.78 E19
2009	3.52 E19
Total	**5.44 E19**

3. Start-up Issues

Physics requirements have been pushing the needed power of the proton beam on target to values around and above 500 kW. The CNGS secondary beam line, starting with the target, has to cope with this situation, which pushes the beam line equipment and instrumentation to the limits of radiation hardness and mechanical stresses during the CNGS operation. The choice of materials, shielding configurations, remote handling capabilities for maintenance and exchange of equipment were carefully designed and optimized. However, the startup issues faced demonstrate the difficulty in the design and operation of such high intensity facilities. Details of major technical issues and their improvements are summarized in [3].

4. CNGS Performance

During the first complete CNGS physics run in 2008 in total 1.78E19 protons on target were cumulated. Detailed results can be found in [4] and [5].

The CNGS physics run in 2009 started on 1st June 2009 and was scheduled until 23rd November 2009. The overall efficiency of the accelerator complex was very good, i.e. 74%, resulting to 3.52E19 integrated number of protons on target which is 10% higher than expected (3.2E19), (see Fig. 2).

For most of the time the facility operated with 2E13 pot/extraction. The overall beam performance and stability of the primary proton beam line was excellent throughout the run: 110 μm r.m.s in the horizontal plane and 50 μm r.m.s in the vertical plane; no active position feedback is necessary.

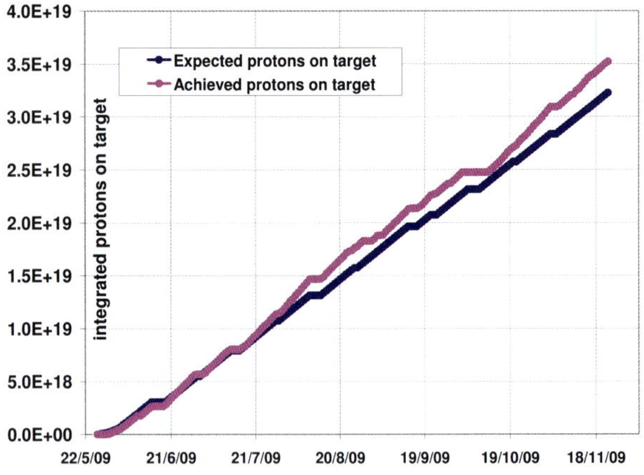

Figure 2: Expected and achieved integrated protons on target for 2009.

The muon detector [6] stations are very sensitive to any misalignment between the proton beam, the target and the horn; for example an 80μm parallel beam shift on the target corresponds to a 5cm shift of the muon profile centroid. This allows to precisely optimizing on-line the secondary particle and neutrino production.

Comparison between the muon profiles measurements and FLUKA [7] simulations shows very good agreement (see Fig. 3). This proves that the CNGS facility is very well understood and special features of the facility are considered: e.g. in the horizontal muon profiles an asymmetry between operating in neutrino mode (focusing of mesons with positive charge) and anti-neutrino mode (mesons with negative charge) has been observed. This can be explained due to the earth magnetic field in the 1000m long decay tube which results in the profile shifts of the observed magnitude [8].

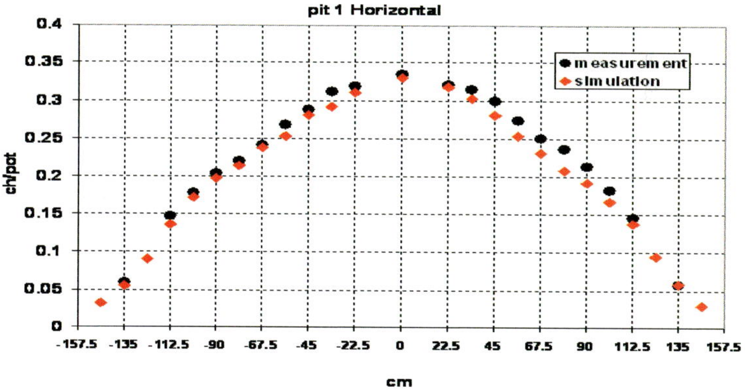

Figure 3: Measured (black circles) and simulated (red rectangles) horizontal muon profile in the first muon detector station. Each point corresponds to one ionization chamber readout.

The observation of horizontal muon detector signals in 2007 non-linear w.r.t. the intensity was attributed to the wire topology of the readout cables. The remedy was to increase the capacitance of all wires to a fixed value [6].

5. Summary

After successfully resolving of the start-up issues since the beam commissioning in 2006, CNGS has started smoothly the physics run in 2008. Since then 5.44E19 protons on target have been cumulated. With the present statistics first tau-neutrino events are expected.

References

1. M. Meddahi *et al*, CERN-AB-2007-039, PAC07, pp692, Albuquerque, USA (2007).

2. R. Acquafredda *et al*, 2009 JINST **4** P04018 (2009).
3. E. Gschwendtner *et al*, 'The CNGS Facility: Performance and Operational Experience', presented at the 11th ICATPP Conference, Villa Olmo, Como, October 2009, (2009).
4. I. Efthymiopoulos *et al*, 'First Year of Physics at CNGS' presented at PAC08, Vancouver (2008).
5. E. Gschwendtner, 'CNGS: First Year of Operation', AB-Seminar, CERN, Nov. 2008 (2008).
6. A. Marsili *et al*, CERN-AB-Note-2008-044-BI (2008).
7. A. Fasso, A. Ferrari, P.R. Sala, Proc, Monte Carlo 2000, Lisbon, 23-26 Oct. 2000, A. Kling *et al*, eds., Springer-Verlag, Berlin, p. 159-164, p. 995-960 (2001).
8. E. Gschwendtner, CERN-AB-Note-2008-050-ATB, POS: NuFact08, pp008, Valencia, Spain (2008).

PARTICLE PHYSICS ENABLED WITH SUPER-CONDUCTING RF TECHNOLOGY — SUMMARY OF WORKING GROUP 1

D. JAFFE

Brookhaven National Laboratory
Upton, New York 11973

R. TSCHIRHART

Fermilab
Batavia, Illinois 60510

The summary of working group WG1 exploring the discovery science enabled with SRF technology is given here.

1. Introduction

Experimental particle physics advances by synthesizing insights won by many techniques. Super-conducting RF technology holds the promise of dramatically extending the energy frontier reach through next-generation colliding beam experiments, and the reach of the intensity frontier with a broad sweep of experiments ranging from the search for an electric dipole moment (EDM) of the neutron to long-baseline neutrino experiments that span the Great Plains of the United States.

To date, colliding beam experiments have taken us to the energy frontier with proton-antiproton collisions at the Fermilab Tevatron and electron-positron interactions at LEP(CERN) and SLC(SLAC). Colliders have enabled measurements of great delicacy at the *B* factories and CLEO(Cornell). Fixed-target experiments have been major contributors to flavor physics, our knowledge of nucleon structure, and our understanding of neutrino properties and interactions. However, many of the accelerator experiments that have helped define the diversity and scale of our program, and have helped drive the past decades of discovery, are coming to an end.

Beyond the Tevatron Collider program, we look forward to the start of the Large Hadron Collider program at CERN, which will place experiment squarely in the heart of the TeV scale essential to the understanding of electroweak symmetry breaking. The U.S. is a key player in the LHC program, as a developer of super-conducting magnets and major detector and computing systems for LHC experiments. The high-energy frontier is part of the U.S. program's heritage, and remains an essential element of its future.

Doing good science and making new scientific tools available is critical to the future of experimental particle physics in the U.S. . This requires the broadband energy frontier assault that the LHC provides, followed by the incisiveness of lepton colliders: the International Linear Collider (ILC) with direct physics reach below 1 TeV and the Compact LInear Collider (CLIC), and Muon-Collider concepts for direct reach beyond the 1-TeV scale. Both the CLIC and Muon-Collider concepts pose extreme technical challenges. Bold new concepts to overcome some of these challenges have evolved in recent years and now need to be tested.

2. The Flavor of New Physics Accessible at the Intensity Frontier

Flavor phenomena have contributed to shaping the structure of the Standard Model of particle physics (SM) as much as any other ingredient. The existence of flavor for quarks and leptons gives the SM its family and generation structure. Assembling quarks in electroweak doublets to suppress flavor-changing neutral currents with kaons led to the prediction of the charm quark. Kaon decays also led to the observation of CP violation, and to the CKM model, which in turn predicted the existence of a third generation. Mixing of neutral B mesons, similar to the role played by neutral kaon mixing in getting the mass range for the charm quark, was the first experimental observation that correctly anticipated the large value of the top quark mass. The dramatic discovery of neutrino masses provided the first incontrovertible evidence that the SM is incomplete and may provide a window to grand unification of forces. It is not surprising that several of the great questions of particle physics have flavor at their core, and it is reasonable to expect that flavor physics will continue playing a crucial role in the progress of the field.

There are several elements that directly associate LHC physics with flavor issues. In the absence of a higgs or some other mechanism of electroweak symmetry breaking (EWSB), quark flavor effects would not even exist. All flavor phenomena in the SM are encoded by a handful of input parameters that lack any dynamical content. But when we move beyond the Standard Model, flavor phenomena can cover a much wider landscape and are even more strongly entangled with the dynamics of symmetry breaking. Flavor changing processes can be mediated by new particles such as charged higgses or supersymmetric partners. New flavors may appear, either in the form of new generations, or as exotic partners of standard quarks (such as Kaluza-Klein excitations in the case of extra-dimensional theories, or the T' and B' quarks in "little higgs" models). New sources of CP violation can arise from couplings of non-minimal higgs sectors, or of superpartners. With all of these new sources of flavor effects, the natural suppression of most flavor violating phenomena in the SM is put into jeopardy, and much larger effects are generally expected from new Terascale physics. In the context of Beyond-SM (BSM) model building, this is a fundamental issue called "the flavor problem".

Equally important is the fact that in several BSM frameworks the parameters of flavor are not just arbitrary inputs but instead are the result of dynamics or symmetries of the underlying theory. Unified theories predict relations between the couplings of quarks and leptons. In supersymmetric models with neutrino masses, a mix of symmetry relations and dynamics connects neutrino mixing and flavor transitions in the charged-lepton sector. In extra-dimensional theories the family replicas can be understood as different branes on which fermions are bound to live, and mixings are tied to the relative positions of these branes in the extra dimensions. In supersymmetric theories, EWSB can be generated dynamically by the large value of the top quark mass, making EWSB, in some sense, a flavor-driven phenomenon. Finally, the numerical coincidence of the top mass value with the scale of EWSB is yet another mysterious hint of a possible direct connection between electroweak symmetry breaking and flavor.

These connections between symmetry breaking and flavor, as well as the flavor mysteries of neutrino masses and the matter antimatter asymmetry of the universe, very strongly suggest that flavor will play a key role in exploring the new physics landscapes unveiled by the LHC. Most new physics manifestations that we can envision will provide new sources of flavor phenomena, accentuating our need to address the flavor problem. The optimal approach to understanding flavor will depend on the details of what is discovered. It is sensible to expect, on the basis of the history of particle physics and of the explicit models of new physics available today, that experiments at the high-energy frontier and flavor experiments at the high-intensity frontier will provide complementary keystones in the coming phases of exploration of the laws of nature. The confluence of discoveries and measurements from many facilities will be needed to achieve a deeper understanding of the great questions of particle physics.

3. Scientific Opportunities at the Intensity Frontier

The "Project-X" concept being developed at Fermilab has the potential to drive world-leading particle physics experiments at the intensity frontier. These experiments include neutrino oscillations over a long baseline, rare processes in quarks and charged leptons that are suppressed or forbidden in the Standard Model, and precision measurement of neutron decay parameters. The scientific reach of this suite of measurements today is limited by available proton beam power. An incomplete list of world-leading experiments that could be mounted at the multi-Megawatt Project-X proton beam facility is presented here:

- **Long-Baseline Neutrino Experiment (LBNE):** A high power proton source in excess of two Megawatts at Fermilab driving massive detectors at the Deep Underground Science and Engineering Laboratory (DUSEL) could map out the fundamental properties of neutrinos and search for matter-antimatter asymmetries among neutrinos. Evidence for matter-antimatter asymmetry in the neutrino sector could help us understand one of the most vexing problems in particle physics today: The clearly apparent asymmetry of matter and antimatter in our world and how that could have evolved from the symmetric state in Big Bang cosmology. Further, precision measurement of neutrino parameters may offer a glimpse of the scale where forces unify (the GUT scale, around 10^{16} GeV/c^2) through various "see-saw" models that inversely leverage the very small neutrino mass scale to scales far beyond the direct reach of conceivable colliders.
- **Muon-to-electron conversion experiments:** The relatively long muon lifetime permits access to very rare phenomena induced through virtual loop process. Dramatic processes such as the coherent conversion of a muon into an electron in the field of a nucleus could be mediated by such BSM physics, in many cases beyond the reach of the LHC. The Project-X beam power can probe rates as small as 10^{-18} of the normal muon nuclear capture rate near a nucleus.
- **Search for the muon electric dipole moment:** The search for electric dipole moments (EDMs) of fundamental particles is a uniquely incisive probe of matter-antimatter asymmetries in particle interactions. Many theories beyond the Standard Model, such as Super-Symmetry can induce EDMs. A search for the muon EDM is uniquely sensitive to generation & flavor dependence of new physics that generates EDMs. Several novel concepts have been developed to deeply probe the muon

EDM, all of which require very bright muon sources and consequently very high power proton drive beams.

- **Search for ultra-rare muon decays:** Ultra-rare muon decays such as muon decaying into a photon and electron or into 3 electrons ($e^+e^-e^+$) are sensitive probes of BSM models and likewise require very bright muon sources.
- **Precision measurement of ultra-rare kaon decays:** Many examples of BSM physics induce substantial enhancements through loop effects to the Standard Model process of $K \rightarrow \pi\nu\nu$ which can be calculated to high accurarcy in the Standard Model. Measuring this highly suppressed process, ~50 parts per trillion in the Standard Model, requires very bright kaon sources and consequently extremely intense proton drive beams with flexible timing characteristics.
- **Precision measurement of interference phenomena in the neutral kaon system:** The neutral kaon system is a very sensitive interferometer by virtue of the tiny mass difference between the K_S and K_L mass states. Measurements in the neutral kaons system today are the most sensitive probe of the microscopic structure of space-time (CPT violation). The proton beam power available at Project-X can drive next generation experiments that can probe inverse mass differences between K^0 and anti-K^0 approaching the Plank scale.
- **Precision measurement of the neutron lifetime:** Precision measurement of the neutron lifetime can be used to determine the fundamental coupling of the up-quark to the down-quark with high accuracy. This determination combined with other precision measurements of quark couplings is sensitive to new families of quarks that may exist beyond the known three generations.
- **Search for the neutron electric dipole moment:** The search for an electric dipole moment of the neutron is broadly recognized as one of the most incisive probes of new physics today. The current search sensitivity of neutron EDMs tightly constrains the character of new physics beyond the Standard Model, and further sensitivity driven by Megawatt class proton beams can probe many new physic models.
- Search for a nuclear electric dipole moment:The search for a nuclear electric dipole moment shares the strong physics motivations with the search of EDM in the neutron and the muon. It tests time-reversal symmetry and probes physics beyond the Standard Model. In particular, it is sensitive to T-violating mechanisms in the strong interaction present among the nucleons. In the nuclear sector, radium-

225 (half-life = 15 d) is a particularly good case for EDM searches due to the octupole deformation of the nucleus. Project-X can become an abundant source of radium-225. Its yield can exceed the current level by 4-5 orders of magnitude, thus offering the opportunity to dramatically improve the EDM search sensitivity.

Working Group 2

SRF Linac for Muon Collider and Neutrino Factory

Conveners

Roland Garoby (Cern)

John Galambos (Oakridge)

Stephen Geer (Fermilab)

PROTON BEAM REQUIREMENTS FOR A NEUTRINO FACTORY AND MUON COLLIDER*

MICHAEL S. ZISMAN
Accelerator & Fusion Research Division, Lawrence Berkeley National Laboratory
Berkeley, CA 94720 U.S.A.

Both a Neutrino Factory and a Muon Collider place stringent demands on the proton beam used to generate the desired beam of muons. Here we discuss the advantages and challenges of muon accelerators and the rationale behind the requirements on proton beam energy, intensity, bunch length, and repetition rate. Example proton driver configurations that have been considered in recent years are also briefly indicated.

1. Introduction

Design and performance evaluations for a Neutrino Factory (NF) and a Muon Collider (MC) have been ongoing for more than a decade. In the case of a Neutrino Factory, the effort is fully international and includes the Neutrino Factory and Muon Collider Collaboration (NFMCC) in the U.S., the UK Neutrino Factory group and the EUROnu design study in Europe, and the Japan Neutrino Factory Working Group in Asia.

Here we focus on the requirements such facilities place on the proton beam parameters, including power, energy, bunch length, repetition rate and bunch train structure.

2. Muon Accelerator Advantages

The interest in muon accelerators is based on their ability to address several of the outstanding particle physics questions that require accelerator facilities. In the neutrino sector the neutrinos result from the decays of either μ^+ or μ^- circulating in a decay ring. These decays produce high-energy electron neutrinos or anti-neutrinos. The kinematics of this decay process are well understood, so there are minimal uncertainties in the spectrum and flux. Moreover, the

* This work is supported by the Director, Office of Science, Office of High Energy Physics, of the U.S. Department of Energy under Contract No. DE-AC02-05CH11231.

neutrinos are produced at high energies, above the tau threshold. Studies [1] indicate that a muon-based NF would have unmatched sensitivity for CP violation, determination of the mass hierarchy, and deviations from unitarity.

At the energy frontier, the fact that the muon is a point particle means that the full beam energy is available for particle production. The heavier mass of the muon compared with that of the electron means that synchrotron radiation is negligible. This allows the use of a circular collider, which makes much more efficient use of the expensive rf equipment, and also nearly eliminates the energy spread at the interaction point arising from beamstrahlung.

3. Muon Beam Challenges

Despite the advantages of muon beams described in Section 2, they are not easily created. There are two main challenges:

1. muons are created as a tertiary beam ($p \rightarrow \pi \rightarrow \mu$)
2. muons have a short lifetime (2.2 μs at rest)

The first issue means that production rates are low. A multi-MW proton source is needed to give the desired number of muons, along with a production target that can handle this power level. Because of the production mechanism, the muons are created with a large beam emittance and a large energy spread. Solenoidal focusing, which operates in both planes simultaneously, is needed to maintain the beam, and emittance cooling is required to produce beam suitable for downstream accelerator systems. Even with such cooling, a large acceptance and rapid acceleration system is needed to bring the muons to the desired energy.

The short lifetime means that all beam manipulations must be rapid, requiring (in the cooling channel) high-gradient rf cavities operating in a strong solenoidal field. In addition, the decay electrons from stored muons in either the NF decay ring or MC collider ring result in a substantial heat load in the mid-plane of the superconducting magnets.

4. Proton Beam Requirements

As indicated in Fig. 1, the present vision is that the NF and MC could share a common front end. For this reason, the NF requirements generally represent those of the MC as well. There are a few differences, however. In particular, a MC prefers a lower repetition rate than does a NF (10–15 Hz compared with ~50 Hz), and the MC prefers a single bunch (or at least a short bunch train), whereas a NF has no a stringent requirement on the bunch train length. Table 1 summarizes the NF proton driver parameters obtained during the ISS [2].

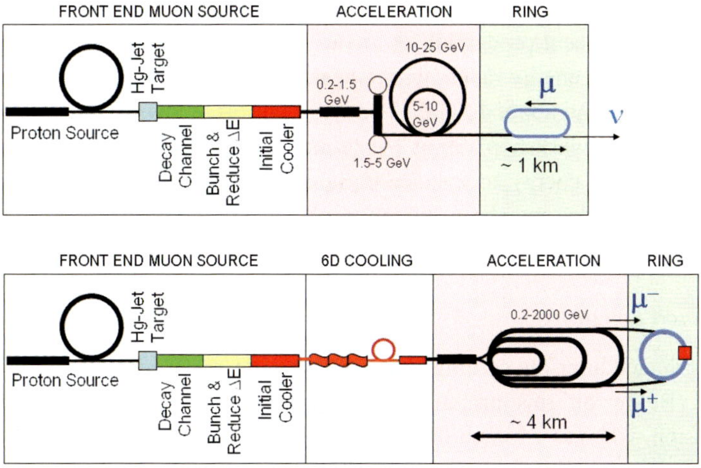

Fig. 1. (top) NF systems (schematic); (bottom) MC systems (schematic). The front end muon source could be the same for both facilities.

Table 1. Proton driver requirements for a Neutrino Factory.

Parameter	Value
Average beam power (MW)	4
Pulse repetition frequency (Hz)	50[a)]
Proton energy (GeV)	10 ± 5
Proton rms bunch length (ns)	2 ± 1
No. of proton bunches	3 or 5
Sequential extraction delay (μs)	≥ 17
Pulse duration, liquid-Hg target (μs)	≤ 40
Pulse duration, solid target (ms)	≥ 20

[a)]For a Muon Collider a lower repetition rate, 10–15 Hz, is required.

For either the MC or NF, the target comprises a free Hg jet within a 20-T solenoid. A tapered series of solenoids brings the field smoothly down to 1.75 T for further transport into the downstream phase rotation and cooling sections. The MERIT experiment [3] serves as a proof-of-principle for the target concept.

4.1. *Beam Energy*

The muons captured by the downstream channel are low energy, with kinetic energies in the range of about 100–300 MeV. Initial studies with MARS14 [2] indicated that the proton energy range for optimal production was 6–11 GeV for μ^- and 9–19 GeV for μ^+. The ISS adopted 10 ± 5 GeV as the range that represented these results. Below 5 GeV, a steep fall-off in production was predicted. More recent production estimates using MARS15 suggested an even narrower optimum energy range, with a peak at 8 GeV and a fall-off by about a factor of two beyond about 40 GeV. The steep fall-off at low energies remained. At the NuFact09 workshop, Strait [4] looked at the HARP data for particles within the NF/MC energy acceptance in the low-energy regime. As indicated in Fig. 2, the data do not show evidence of the steep fall-off at low energy predicted by MARS.

4.2. *Bunch Length*

When evaluated after the cooling channel, there is a preference for short proton bunches. Below an rms proton bunch length of 1 ns, there is no loss in transmission, whereas a bunch length of 3 ns results in a 10% loss in intensity. For this reason, the ISS [2] and the follow-on IDS-NF [5] adopted an rms bunch length requirement of 2 ± 1 ns, as listed in Table 1. For a linac solution, downstream accumulator and compressor rings would provide this structure.

4.3. *Repetition Rate*

For the baseline Hg-jet target, the maximum repetition rate is limited by disruption of the target material. At high magnetic fields, the disruption length is reduced, and is typically about 0.2 m at the intensities of interest to a NF. For a jet velocity of 15 m/s, the disruption recovers in about 15 ms, which would permit a repetition rate of 70 Hz. Present NF designs call for a repetition rate of about 50 Hz.

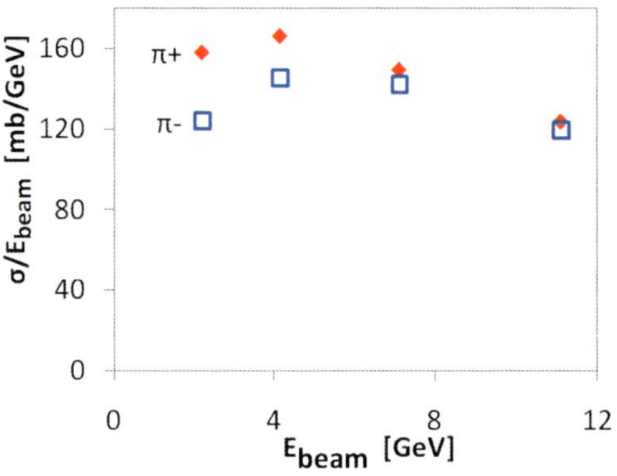

Fig. 2. HARP results in the acceptance of relevance to a NF or MC.

The minimum repetition rate will be limited by the space-charge tune shift in the compressor ring, which will define how short a bunch the ring can deliver. For the MC, where fewer bunches are desirable, it may be necessary to provide a workaround, e.g., using multiple bunches in the compressor ring with separate transport lines that serve as "delay lines" to deliver the separately extracted bunches to the target simultaneously.

4.4. *Bunch Train Length*

As mentioned earlier, the NF does not have a strong constraint on the length of the 201-MHz bunch train, and trains of up to ~70 bunches have been considered. To make the front end more compatible with the MC, Neuffer [6] has developed a shorter bunching and phase rotation section that reduces the bunch train length to ~20 bunches. This is helpful, but will not obviate the need for some form of further bunch merging to obtain a single muon bunch of each charge in the collider ring.

5. Implementation Schemes

There are a number of ways to provide the required proton parameters, and the choice of which approach to adopt is thus a site-specific one.

Rees and Prior [2] have developed a 10 GeV system with a 180 MeV linac followed by a 3-GeV rapid cycling synchrotron and a 10 GeV non-scaling FFAG ring. J-PARC [7] has already successfully commissioned a proton source with a linac followed by two RCS rings, the first providing 3 GeV and the second up to 50 GeV. Fermilab's Project X [8] has considered several different configurations. Their initial configuration was based on an 8 GeV pulsed superconducting linac. Other variants, e.g., an 8 GeV CW linac and a lower energy linac followed by an RCS to reach 8 GeV, have also been explored. A 5 GeV superconducting linac (the SPL [9]) is being discussed at CERN.

6. Summary

As of today, the proton beam requirements for either a NF or a MC are reasonably well understood. These include a beam energy of 5–15 GeV, a beam power of ~4 MW, a bunch length at the target of 1–3 ns, and a repetition rate of 10–50 Hz. These parameters are considered to be achievable, and there are several approaches that should be able to deliver them. Because the requirements for the two types of facility are very similar, it is expected that a single implementation can satisfy either one.

It is worth noting that the facility to house a 4 MW target and beam dump is a substantial one, and requires a significant amount of engineering to develop. Radiation shielding, equipment and air activation, and remote handling and repair are just some of the issues that must be addressed to develop a 4 MW target hall. In practice, it would be prudent to consider designing for even higher beam power to allow for future upgrades.

Acknowledgments

I would like to thank my colleagues in the NFMCC, the Fermilab Muon Collider Task Force (MCTF), the International Scoping Study (ISS) and its follow-on International Design Study of a Neutrino Factory (IDS-NF) for their contributions to the concepts and designs of the muon-based facilities described here. I also wish to thank both DOE and NSF for having the wisdom to provide continued support for this challenging endeavor.

References

1. S. F. King, K. Long, Y. Nagashima, B.L. Roberts, O. Yasuda (Eds.), RAL-TR-2007-019, December 2007; Rep. on Prog. in Physics 72, 10621, 2009; http://www.iop.org/EJ/article/0034-4885/72/10/106201/rpp9_10_106201.pdf.
2. M. Apollonio *et al.* (ISS Accelerator Working Group), "International scoping study of a future Neutrino Factory and super-beam facility: Accelerator design concept for future neutrino factories, (ed. M. Zisman), RAL-TR-2007-23, December 2007; arXiv:0802.4023v1 [physics.acc-ph], 27 February 2008; JINST 4, P07001, 2009; http://www.iop.org/EJ/article/-search=67573990.1/1748-0221/4/07/P07001/jinst9_07_p07001.pdf.
3. H. G. Kirk *et al.*, "The MERIT high power target experiment at CERN," in Proc. EPAC 2008, p. 2886.
4. J. Strait, in Proc. NuFact09, Chicago, July 20–25, 2009, to be published; see also http://nufact09.iit.edu/wg3/wg3_strait-marsvsharp.ppt.
5. IDS Web page: https://www.ids-nf.org/wiki/FrontPage.
6. D. Neuffer, "Exploration of the 'High-Frequency' Buncher Concept," NFMCC note MUC-269, http://nfmcc-docdb.fnal.gov/cgi-bin/RetrieveFile?docid=269&version=1 &filename=muc0269.pdf
7. See http://hadron.kek.jp/~accelerator/TDA/tdr2003/index2.html.
8. V. Lebedev, in Proc. NuFact09, Chicago, July 20–25, 2009, to be published; see also http://nufact09.iit.edu/wg3/wg3_lebedev-projectxupgrades.pdf
9. F. Gerigk (ed.), "Conceptual Design of the SPL II – A High Power Superconducting H^- Linac at CERN," CERN-2006-006.

PROTON BUNCHING OPTIONS*

R. B. PALMER*

Brookhaven National laboratory,
Upton, New York, 11786, USA
**E-mail: palmer@bnl.gov*

Muon Colliders [1] need intense, very short, proton bunches. The requirements are presented and a number of possible bunching systems discussed. The best solution uses a small super-conducting buncher ring with 6 bunches that are taken though separate transports and combined on the target.

Keywords: Muon Collider; proton driver; space charge tune shift

1. Introduction

Because a muon collider luminosity depends on the square of the bunch charges, intense bunches of 2 10^{12} muons are required. To generate such bunches needs intense proton bunches. Fig.1 shows the relative muon fluxes vs. the proton energy used to make them. Pi production was from MARS 15 [2]. Muons were from an ICOOL [3] simulation [4] of the front end of a neutrino factory including pion decay, muon phase rotation, 80 m of transverse cooling, and accelerator acceptances of 30 mm transverse, 150 mm longitudinal. Since a muon collider is assumed to have a similar front end, this is a good measure of relative efficiency. This efficiency is maximum at around 8 GeV. At this relatively low energy, the required number of protons is very large ($\approx 200\ 10^{12}$), and the space charge tune shift with the required 2 ns bunch length, can be a problem.

Space Charge Tune Shift [5] is given by

$$\Delta\nu \;=\; F_{dist} \left(\frac{2\pi R}{\sqrt{2\pi}\ \sigma_z} \right) \frac{N_p\ r_o}{2\pi\ \epsilon_N\ \beta_v \gamma^2}$$

*Work supported by US Department of Energy under contracts AC02-98CH10886 and DE-AC02-76CH03000

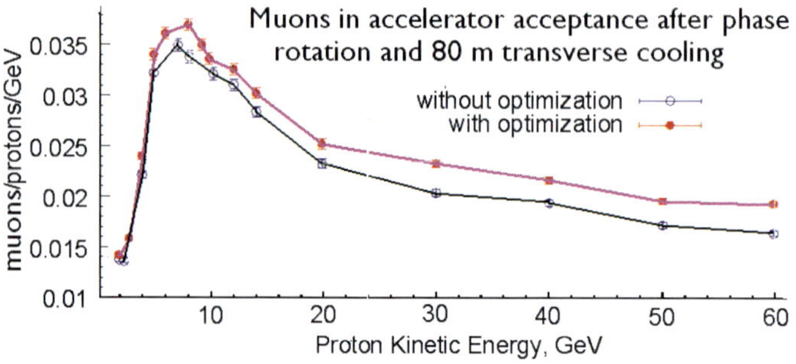

Fig. 1. Relative pion production vs. proton energy.

For Gaussian beams $F_{dist} = 3$, ϵ_N is the normalized (95%) emittance as used for protons at FNAL. The rms normalized emittance $\epsilon_\perp = \epsilon_N/6$.

2. Six Buncher Cases Considered

The above formula is applied to the following numbered cases with parameters given in Table 1.

Table 1. Parameters of bunchers.

		1 Booster at inj	2 Booster at 8 GeV	3 SC	4 24 GeV	5 FFAG	6 6 bunch
E	GeV	0.4	8	8	24	8	8
Circ	m	474	474	200	561	339	200
N_p	10^{12}	0.06	200	200	96	200	200/6
σ_z	m	1.2	.66	.66	.66	.66	.66
σ_θ	mrad	1.2	1.2	2.5	0.83	25	2.5
ϵ_N	μm	12	112	200	200	2000	200
$\epsilon_\perp$	μm	2	19	33	33	330	33
nb		84	1	1	1	1	6
$\Delta\nu$		0.4	4.0	0.95	0.17	0.17	0.16

(1) To check the formula we consider the FNAL Booster at its 400 MeV injection energy. It gives a tune shift of 0.4, as published [6].
(2) This Booster, bunching 200 10^{12} protons to 2 ns at 8 GeV, assuming the full geometrical emittance as used at injection (normalized 95% emittance at 8 GeV of 112 μm), gives a space charge tune shift of 4: far above the value of 0.2 considered maximum for such an intense beam.

(3) A ring using super-conducting magnets would have a smaller circumference e.g. ≈ 200 m (instead of 474 m) and could have a larger acceptance: e.g. a 95% emittance of 200 μm (instead of 112 μm). But it's tune shift would still be an unacceptable 0.95.

(4) Since $\Delta\nu \propto 1/\gamma^2$, the tune shift can be lowered to a reasonable value by raising the proton energy to 24 GeV. From Fig.1 the needed proton bunch intensity is now 96 10^{12}. Assuming the same buncher ring normalized acceptance and average bending field as case # 3, then the tune shift for is a reasonable 0.17. But because pion production per GeV is now less, the required proton power is a factor of $\approx$1.7 higher: Not a good solution.

(5) Instead of raising the energy, one can try using a very large acceptance ring, such as the 5-10 GeV FFAG designed for muon acceleration in Study 2a [8]. That ring has a 339 m circumference and a normalized muon acceptance of 30 mm. With a 95% emittance one half this, the normalized proton emittance $\epsilon_N = 30,000/2 \times 106/970 \approx 2000\ \mu m$. The tune shift is now a reasonable 0.17. But when this huge emittance is focused down to one third of the 5 mm target radius ($\sigma_r = r_{\text{target}}/3 \approx 1660\ \mu$m) then three times the rms angular spread is:

$$3\sigma_\theta = 3\ \frac{\epsilon_\perp}{\beta\gamma\ \sigma_r} = 3\ \frac{330}{8\ 1660} = 75 \quad \text{mrad}$$

over twice the crossing angle between the beam and jet, which is not viable.

(6) But if one can have multiple bunches in the ring and bring them simultaneously onto the target, then there appears to be a good solution. Consider the same ring as case #3 but with 6 bunches. The tune shift is now an acceptable 0.16. The bunches can be extracted into transports of differing lengths (trombones) [7] to bring them all onto the target at the same time: see Fig.2a. Since it is assumed that the beam intersects the mercury jet target from the side at an angle $\approx$ 33 mrad, it should be possible to bring multiple beams in from multiple azimuths, all at the same angle to the jet, see Fig.2b. The three sigma angular spreads are now 7.5 mrad, small enough to keep the transports for the beams well separated from the jet and one another.

3. Conclusion

Pion, and thus muon, production is predicted to have a maximum for 8 GeV protons. The muon collider then requires 200 10^{12} protons/bunch with $\sigma_t \approx$

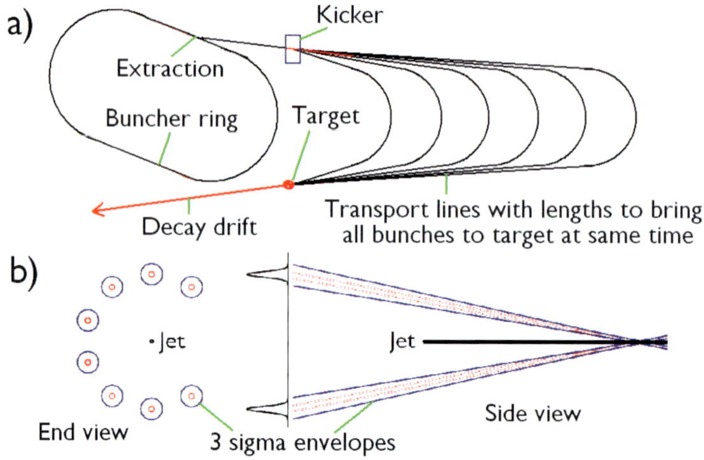

Fig. 2. a) Trombone transport lines to bring all bunches to the target at the same time; b) Multiple beam target geometry.

2 ns. The space charge tune shifts of such bunches in an FNAL Booster-like ring is excessive.

The space charge is reduced if higher bending fields allow a smaller circumference ring (474→200 m), and if the acceptance is increased ≈ 1.8×, but this is not enough unless the proton energy is increased to 24 GeV, with a resulting increase in required power of 1.7. The best solution is to use multiple bunches in the ring, extracted into different transports that bring them all to the target at the same time.

References

1. R. B. Palmer,*Muon Collider*, in these proceedings.
2. *MARS15* http://www.physics.princeton.edu/mumu/target/mokhov/mokhov_102009b.pdf
3. *ICOOL* http://pubweb.bnl.gov/users/fernow/www/icool/readme.html
4. X. Ding et al,*Optimized parameters for a mercury jet target*, Proc PAC, Vancouver, BC,Canada (2009)
5. S Y Lee, *Accelerator Physics*, World Scientific, p110
6. *Booster*, http://www-bd.fnal.gov/pdriver/booster/database/parameter_list.pdf
7. C. Ankenbrandt, http://indico.fnal.gov/getFile.py/access?contribId=65 &sessionId=28&resId=0&materialId=slides&confId=2854
8. J.S. Berg et al, *Cost-effective design for a neutrino factory*, Phys. Rev. ST Accel. Beams **9** (2006) 011001.

CW SRF H⁻ LINAC AS A PROTON DRIVER FOR MUON COLLIDERS AND NEUTRINO FACTORIES*

MILORAD POPOVIC[1†], CHARLES M. ANKENBRANDT[2], ROLLAND P. JOHNSON[2]

[1]*Fermilab, P.O. Box 500, Batavia, IL, 60510, USA*

[2]*Muons, Inc., 552 N. Batavia Ave., Batavia, IL, 60510, USA*

We describe a Project-X proton driver based on a CW Superconducting RF Linac with final energy as high as 8 GeV. This machine would have the potential to produce multi-MW H^- beams to drive the Fermilab neutrino programs, rare kaon and muon decay experiments, muon cooling R&D programs, neutrino factories, and muon colliders. Keys to a CW machine to suit these uses include ways to generate the desired bunch trains and ways to accumulate many protons in an intermediate storage ring before they are bunched and directed to a target. Enhanced carbon foil techniques can allow accumulation of intense proton beams from a CW linac, which we propose to be extended from 3 GeV to as much as 8 GeV for the most efficient muon production for colliders and neutrino factories and to replace the Booster for improved Main Injector operation.

1. Introduction

One of the possible realizations of Project X [1] at Fermilab is a CW Superconducting RF Linac accelerating H^- beam to 3 GeV. The initial mission of the linac would be to provide intense kaon and muon beams for rare-process experiments. Such experiments typically require high macroscopic duty factors; a variety of bunch lengths, intensities, and patterns are required. Elaborate chopping and beam switching techniques are envisioned to satisfy these requirements simultaneously for multiple experiments. The chopping would occur at high enough frequencies that the high-Q RF cavities would effectively average over the beam microstructure. The beam needed for these experiments, with average beam current as high as 1 mA and kinetic energy as high as 3 GeV, would, with additional acceleration, also provide the basis for the higher-energy, multi-MW beams that are needed for neutrino factories and muon colliders. Additional acceleration beyond the energy of the CW linac of Project X would also allow beam to be delivered to the Main Injector.

---

* Supported in part by Fermilab Research Alliance, LLC under Contract No. DE-AC02-07CH11359

† popovic@fnal.gov

2. Low Energy Linac Front End

The proposed front end for Project X based on a CW SRF Linac assumes a 10 mA H^- source with a warm CW RFQ. The RFQ would operate at 325 MHz or 162.5 MHz to allow the chopper to reject individual bunches. To reduce residual radio-activation, it has been suggested that the output energy of the RFQ be less than 2.5 MeV. Since the required beam peak current is low, the input energy of the RFQ can be 30 keV. Except for the continuous beam requirement, these parameters are not very demanding for the RFQ. The input energy and relatively low peak current also make chopping of input beam less difficult. SNS operation has demonstrated that an ion source extraction system combined with electrostatic focusing and a very short LEBT can preserve the beam emittance even with a peak current of 40 mA. The present SNS LEBT is 12 cm long and allows chopping 200 ns of 40 mA beam at 1 MHz rep rate.

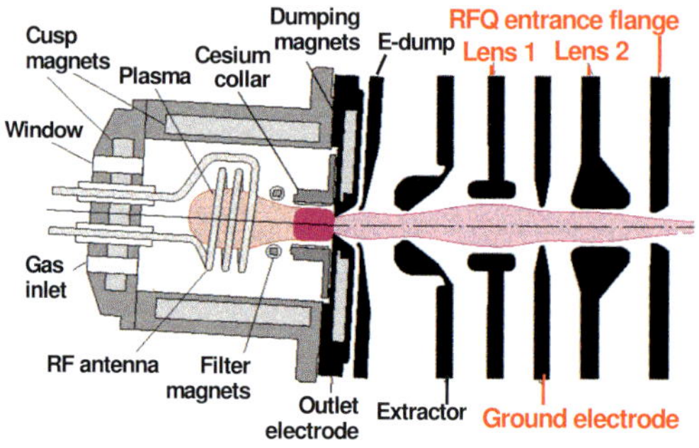

Figure 1: SNS Ion Source with LEBT.

Lowering the peak current to 10 mA reduces space charge effects, removes the need for beam neutralization, and allows a longer electrostatic focusing system with less demanding peak voltages for chopping. It is assumed that the very different beam structure requirements for different experiments can be satisfied by chopping at low energy and splitting with transversely-kicking RF to multiple experiments downstream of the linac.

3. CW Linac from 3 GeV to as High as 8 GeV

It is likely that the 3 GeV Linac, after the Front End and the MEBT, will be based on 325 MHz spoke cavities up to ~200 MeV, continued with 650 MHz beta=0.6 five-cell bell-shaped cavities up to ~500 MeV, followed with 650 MHz beta=0.9 cavities up to about 2 GeV and 1.3 GHz ILC-like cavities up to 3 GeV. It is natural to assume that the linac extension beyond 3 GeV also will be based on the same 1.3 GHz technology extended to a final energy as high as 8 GeV.

4. Beam Delivery and Beam Accumulation

A currently open issue is whether the linac extension should be pulsed or CW. There are various pros and cons. A CW extension would seem reasonable because no modifications to the original 3 GeV CW linac would be needed, no additional R&D would be needed for the beta=1 cavities, and there is not yet a convincing method to make the transition from CW to pulsed mode that conserves the potential beam power enabled by the CW portion.

Muon colliders and neutrino factories need a relatively small number (as few as 10 for a collider, as many as 300 for a factory) per second of intense, short (<~3 ns rms) proton bunches arriving at a pion production target. One or more "post-processing" rings will undoubtedly be necessary to provide the required bunch structures. The crux of this paper is to consider whether and how a CW linac can be used to fill an accumulator ring for these purposes. There are two issues, the long accumulation time and, if a classical stripping foil is used, foil heating.

Another likely use of a higher energy linac would be to fill the Main Injector for its neutrino program to allow the aging Booster synchrotron to be retired. This case is considered first as an example of how to cope with long accumulation time and foil heating. We assume protons are accumulated in the Recycler from a 1 mA CW linac at 8 GeV using existing foil stripping technology. For 4 mm linac beam radius, 600 $\mu g/cm^2$ carbon foil, and assuming that on average every proton comes back onto the foil every 5th turn, we calculate that the temperature of the foil can be kept below 1800 degrees, and 1.5×10^{14} protons can be accumulated in the Recycler in less than 100 ms. Techniques discussed below can reduce the time to inject the whole beam in 2200 turns into the Recycler in 24 ms. Figure 2 shows the foil heating and cooling for the case of injecting in the Recycler. Similar considerations would apply for the case of injecting directly into the Main Injector from a CW linac of less than 8 GeV.

Injection into an accumulation ring for the neutrino factory and/or muon collider case is considerably more challenging, requiring more turns to be injected, because each of these applications requires about 4 MW of beam power and because the accumulation ring must be considerably smaller in circumference than the Recycler (or Main Injector) to alleviate space-charge effects in the subsequent bunch-shortening ring. However, there are also a few factors that will make it easier. The aperture of the ring will be considerably larger than the Recycler in order to accommodate the very large beam sizes that are required in any case to make space charge effects tolerable. The added phase space for the circulating beam, combined with the small emittance of the injected beam, makes it less likely that circulating protons will strike the foil.

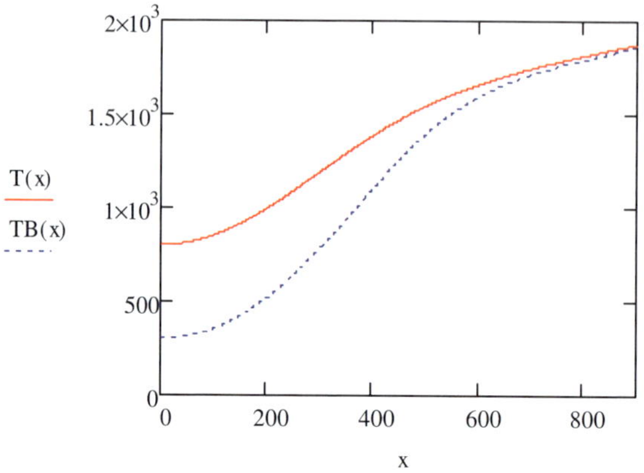

Figure 2: Foil Temperature vs. number of injected turns for initial foil temperatures of 300K (TB-blue dotted) and 800K (T-red solid).

And the smaller CW injected beam current (~ 1 mA or less) means that the foil does not overheat where the incoming beam strikes it. Basically, stretching out the injection process in the CW case allows the foil to radiate considerable power during the injection process, whereas in the pulsed case the injection time is too short to allow much radiative cooling during the incoming beam pulse.

The above factors are probably not sufficient to make a convincing case that foil-stripping injection is feasible from a high-power CW linac. Fortunately, four new concepts have been developed that make it plausible that injecting as many as 10^5 turns into a ring using a stripping foil will be feasible. Of course detailed simulations of the whole process will be necessary to help decide the issue.

Resonant Foil Bypass: The first new idea was presented at the Workshop on Applications of High Intensity Proton Accelerators at Fermilab [2], and dubbed "resonant foil bypass" by the organizers of the parallel session. Basically, the idea is to modulate the incoming beam at a relatively high frequency (so high that the high-Q SRF cavities average over the gaps) using the chopper, and to move the closed orbit synchronously at the foil so that it is closest to the foil only when the incoming beam is "on". In other words, the idea is to run the CW linac with pulsed beam, albeit beam that is pulsed at a high frequency. If the average current during the pulses is 10 mA and the macroscopic average current is 1 mA, then the probability of a circulating proton hitting the foil might be reduced by about an order of magnitude by this means.

Longitudinally Segmented Stripping Foils: The second idea is to take advantage of the facts that the optimum stripping foil thickness for a multi-GeV beam is about 600 μg/cm^2 and that 200 μg/cm^2 thick stripping foils have proven to be quite durable. Thus the stripping foil can be segmented longitudinally into (say) 3 foils, so that there are six surfaces to dissipate the heat by radiation instead of two. The use of at least two foils has been advocated in the past to mitigate the effects of any "pinholes" in the foils.

Rotating Stripping Foils: The third idea is to rotate a circular foil rapidly, with an annulus having a radial extent of about 1 cm exposed in one corner of the aperture. That will spread out the energy deposition over an effective area of several square cm. Mechanical concepts for mounting and supporting such a rotating foil are being developed.

In-ring Debuncher Cavities that also Decelerate: The fourth idea is a so-called in-ring debuncher. Conventionally, debunching cavities are located after a long drift downstream of the linac to reduce the momentum spread of the linac beam before it enters the ring. It is possible that such cavities could instead be located in the ring itself, almost a full turn downstream of the injection region. If they are operated in such a way as to decelerate the incoming beam as well as to rotate it in longitudinal phase space, and if the foil is located in a dispersive region, then this can also reduce the probability that circulating beam strikes the foil on subsequent turns. If the RF frequency of the debunching cavities is not an integer multiple of the revolution frequency, then they will not form buckets that would capture the beam. Simulation will show whether the quasi-random kicks on subsequent turns are tolerable. With luck, they may serve to stabilize the beam rather than to disturb it.

Conclusions

We have presented a concept of a multi-MW CW linac with energy up to 8 GeV that could serve as a natural upgrade to the 3 GeV CW SRF Project X linac that is now under active consideration. We have shown that injection of 1.5×10^{14} particles in 100 ms in the Recycler can be done using conventional foil stripping, keeping the foil temperature below 1800K. Four new concepts were presented that may allow injection of very large numbers of turns from a CW linac into a proton accumulation ring. The accumulated proton beam can then be transferred to another ring where the intense bunches are shortened by a rotation in longitudinal phase space and targeted to create intense pion and muon beams for a neutrino factory and/or a muon collider.

References

[1] Fermilab Project X http://projectx.fnal.gov/

[2] C. Ankenbrandt, "Accumulation and Bunch Rotation Schemes", Workshop on Applications of High Intensity Proton Accelerators, October 19-21, 2009, Fermi National Accelerator Laboratory, Batavia, IL, USA. http://conferences.fnal.gov/App-Proton-Accelerator/index.html

RAPID CYCLING SYNCHROTRON OPTION FOR PROJECT X*

WEIREN CHOU†

Fermi National Accelerator Laboratory

P.O. Box 500, Batavia, Illinois 60510, USA

This paper presents an 8 GeV Rapid Cycling Synchrotron (RCS) option for Project X. It has several advantages over an 8 GeV SC linac. In particular, the cost could be lower. With a 2 GeV 10 mA pulsed linac as injector, the RCS would be able to deliver 4 MW beam power for a muon collider. If, instead, a 2 GeV 1 mA CW linac is used, the RCS would still be able to meet the Project X requirements but it would be difficult for it to serve a muon collider due to the very long injection time.

1. Introduction

Work on an RCS as an option for a proton driver for a muon collider began around 1998. Several laboratories including Fermilab, BNL, KEK and RAL carried out studies and published design reports. At Fermilab, two proton driver studies, with acronyms PD1 and PD2, were initiated in the period from 1998 to 2002. PD1 was a 16 GeV RCS and PD2 an 8 GeV RCS. Both studies were documented [1,2]. In 2004, Fermilab decided to abandon the synchrotron approach and opted for an 8 GeV SC linac, also a proton driver. The principal reason for this choice was its synergy with the ILC, the ultimate goal of Fermilab at that time. Unfortunately, the proton driver proposal was rejected by DOE because of the "fast track" strategy for the ILC. In 2007, however, when the ILC cost estimate was published and it was seen to be about twice the previous TESLA estimate, DOE lost enthusiasm for this project. Fermilab decided to go back to the proton driver, which was renamed Project X. Furthermore, DOE allowed decoupling between Project X and the ILC. Thus, after a complete circle the RCS option is back on the table. In the meantime, Japan has built a 1 MW 3 GeV RCS as part of the J-PARC project.

* This work is supported by Fermi Research Alliance, LLC under Contract No. DE-AC02-07CH11359 with the United States Department of Energy.

† E-mail: chou@fnal.gov

There are two fundamental requirements of a proton driver: high beam power and short bunch length. Both are achievable by either an RCS or a linac. The main advantage of an RCS is the lower cost. A preliminary comparison shows the cost of an 8 GeV RCS could be 40% lower than an 8 GeV SC linac ($0.93B *vs.* $1.5B). Moreover, since the injection energy of an RCS is low, the stripping foil would be easier, stripping efficiency higher and more beam loss at injection could be tolerated. The transport of H^- ions in the beam line would be easier because stronger bends could be used and there would be no need for a cryogenic beam pipe. Existing enclosures could be reused. On the other hand, however, a linac is simpler to design, build and operate than a synchrotron and has higher reliability. The hardware of an RCS is more challenging (large aperture magnets, rapid cycling power supplies, high power tunable RF system, field tracking during cycle, eddy current in beam pipe, etc.).

2. RCS for Project X

Compared to previous RCS studies, there are several new conditions that would make a better RCS possible for Project X.

- Elimination of the Main Injector (MI) injection front porch: This is piggy-backing on the NOvA project, which uses the Recycler as an accumulator.
- Free housing: The antiproton source enclosure would be available after 2011 and could house an 8 GeV RCS (see below).
- Higher injection energy: A 2 GeV SC linac (*vs.* 600 MeV in PD2) as an injector of the RCS would reduce the RCS aperture requirement.

Fig. 1 shows an 8 GeV RCS located in the present antiproton source enclosure. It has the same size and shape as the Debuncher [3]. The triangular lattice shown in Fig. 2 is transition-free thanks to its high γ_t (18.6). This is a simple doublet lattice and uses only one type of bending magnets and one type of quadrupoles. The missing magnet scheme in mid-cell provides zero-dispersion straight sections without dispersion suppressors [4].

3. RCS for the Muon Collider

To use an RCS as a proton driver for a muon collider, the main requirements are: 4 MW beam power and 3 ns bunch length. We consider two scenarios.

3.1. *Pulsed Linac*

Table 1 lists the parameters when a 10 mA pulsed linac is used as the RCS injector. In order to reduce the bunch number from 90 to 5, one may add a compressor ring, which would coalesce 18 bunches into one and form 5 super-

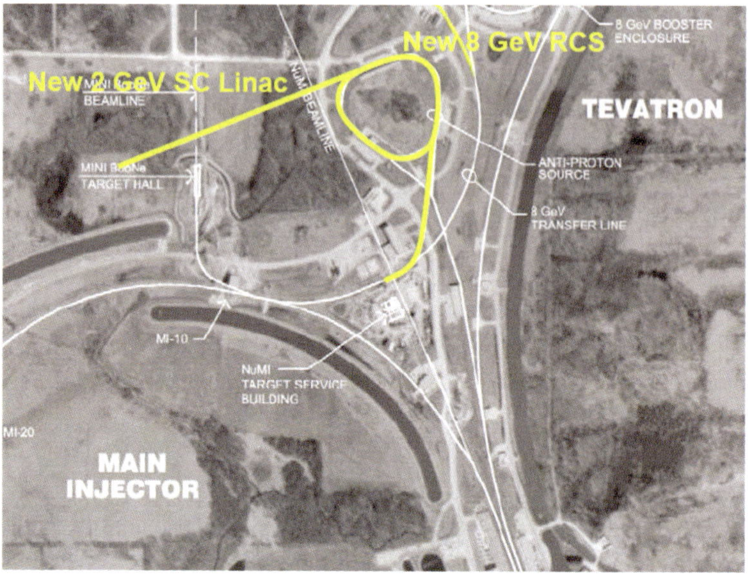

Figure 1. Layout of a 2 GeV SC linac and an 8 GeV RCS.

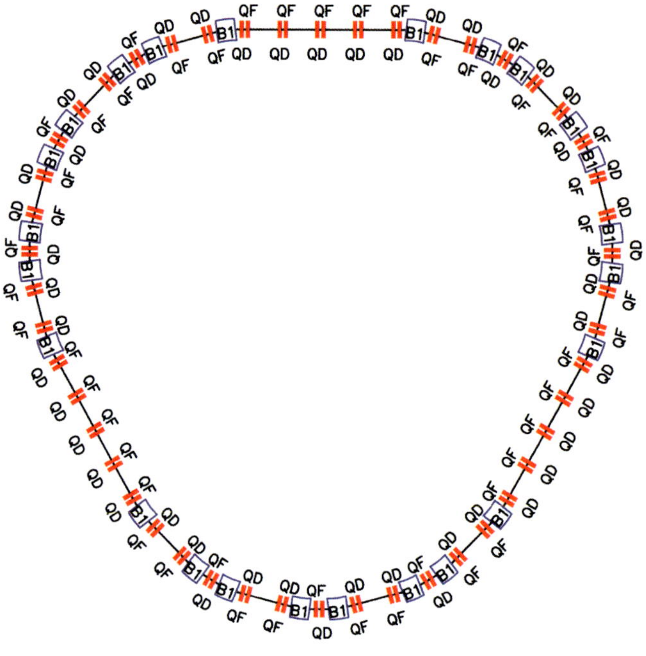

Figure 2. 2D plot of a triangular transition-free lattice.

bunches, each of 2×10^{13} protons (20 TP). These super-bunches would be phase-rotated to 3 ns in length, a compression ratio of 6. The bunch spacing can be controlled by using RF barriers.

3.2. *CW Linac*

Table 1 also lists the parameters when a 1 mA CW linac is used as the RCS injector. It is seen that the injection time becomes very long (16 ms), which would lead to two major problems. One is the excessive heating on the stripping foil due to the large number of hits per particle. Another is 30 Hz operation of the RCS would be questionable because the injection takes half cycle time.

It should be pointed out that a CW linac serving the RCS would be fine for Project X, since the RCS needs only to deliver 0.5 MW, not 4 MW.

Table 1. Main parameters of a 4 MW RCS with a pulsed or CW linac.

	Pulsed Linac	**CW Linac**
Linac		
Rep rate (Hz)	30	CW
Kinetic energy (GeV)	2	2
Peak current (mA)	10	1
Pulse length (μs)	1600	16000
H^- per pulse	1×10^{14}	1×10^{14}
Beam power (MW)	1	1
RCS		
Rep rate (Hz)	30	30
Circumference (m)	505.3	505.3
Extraction energy (GeV)	8	8
Protons per bunch	1.1×10^{12}	1.1×10^{12}
Number of bunches	90	90
Protons per cycle	1×10^{14}	1×10^{14}
Norm. transverse emit. (mm-mrad)	40π	40π
Longitudinal emit. (eV-s)	0.2	0.2
Injection time (μs)	1600	16000
Injection turns	950	9500
Maximum Laslett tune shift	0.15	0.15
Extraction bunch length (rms, ns)	1	1
RF frequency (MHz)	53	53
Beam power (MW)	4	4

References

1. "The Proton Driver Design Study," *Fermilab-TM-2136* (December 2000).
2. "Proton Driver Study II – Part 1," *Fermilab-TM-2168* (May 2002).
3. W. Chou, "A Hybrid Design of Project X," *PAC2009 Proceedings*, May 4-8, 2009, Vancouver, Canada.
4. W. Chou, "A Simple Transition-free lattice of an 8 GeV Proton Synchrotron," *PAC2009 Proceedings*, May 4-8, 2009, Vancouver, Canada.

LINAC-BASED PROTON DRIVER FOR A NEUTRINO FACTORY

R. GAROBY, E. BENEDETTO, M. AIBA, M. MEDDAHI
CERN, Geneva, Switzerland

A Neutrino Factory Proton Driver based on a superconducting proton linac has been designed in the CERN context. The 5 GeV/4 MW H^- beam from the linac is accumulated using charge exchange injection in a fixed-energy synchrotron and afterwards transferred to a compressor ring, where bunch rotation takes place. The lattices of the accumulator and compressor are described, as well as magnet technology and RF manipulations. Critical issues related to charge-exchange injection, space-charge effects in the compressor and beam stability in the accumulator, are addressed. The analysis is focused on the baseline scenario, which provides 6 bunches on the target. Results of preliminary analysis of options with less bunches (three and one) are also presented.

1. Introduction

According to the specification assembled in the context of the International Scoping Study [1], the Proton Driver for a Neutrino Factory must deliver 4 MW of beam power at 5-15 GeV, in a small number of short bunches (~2 ns rms) onto a production target. The CERN–proposed scenario [2] is based on the 5 GeV High Power version of the Superconducting Proton Linac (HP-SPL) which can deliver 10^{14} protons at a repetition rate of 50 Hz [3]. Two additional fixed-energy rings are needed to transform the long linac pulse into the required bunches on the target [2, 4, 5].

2. Principle

In the first ring (the accumulator), the chopped linac beam is accumulated in a few long bunches, using charge-exchange injection. The accumulator is isochronous to preserve the time structure of the linac beam, and it has no RF system to minimize impedance. Once accumulation is finished, bunches are transferred one by one to the compressor ring where bunch rotation takes place with the energy stored in the RF cavities. The ratio between the two rings is such as to guarantee the arrival of the successive bunches at the correct location in the compressor. This principle is described in Ref. [4] and sketched in Fig. 1 in the case of 6 bunches.

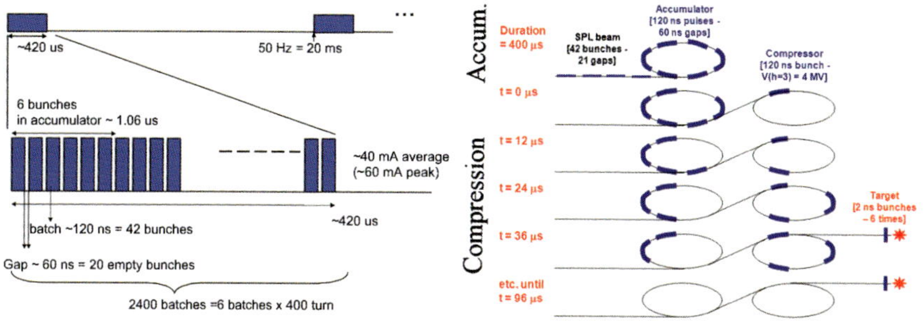

Figure 1. Principle of bunch generation with a Linac-based Proton Driver.

3. Accumulator and Compressor Design

Figure 2 shows the lattices of both rings in the 6 bunches case. In contrast with the accumulator which is isochronous and without RF, the compressor has large slippage factor and RF voltage for a fast phase rotation. Table 1 summarizes the main parameters of both rings. The transverse emittance of 3π mm.mrad results from the competing requirements of injection foil heating, aperture, space charge and beam size on target, while the longitudinal parameters are set by phase rotation and required bunch length on target, leading to a momentum spread of 0.863×10^{-3} rms and a total bunch length of 120 ns before rotation. Variable chopping of the linac beam is used to obtain a flat profile, with smooth edges to minimize longitudinal space-charge effects.

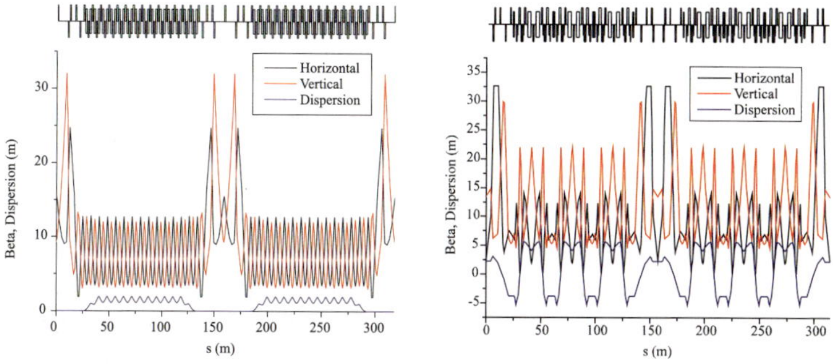

Figure 2. Accumulator and compressor lattices for the 6-bunches scenario.

4. Critical Issues

Two–dimensional painting is used in the charge exchange injection to keep the stripping-foil temperature below 2000 K. A moving vertical bump of the H^- beam is also necessary in the case of a single bunch [5]. The minimum emittance achieved in this case is 67% larger (5 instead of 3 π mm.mrad).

Table 1. Main parameters of the accumulator and compressor rings.

Ring	Parameter	6 bunches case	3 & 1 bunch cases
Accumulator	Circumference (m)	318.5	185.8
	Nb. of turns of accumulation	400	640 / 1920
	Working point (H/V)	7.77 / 7.67	7.37 / 5.77
	Type of magnets	NC	SC
Compressor	Circumference (m)	314.2	200
	Nb. of turns for compression	36	86
	RF voltage (MV)	4	1.7
	Transition gamma	2.3	2.83
	Working point (H/V)	10.79 / 5.77	4.21 / 2.74
	Type of magnets	SC (bendings)	NC
	Time interval between bunches (μs)	12	30

Results of ORBIT [7] simulations of bunch rotation (Fig. 3) show that transverse space-charge is tolerable, partly because of the rapidity of the rotation and partly because of the large horizontal beam size due to dispersion which keeps the Laslett tune shift small enough to avoid integer resonance crossing.

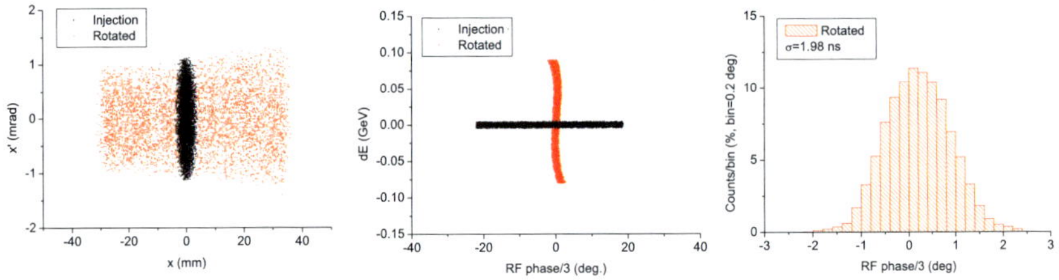

Figure 3. Phase space plots before and after bunch rotation.

Because of the absence of synchrotron motion in the isochronous accumulator, beam stability in the longitudinal and transverse planes deserved study. Analytical and computer tracking results (using HEADTAIL [8]) show that the beam can remain stable in all planes applying acceptable practical constraints [6]. The narrow-band component of machine impedance can be neglected due to the absence of RF cavities and the resistive-wall is not an issue because the rise-time $\tau \sim 8.2$ ms is long compared to the accumulation time. The effect of the broad-band impedance (BB) was analyzed with simulations, assuming the full intensity during all of the accumulation. Figure 4 (left) shows the transverse beam size evolution for different values of chromaticity, considering a BB impedance of 1 MΩ/m, 1 GHz and $Q_R = 1$. The instability can be cured with a tune spread of $\Delta Q \sim 0.02$ either from chromaticity or from octupoles detuning with amplitude. Threshold scales linearly with impedance.

As shown in Fig. 4 (right), a longitudinal broad-band impedance of a few Ω is tolerable. Electron-clouds should not be an issue, because the long and flat bunch profiles are not favorable to their build-up. Similar numbers are obtained for the 3-bunches accumulator. The single bunch scenario is expected to lead to three times more stringent values, since the bunch intensity is triple.

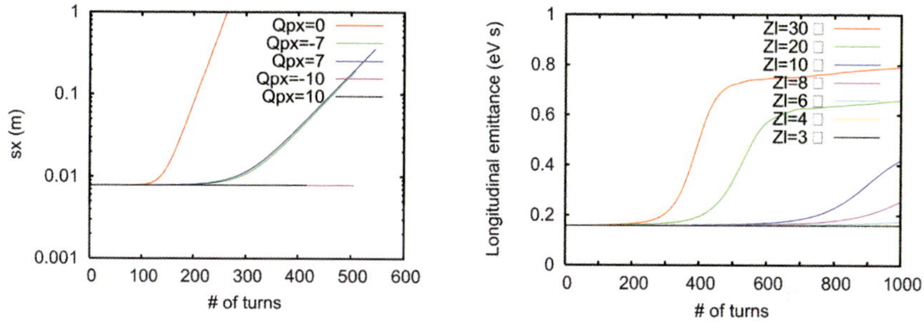

Figure 4. Left: Horizontal beam size evolution, as a function of chromaticity for $Z_{\perp}$=1MΩ/m, Q_R=1, f_R=1 GHz. Right: Longitudinal emittance evolution for different values of $Z_{//}$.

5. Conclusion

A 4 MW beam with the time structure required by a neutrino factory can be obtained with a linac-based proton driver. More work is now required for minimizing beam-loss and cost, in close connection with the design of equipment like RF, collimators and injection and extraction systems.

Acknowledgments

We are grateful to C. Carli, S. Cousineau, F. Gerigk, M.Martini, E.Metral, W. Herr, T. Nakamoto, G. Rumolo and A. Shishlo for their suggestions.

References

1. M. Apollonio, *et al.*, IOP Electronic journals, 2009_JINST_4_P07001.
2. R. Garoby, in Proc. NuFact'06, Irvine, CA (2006).
3. R. Garoby, sLHC-PROJECT-Report-0015 (2009).
4. M. Aiba, CERN-AB-2008-060 BI (2008).
5. M. Aiba, CERN-AB-Note-2008-048-BI (2008).
6. E. Benedetto, CERN-BE-2009-037 and CERN-NUFACT-Note-156 (2009).
7. J. Galambos, *et al.*, SNS/ORNL/AP TN 0011 (1999).
8. G. Rumolo, F. Zimmermann, CERN-SL-Note-2002-036 (2002).

PION PRODUCTION FOR NEUTRINO FACTORIES AND MUON COLLIDERS*

N. V. MOKHOV†, K. K. GUDIMA, J. B. STRAIT, S. I. STRIGANOV

Fermilab, MS 220, P.O. Box 500
Batavia, IL 60510, U.S.A.

Optimization of pion and muon production/collection for neutrino factories and muon colliders is described along with recent developments of the MARS15 code event generators and effects influencing the choice of the optimal beam energy.

1. Target System

Neutrino factory and muon collider (NFMC) projects are feasible only if adequate numbers of muons can be produced and collected. In the baseline front-end, a proton beam hits a thick target in the center of a high-field solenoid, followed by a matching section and a solenoidal decay channel which collects muons resulting from pion decays. A list of targetry issues includes [1]: maximal production of pions resulting in the highest number of useful muons at the downstream end of the system, background particles transported down the beamline, target survivability, beam window fatique and structural integrity, superconducting (SC) coil quench stability, heat loads on target and cryogenic system, radiation damage and activation of materials near the beam, spent proton beam handling, and numerous shielding issues from prompt radiation to ground-water activation. "Useful" muons are those captured by a phase rotation system downstream of the decay channel, with a kinetic energy in the range $40 < T_\mu <$ 180 MeV.

All these issues have been addressed in detailed simulations. Thorough MARS optimizations of the target system have been performed [1, 2] for 4 to 30 GeV proton beams. The findings included: a gain from high-Z target materials, especially at the high beam energies; a 1-m long hybrid solenoid at $B_z \times R_a^2=1125$

* Work supported by Fermi Research Alliance, LLC under contract No. DE-AC02-07CH11359 with the U.S. Department of Energy.

† E-mail: mokhov@fnal.gov

T×cm^2 with B_z=20T and aperture R_a=7.5cm followed by a matching section (to ~20 m) and decay channel (R=30 cm, B=1.25T) to 50-100 m with SC coils protected by water-cooled tungsten-carbide balls; open mercury jet target (R=5 mm) tilted at 100 mrad with a 2λ (~30 cm) beam-jet interaction region; proton beam at 67 mrad with an RMS beam spot size of 1.5 mm; pion/muon yields Y at the end of the decay channel growing with the beam energy E_p, while Y/E_p having a broad maximum at E_p ~ 6 GeV.

2. MARS Event Generators

The MARS code [3] was and still is the main tool to study the NFMC targetry issues listed above. Particle production in the code is modeled with the built-in event generators. In the current version, MARS15, as a default, an inclusive phenomenological model is used from 3-5 GeV to tens of TeV supplied at lower energies with an exclusive Cascade-Exciton Model code CEM03 [4], combined with the Fermi break-up model, coalescence model, and Generalized Evaporation-fission Model (GEM2). A recent multi-fragmentation extension is available. Alternatively, the Los Alamos Quark-Gluon String Model code, LAQGSM [4], is used for photon, hadron and heavy-ion projectiles from a few MeV/A to about a few TeV/A. It shares with CEM03 some modules at E < 3-5 GeV. This option provides a power of full theoretically consistent modeling of exclusive and inclusive distributions of secondary particles, spallation, fission, and fragmentation products. It includes recently improved unified algorithms at intermediate (2 to 10 GeV) and low (< 20 MeV) projectile energies. Among other things, the code has been switched to Δ and N* resonance production and interaction at E < 4.5 GeV with pions produced in Δ-decays. Results of modeling are in an agreement with pion production data, specifically at energies 2 < E < 15 GeV (see, for example, recently published papers on BNL-E910 and HARP [5] experiments).

3. Front-End Channel Acceptance and Pion Production

The MARS calculated acceptance of the front-end channel described above (see also Ref. [1, 2]) is shown in Fig. 1. It is defined as the probability that a pion of a given angle and momentum at the downstream end of the target yields a surviving muon (or pion) at the end of the channel (z=50 m). To be accepted and efficiently captured by the downstream RF channel, the muon must have a kinetic energy in the range $40 < T_\mu < 180$ MeV. One sees that the majority of the useful pions has the momenta $0.075 < p < 0.475$ GeV/c and is produced in the forward direction ($\theta < \pi/2$). For protons interacting with a mercury nucleus (thin target), a pion (kaon) yield Y in this phase space grows rapidly with proton energy starting from the threshold (~0.2 GeV) to about 2 GeV, slowing down at higher energies and reaching a logarithmic growing mode at about 15 GeV. As a result, the yield over proton energy, Y/E_p, has a maximum at ~2 GeV (Fig. 2).

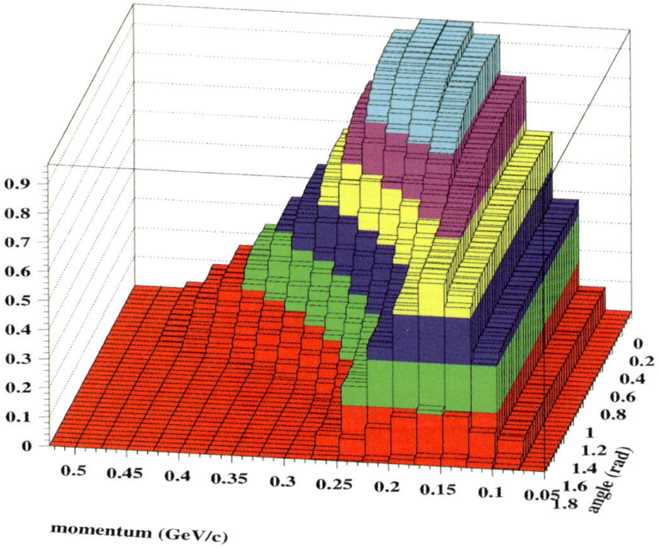

Figure 1. Front-end channel acceptance.

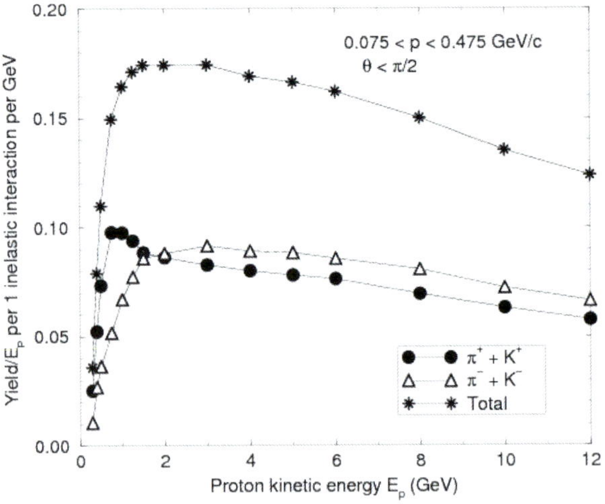

Figure 2. Pion/kaon yield Y/E_p for p+Hg interactions as calculated with MARS15(LAQGSM).

4. Pion Production and Optimal Beam Energy for NFMC

In a realistic thick target, absorption of low-energy pions in target material and secondary interactions (showering) at high energies move the maximum of Y/E_p to a larger E_p. As a result, the Y/E_p (E_p) dependence becomes pretty flat at $2 < E_p < 15$ GeV. The total acceptance-weighted yield $Y/E_{p,}$ modeled with MARS15(LAQGSM) for a 2λ thick mercury target, is constant 0.15±5% in this energy range with a broad maximum around E_p=7 GeV. These results are shown in Fig. 3 in comparison with HARP data for a thin tantalum target [5]. The later are acceptance-weighted and corrected for the unmeasured angular region, $\theta <$ 350 mrad (via quadratic extrapolation to θ=0 of HARP group data, 12-19% effect), and MARS-calculated thick target effect.

One can conclude that any beam energy in the 4-12 GeV range represents a good choice for the proton driver for a neutrino factory or muon collider.

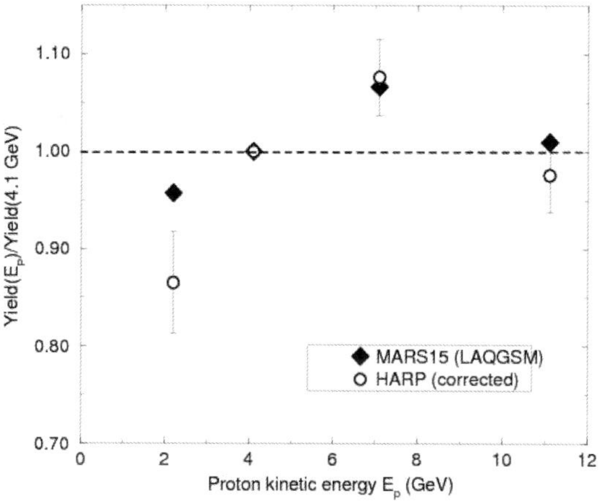

Figure 3. Beam-power normalized muon yield at z=50 m with respect to that at E_p=4.1 GeV as calculated with MARS15(LAQGSM) for mercury and measured in Ref. [5] for tantalum.

References

1. N.V. Mokhov, *Nucl. Instr. Meth. Phys. Res.*, **A472**, 546 (2001).
2. N.V. Mokhov, *Proc. 2001 Part. Accel. Conf., 745-747, Chicago, June 2001.*
3. N.V. Mokhov, Fermilab-FN-628 (1995); N.V. Mokhov, S.I. Striganov, in *Proc. of Hadronic Shower Simulation Workshop*, Fermilab, *AIP Conf. Proc., **896**, pp. 50-60 (2007)*; http://www-ap.fnal.gov/MARS/.
4. S.G. Mashnik et al., LANL Report LA-UR-08-2931 (2008); arXiv:0805.0751v1 [nucl th] 6 May 2008.
5. M. G. Catanesi et al. Phys. Rev. **C 77**, 055207 (2008).

PROTON BUNCH COMPRESSION STRATEGIES*

VALERI LEBEDEV†
Fermilab, P.O. Box 500
Naperville, IL 60563, USA

The paper discusses main limitations on the beam power and other machine parameters for a 4 MW proton driver for muon collider. The strongest limitation comes from a longitudinal microwave instability limiting the beam power to about 1 MW for an 8 GeV compressor ring.

1. Introduction

The muon collider luminosity is proportional to the square of the number of muons delivered to a collider and consequently to the square of proton driver power. In the energy range of 8 to 20 GeV the efficiency of muon production per unit power of the proton beam has a weak dependence on the beam energy and therefore the major requirement for the proton driver is the total beam power. Presently, there are a few proposals for a muon collider [1]. All of them require a proton driver with 4 MW beam power and the rms bunch length at the muon production target below 2-3 ns. While the repetition rates of different proposals vary, in this paper we assume the repetition rate of 15 Hz.

At the very least the proton part of the muon collider consists of two pieces. The first one is the proton driver which can be based on a rapid cycling synchrotron (RCS) or superconducting (SC) linac. The second one is a compressor ring. In order to simplify the requirements of the RF system some proposals suggest an intermediate ring to accumulate the strip-injected H^- beam accelerated in the SC linac – the Accumulator [2]. In this paper we do not discuss limitations coming from the beam injection and RF systems and therefore do not separate the Accumulator and Compressor rings. In this case the process consists of two major steps: (1) the injection of the beam to the

* This work is supported by the U.S. Department of Energy under contract No. DE-AC02-76CH03000
† val@fnal.gov

compressor ring, and (2) the longitudinal bunch compression performed by the bunch rotation in the phase space excided by high voltage RF system.

The main beam power limitations come from the longitudinal microwave instability, the betatron tune shift due to beam space charge and the beam size limitation on the target. There are also a number of engineering limitations which are also important but will not be discussed in here in any detail.

2. Longitudinal Microwave Instability

First, we consider the beam stability criterion for the continuous beam. Let the beam longitudinal distribution be described by function $f(p)$ which is normalized to unity:

$$\int f(p)dp = 1 \ , \tag{1}$$

where p is a momentum deviation from its mean value, p_0. We additionally rescale the argument of the distribution function by the rms momentum spread so that the new and old distributions are related as following:

$$f_0(p) = \frac{1}{\sigma_p}\psi_0\left(\frac{p}{\sigma_p}\right) \quad , \quad \sigma_p{}^2 = \int p^2 f_0(p)dp \quad . \tag{2}$$

This results in the following properties of the new distribution function:

$$\int \psi_0(x)dx = 1 \quad , \quad \int x^2\psi_0(x)dx = 1 \quad , \tag{3}$$

where $x = p/\sigma_p$.

The coherent frequency, $\delta\omega$, of the mode n can be found from the well-known dispersion equation [3]:

$$1 - iA_n \int_{\delta\to+0} \frac{d\psi_0/dx}{y + x - i\delta\operatorname{sign}(n)}dx = 0, \tag{4}$$

where

$$A_n \equiv \frac{eI_0}{2\pi cp_0\beta\eta}\left(\frac{p_0}{\sigma_p}\right)^2\left(\frac{Z_n}{n}\right), \quad y = \frac{\delta\omega}{n\omega_0\eta}\left(\frac{p_0}{\sigma_p}\right) , \tag{5}$$

I_0 is the beam current, e is the particle charge, Z_n is the longitudinal impedance on the n-th harmonic of the revolution frequency, $\eta = \alpha - 1/\gamma^2$ is the ring slip factor, and $c\beta$ is the beam velocity. At the stability boundary the imaginary part of coherent frequency is equal to zero, Im($\delta\omega$) = 0. Using this condition in Eq. (4) one obtains a parametric equation for the stability boundary:

$$A(y) = \left(i \int_{\delta \to +0} \frac{d\psi_0 / dx}{y + x - i\delta \mathrm{sign}(n)} dx = 0 \right)^{-1} . \qquad (6)$$

Numerical integration of Eq. (6) was carried out for the following set of functions:

$$\psi_k(x) = \frac{1}{b_n} \exp\left(-\frac{x^{2k}}{s_k} \right) , \qquad (7)$$

where coefficients b_k and s_k are determined by Eqs. (3). For the first four functions, k=1…4, they are:

$$b_k = \left\{ \sqrt{2\pi}, \frac{\pi^{3/2}}{\sqrt[4]{2}\,(\Gamma(3/4))^2}, \frac{2^{3/2}\pi^{5/4}}{3(\Gamma(5/6))^{3/2}}, \frac{(\Gamma(1/8))^{3/2}}{4\sqrt{\Gamma(3/8)}} \right\} \approx \{2.507, 3.118, 3.288, 3.358\},$$

$$s_k = \left\{ 2, \frac{\sqrt{2}\pi}{(\Gamma(3/4))^2}, \left(\frac{2\sqrt{\pi}}{\Gamma(5/6)} \right)^3, \left(\frac{(\Gamma(1/8))}{\Gamma(3/8)} \right)^4 \right\} \approx \{2, 8.754, 30.97, 102.04\}. \qquad (8)$$

The function $\psi_1(x)$ describes the Gaussian distribution. Figure 1 presents plots for the first four functions and corresponding stability boundaries.

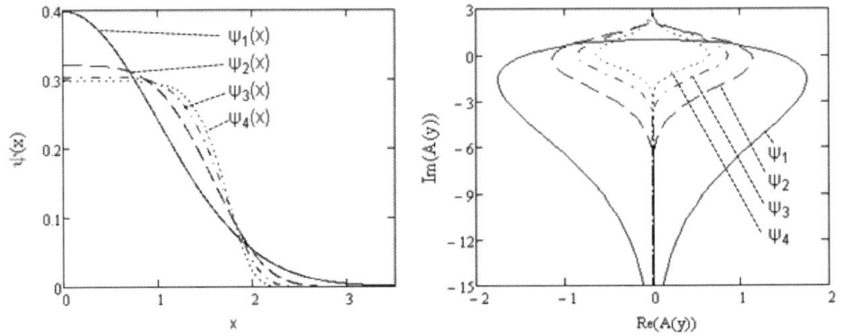

Figure 1. Plots for functions $\psi_k(x)$ (left) and corresponding stability boundaries (right).

In the energy range necessary for the proton driver the impedance is dominated by the beam space charge and $Z \propto i$. Consequently, below the transition ($\eta < 0$) $A_n \propto Z\eta \propto -i$. One can see from the Figure 1 that for Gaussian beam (k = 1) the stability region below transition is significantly higher than above. However this difference is much less significant for distributions with steeper edge (k > 1). Note that these distributions better represent a longitudinal distribution obtained by painting small emittance linac beam into a much larger ring acceptance. Consequently, operation above critical energy looks more preferable because significantly larger value of slip factor can be achieved. In this case the stability criterion is:

$$|A_n| \equiv \left| \frac{eI_0}{2\pi c p_0 \beta\eta} \left(\frac{p_0}{\sigma_p} \right)^2 \left(\frac{Z_n}{n} \right) \right| \le B_k \quad , \tag{9}$$

For the first four functions $\psi_k(x)$ the coefficients B_k are: $B_k = \{1, 2.189, 2.665. 2.747\}$ (these numbers correspond to y coordinates of top points at the stability diagram of Figure 1.) In further consideration we will assume $B_k \approx 2$.

The space charge impedance for the round beam with Gaussian distribution with rms size $\sigma_\perp$ in the round vacuum chamber of radius a is

$$\frac{Z_n}{n} \approx i \frac{Z_0}{\beta\gamma^2} \ln\left(\frac{a}{1.5\sigma_\perp} \right) , \tag{10}$$

where $Z_0 \approx 377\ \Omega$ is the impedance of free space, and β and γ are the relativistic factors. In the energy range of 8 to 20 GeV $|Z_n/n|$ is changing from ~10 to ~2 Ω. An achievement of high bunch density implies good electrodynamics of vacuum chamber. In this case the impedances due to vacuum chamber inhomogeneities and a finite value of its wall resistivity do not exceed 1 Ω for frequencies above the particle revolution frequency of ~1 MHz. Substituting Eq. (10) into Eq. (9) and equaling $B_k = 2$ results in the stability condition:

$$\frac{r_p N}{\beta^2\gamma^3\eta(\sigma_p / p)^2 L_b} \ln\left(\frac{a}{1.5\sigma_\perp} \right) \le 1 \quad , \tag{11}$$

where r_p is the classical proton radius, N is the number of particles in the beam and L_b is the bunch length. Here we imply the uniform particle density along the bunch. It happens naturally if the beam is adiabatically debunched after arriving from the RCS. As can be seen from Eq. (11) the stability threshold does not depend on the mode number because the longitudinal impedance, Z_n/n, does not change up to very high frequency, $f_{max} \approx c\gamma/(4a)$. For the energy range of 8 to 20 GeV it corresponds to frequency range of ~20 – 40 GHz. Above this frequency Z_n/n starts to decrease.

Well above the instability threshold the instability growth rate can be easily calculated and is equal to:

$$\lambda_n \approx n\omega_0 \sqrt{\frac{2r_p N\eta}{\beta^2\gamma^3 L_b} \ln\left(\frac{a}{1.5\sigma_\perp} \right)} \quad , \tag{12}$$

where we assume that the space charge makes a dominant contribution to the longitudinal impedance. As one can see from the above equation the growth rate is proportional to the harmonic number and the instability can be extremely fast at the high frequency end. Although a low frequency part, ≤0.5 GHz, can in principle be stabilized by a longitudinal damper the high frequency part cannot be stabilized; and therefore one should not exceed the instability threshold. Note that Eq. (11) and (12) were derived for the continuous beam. However at sufficiently high frequencies, where the growth rate becomes larger than the synchrotron frequency, these equations correctly describe the threshold and growth rate. In this high frequency region this instability is called the microwave instability.

3. Mitigation of the Microwave Instability

Expressing the number of particles through other machine and beam parameters in Eq. (11) one obtains the beam power limitation imposed by the longitudinal instability:

$$P_{\max} \le f_{rep} m_p c^2 (\gamma - 1) \frac{\beta^2 \gamma^3 \eta (\varepsilon_L / p_0)^2}{F_\varepsilon^{\ 2} r_p L_b \ln(a/(1.5\sigma_\perp))} , \tag{13}$$

where f_{rep} is the repetition rate, m_p is the proton mass, $\varepsilon_L = F_\varepsilon\, \sigma_p\, L_b$ is the 95% beam longitudinal emittance, and F_ε is a form-factor binding up the bunch length, the rms momentum spread and the longitudinal emittance. For the first four functions ψ_k the form-factors are: $F_\varepsilon \approx \{1.96, 1.80, 1.75, 1.71\}$.

The longitudinal emittance is approximately conserved during the bunch rotation, and, as can be seen from Eq. (13), the beam power is limited by the initial bunch length, L_b. The bunch rotation shortens the bunch and, consequently, widens the stability margin. Note that the momentum acceptance of the compressor ring is much larger than the initial momentum spread and therefore the instability does not cause the beam loss. It rather results in a fast increase of the momentum spread to its value determined by Eq. (11). Consequently, the bunch rotation results in a longer bunch on the target.

The required value of longitudinal emittance is set by the rms bunch length on the target, L_t, and the maximum momentum spread, $\Delta p_{\max}$. The bunch length on the target is determined by requirements of the muon cooling channel and, as it was already quoted, is ~60 cm. The relative momentum spread is set by nonlinear chromatic effects in the compressor ring and the chromaticity of the beam focusing on the target. In our considerations we will assume the momentum aperture ±2.5% – same as for the Debuncher – the Fermilab ring with largest momentum and transverse acceptances. It results in $\varepsilon_L/cp_0 \approx 90$ ps, which, in the first approximation, does not depend on the compressor ring energy.

To maximize the slip factor a sufficiently small transition energy should be chosen. In this case $\eta \approx \alpha \approx D/R_0$, where R_0 is the mean ring radius and D is the effective dispersion. On other hand $\Delta p_{max} / p_0 \approx a_{max} / D$, where a_{max} is the maximum beam size in the dipole. Combining one finally obtains: $\eta \Delta p_{max} / p_0 \approx a_{max} / R_0$. Taking this into account we can write an estimate for power limitation resulting from Eq. (13) in the following form:

$$P_{\max} \approx \frac{1}{2}\frac{f_{rep} m_p c^2 (\gamma-1)\gamma^3 \beta^2 L_t^{\,2} a_{\max}}{2\pi r_p R_0^{\,2} \eta_{fill} \ln(a/(1.5\sigma))} \frac{\Delta p_{\max}}{p_0} , \qquad (14)$$

where $\eta_{fill} = L_b / (2\pi R_0)$ is the part of the machine circumference taken by the beam after injection, and the factor 1/2 at the right-hand side corrects for real machine optics and difference between rms and maximum momentum spreads on the target (see details below). Note that in difference to Eq. (13) this equation is approximate and should be used cautiously. However it correctly reflects the dependence of beam power on main machine and beam parameters. For an 8 GeV compressor ring with average radius R_0 = 42 m and aperture a_{max} = 10 cm one obtains $P_{max} \approx 1.2$ MW. One can see that the maximum power has very steep dependence on the beam energy. Taking into account that $R_0 \propto \gamma$ and other parameters have comparatively weak dependence on the beam energy one obtains that the beam power is proportional to γ^2.

4. Limitations on the Beam Emittance

For a round Gaussian beam and the smooth focusing approximation the betatron tune shift due to beam space charge is equal to:

$$\delta\nu_{SC} = \frac{r_p R_0}{2\beta^2 \gamma^3 \varepsilon_\perp} \frac{dN}{ds} \qquad (15)$$

where dN/ds is the bunch longitudinal density, and $\varepsilon_\perp$ is the transverse rms beam emittance. The tune shift achieves its maximum at the end of bunch compression and imposes serious limitations on the machine parameters. Assuming a Gaussian longitudinal distribution at the end of bunch compression one can obtain the power limitation imposed by the tune shift:

$$P_{\max} \le f_{rep} m_p c^2 (\gamma-1) \frac{2\sqrt{2\pi}\beta^2 \gamma^3 \varepsilon_\perp L_t}{r_p R_0} \delta\nu_{SC} . \qquad (16)$$

A conservative estimate based on the experience accumulated in Fermilab and other laboratories requires $\delta\nu_{SC}$ being less or about 0.05. In this case the beam loss is expected to be small – the primary requirement for a MW scale machine. More detailed study is required to see a feasibility of larger tune shift achievement without an increase of the beam loss.

An increase of the transverse emittance is a natural way to mitigate the space charge effects. However this increase is limited by a necessity to focus the beam on the muon production target. An optimal rms beam size on the target is 2.5 mm [4]. The length of proton beam interaction with the target is comparatively large. That limits the beam angular spread on the target and, consequently, the beta-function on the target, β^*. The pion production suffers if $\beta^* \leq 20$ cm. That determines the maximum acceptable emittance to be ~30 mm mrad independently on the beam energy. Taking into account that R_0 is proportional to the beam momentum results in that P_{max} is proportional to $\sim\gamma^3$.

5. Compressor Ring and its Tentative Parameters

The beam power limitations imposed by the microwave instability and by the betatron tune shift due to beam space charge become more stringent with an increase of average machine radius. Therefore the compressor ring considered here is built with high field superferic magnets. That allows one to combine large magnetic field and a large radial aperture.

Table 1. Tentative machine and beam parameters.

Beam energy (kinetic)	8 GeV
Circumference	223 m
Number of particles	$5.2\cdot10^{13}$
Duty factor at injection, $L_b/2\pi R_0$	0.2
Repetition rate	15 Hz
Beam power	1 MW
Transition energy, GeV	2.45
Betatron tunes	5.59
Dipole magnetic field	3 T
Dipole length, m	3.1
Acceptance	140 mm mrad
Rms emittance	8.8 mm mrad
Space charge betatron tune shifts, $\delta\nu_x/\ \delta\nu_y$	0.057/0.087
Harmonic number	1
RF voltage for bunch rotation	2.5 MV
Duration of bunch rotation	146 turns
Rms momentum before rotation	$5.4\cdot10^{-4}$
Longitudinal impedance, Z_n/n	10 Ω
Z_n/n at the stability boundary at injection	20 Ω

The compressor ring energy is chosen to be 8 GeV to be matched to the energy of present Fermilab Booster as well as to the energy of new Fermilab proton source which is expected to start operations before 2020. The compressor ring represents a racetrack with FODO focusing and the dispersion zeroed in the two straight sections by missed dipoles. Figure 2 presents the beta-functions, the dispersion and the beam envelopes for a quarter of the ring and Table 1 presents the beam and machine parameters.

Figure 3 presents the longitudinal phase space and the longitudinal bunch density at the end of bunch rotation. To mitigate the microwave instability the bunch length before the rotation is short and, consequently, a voltage of the first RF harmonic creates sufficiently linear bunch rotation. Effect of the beam space charge on the rotation is also small. One can see that after rotation the rms bunch length is 65 cm – close to the required value. The factor of 2 margin for the microwave instability presented in Table 1 was computed for uniform initial longitudinal distribution. However if the strip-injection is used the beam accumulation occurs during many hundreds of turns. Consequently, the synchrotron motion results in a particle density increase in the bunch center and reduction of the stability margin.

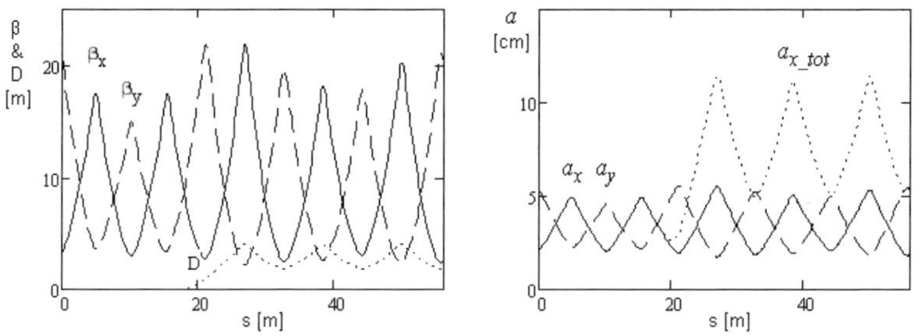

Figure 2. Twiss parameters (left) and beam envelopes (right) for a quarter of compressor ring. Beam envelopes are build for $\varepsilon_x = \varepsilon_y = 140$ mm mrad (4σ), and $\Delta p/p_0 = \pm 0.025$.

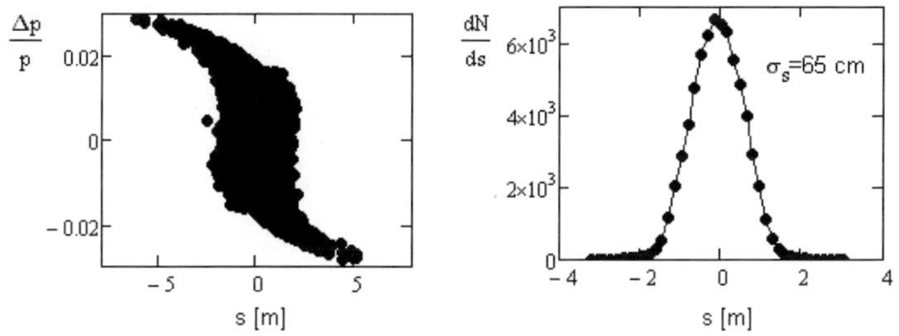

Figure 3. Results of numerical simulations of bunch compression: left – macro-particle positions in the phase space after bunch compression, right – corresponding longitudinal particle density.

6. Discussion

As can be seen from Table 1 both the microwave instability and the betatron tune shift due to beam space charge limit the beam power to ~1 MW for 8 GeV muon collider proton driver if a single bunch at 15 Hz is used. But both limitations diminish quickly with beam energy and become insignificant above ~16 GeV. Thus there are two ways to achieve 4 MW power. The first one is an energy increase and the second one is an increase of number of bunches and merging them on the target [5]. Both possibilities are feasible. An accurate engineering analysis is required to make the final choice.

Acknowledgments

The author would like to thank S. Nagaitsev, C. Ankenbrandt, R. Palmer and M. Popovic for useful discussions devoted to different aspects of the muon collider proton driver.

References

1. R. B. Palmer, "Muon Collider," *This conference.*
2. R. Garoby, *et.al.*, "Linac-based proton driver for a neutrino factory, " *This conference.*
3. A. Hofmann, "Landau damping,", in "*CERN Accel. School," CERN 95-06*, 275 (1995).
4. M. S. Zisman, "Proton beam requirements for a neutrino factory and muon collider, " *This conference.*
5. R. B. Palmer, "Proton bunching options," *This conference.*

ACCELERATOR TEST FACILITY FOR MUON COLLIDER AND NEUTRINO FACTORY R&D*

VLADIMIR SHILTSEV†

Fermilab, Accelerator Physics Center, MS221
Batavia, IL, 60510, USA

Over the last decade there has been significant progress in developing the concepts and technologies needed to produce, capture, accelerate and collide high intensity beams of muons. At present, a high-luminosity multi-TeV muon collider presents a viable option for the next generation lepton-lepton collider, which is believed to be needed to fully explore high energy physics in the era following LHC discoveries. This article briefly reviews the needs and possibilities for a Muon Collider beam test facility to carry out the R&D program on the collider front-end and 6D cooling demonstration experiment.

1. Introduction

1.1. *Possible Timeline of the Muon Accelerator R&D in the US*

At present, two teams of accelerator scientists associated into the US-wide neutrino factory and Muon Collider Collaboration (NFMCC) and Fermilab's Muon Collider task Force (MCTF) actively work on the feasibility proof of a Muon Collider (MC) needed to fully explore the physics responsible for electroweak symmetry breaking that requires a center-of-mass energy ($\sqrt{s}$) of a few TeV and a luminosity in the 10^{34} – 10^{35} cm^{-2}s^{-1} range [1]. The NFMCC and MCTF have recently jointly proposed a US National Muon Accelerator R&D program with major goals of (i) delivering a Muon Collider Design Feasibility Study Report (MC DFSR) - interim by the end FY14 and final (with the MC cost range) by the end of FY16; (ii) 4-Dimensional Muon Ionization Cooling Experiment (MICE at RAL, UK) completion by the end of FY13; (iii) completion – by the end of FY12 - of a program of RF studies to provide input

* This work is supported at the Fermi National Accelerator Laboratory, which is operated by the Fermi Research Association, under contract No. DE-AC02-07CH11359 with the U.S. Department of Energy.

† shiltsev@fnal.gov

for down-selection of 6D cooling channel by the end; (iv) participation in an International Design Study and completion of NF-RDR in early FY14; (v) a section (unit) of a 6D muon cooling channel built and tested without beam by the end of FY16. It is thought that around 2013-2014 all of the Neutrino Factory proof-of-principle tests will be completed and if the community wishes to proceed, after a few years of additional R&D, the NF construction could start as early as the late 2010's. Around the same time, the need and feasibility of a Muon Collider will be well understood and – if the MC path found attractive for HEP community – a series of demonstration experiments with muon beams on the 6D cooling, production and collection will need to be carried out that will take (estimated) 5-7 years starting 2015-2016. In case of success at that stage and if the HEP community wishes to go down this path, a MC construction start in the early to mid-2020s is plausible.

This paper addresses the requirements and possibilities for a Muon Collider beam test facility to carry out the R&D program on the collider front-end and 6D cooling demonstration experiment (the Phase II of the Muon Accelerator R&D program).

2. Facilities for Muon Accelerators R&D

2.1. *Existing Facilities, their Timeline and Limitations*

At present, two facilities are dedicated to the MC/NF accelerator R&D. The Mucool Test Area (MTA) at Fermilab has cryogenic capabilities, RF power at 201 MHz & 805 MHz, Liquid H_2 absorber filling capability, 5 T SC Solenoid with 30 cm bore (so, a 805 MHz Cavity fits inside) and beamline which can deliver 400MeV/c protons from Fermilab's Linac to the experimental hall about 8x20 m larger. The MTA facility has an established program of RF cavity studies and SC coils test for the next 4-5 years and keeping the activities there uninterrupted is critical for the success of the MICE. MICE facility at RAL has ~40x12 m hall which will be fully occupied by the 4D cooling demo experiment up until 2013-14 (so called Steps I-V) and, quite probably beyond 2014 for Step VI and a possible test of wedge absorbers. The hall can fit all the spectrometers, two or three liquid H2 absorbers, RF cavities and bucking coils needed for the experiment. Very low intensity muon beam generated by 800 MeV protons in ISIS hitting a target and delivered to the MICE hall via a beamline.

2.2. *Requirements for the Phase II Muon Accelerator R&D Facility*

Main requirements for the Phase II Muon R&D facility are: (i) it is better be available sometime after 2012-13, when the 6D cooling technology is specified so the first unit assembly and test can be started there; (ii) it should offer a low intensity muon beam for experimental studies till ~2018; (iii) it has to have modest incremental cost and enough space for the 6D cooling demonstration experiment; (iv) it has to be upgradable/expandable to take a medium to high intensity Project-X beam and generate high intensity muon bunches for consecutive R&D program until ~2024; (v) after a proper (may be, major) upgrade it should be able to be used as an operational MC or NF Front End facility.

The dimensions of the facility should be of about 80-120 meters long by 8-12 meters wide to host 6D cooling section (e.g. per Ref. [2] some 20-fold reduction of 6D emittance is expected over 120 m of the FOFO snake channel - one of the possible cooling channels schemes under consideration now). About the same length is needed for a MC/NF factory front-end facility which includes a target, drift section, buncher and rotator [3]. A short bunch (ideally, shorter than 10 ns long) of 10^7-10^{10} of ~200 MeV muons every 1-10 sec should suffice for the initial beam studies.

Several possibilities for such a facility were considered and found not adequate – the MTA (busy with its own test program, too small and wrong beam structure), upgraded MICE (not available until the experiment is over, too small, wrong time structure and very low intensity), a new facility adjoin to the mu2e target and experiment building (wrong beam structure, quite longer timescale than needed, uncertainty with location which may be not compatible with the Project-X operation). The former KTeV hall and target area looks an ideal candidate.

2.3. *KTeV Hall and Target Area*

The KTeV experimental hall consists of three areas with dimensions 35m x 4x3m; 40 m x7x6 m; 45m x17x12m. The building has spacious areas for a control room and power supplies; a 25-ton crane, larger LCW water

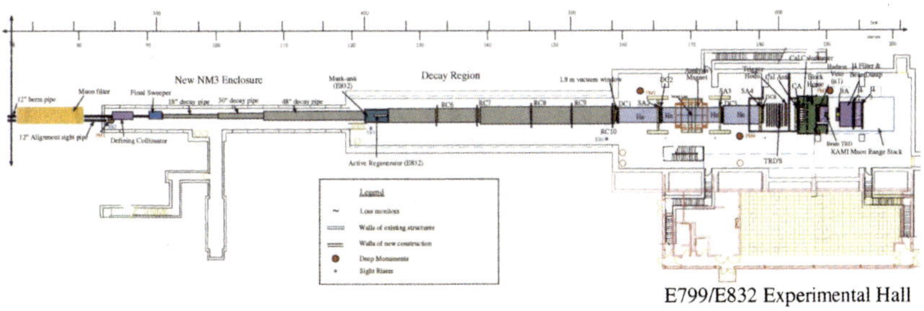

Figure 1. KTeV experimental hall area as of 2009.

infrastructure and a lot of AC power available. The experimental hall is preceded with a target area where 120 GeV high intensity proton beam can be delivered from Fermilab's Main Injector via one of the existing external beamlines. In the current plan of the Fermilab experiments, E906 (Drell-Yan) experiment will take place in the KTeV hall till 2012 (with a possible 1 year extension into 2013). So a possible strategy for the KTeV based Muon Accelerator R&D facility could be: (i) start studies with muons generated by the 120 GeV beam from MI and delivered to the KTeV hall; (ii) build a 8 GeV beam line (MI → KTeV hall); (iii) get Pr-X 8 GeV beam at a low repetition rate, generate muons in the KTeV target area; (iv) build a ~8GeV buncher ring for production of high intensity and very short proton bunches out of the Project X beam, deliver short proton bunches to the KTeV facility; (v) build Muon R&D/Front-end test facility in front of the KTeV building; (vi) transform it and the KTEeV hall into the MC or NF front end facility.

Acknowledgments

I acknowledge useful discussions with S. Geer, E. Ramberg, D. Neuffer, H. White, M. Syphers of Fermilab and R. Abrams of Muons, Inc. on the subject of the paper.

References

1. S. Geer, *Annu. Rev. Nucl. Part. Sci.* **59**:347–65 (2009).
2. Yu. Alexahin, "Helical FOFO snake for 6D ionization cooling of muons. Preprint FERMILAB-CONF-09-536-APC (2009).
3. C. Ankenbrandt *et al., Phys. Rev. ST Accel. Beams* **12:**070101 (2009).

THE SUPERCONDUCTING RF LINAC FOR MUON COLLIDER AND NEUTRINO FACTORY — SUMMARY OF WORKING GROUP 2

J. GALAMBOS
Oak Ridge National Laboratory
Oak Ridge, Tennessee 37831

R. GAROBY
CERN, Geneva, Switzerland, CH-1211

S. GEER
Fermilab, Batavia, Illinois 60510

The summary of working group WG2 exploring the Muon Collider and Neutrino Factory cience enabled with SRF technology is given here.

1. Introduction

Project-X is a proposed project to be built at Fermi National Accelerator Laboratory with several potential missions [1]. A primary part of the Project-X accelerator chain is a Superconducting linac, and In October 2009 a workshop was held to concentrate on the linac parameters [2]. The charge of the workshop was to "..focus only on the SRF linac approaches and how it can be used...". The focus of Working Group 2 of this workshop was to evaluate how the different linac options being considered impact the potential realization of Muon Collider (MC) and Neutrino Factory (NF) applications. In particular the working group charge was, "to investigate the use of a multi-megawatt proton linac to target, phase rotate and collect muons to support a muon collider and neutrino factory". To focus the working group discussion, three primary questions were identified early on, to serve as a reference:

1. What are the proton source requirements for muon colliders and neutrino factories?
2. What are the issues with respect to realizing the required muon collider and neutrino factory proton sources?

a. General considerations.
b. Considerations specific to the two linac configurations identified by Project-X.

3. What things need to be done before we can be reasonably confident that ICD1/ICD2 can be upgraded to provide the neutrino factory / muon collider needs?

A number of presentations were given, and are available at the workshop website [2]. This paper does not summarize the individual presentations, but rather addresses overall findings as related to the three guiding questions listed above.

2. Baseline Linac Configurations

Project-X has already been under study for some years and studies have led to two primary accelerator configurations: "ICD-1" and "ICD-2" [3]. The Initial Configuration Document (ICD)-1 design includes a pulsed (5 Hz) 8 GeV, 20 mA, 1.25 msec pulse length linac. The ICD-2 incorporates a CW, 1 mA, 2 GeV linac. These two linac configurations serve as a primary basis of comparison for the initial acceleration component for muon collider and neutrino factory applications for Project-X.

3. Neutrino Factory/Muon Collider Requirements

The NF and MC proton driver beam requirements are summarized in Table 1 (taken from Ref. 4). The power requirement is based on MC luminosity specifications and neutrino delivery rates at far detectors. The required beam energy has a broad optimum centered around 10 GeV, with a steep fall-off

Table 1. Proton driver requirements for a Neutrino Factory.

Parameter	Value
Average beam power (MW)	4
Pulse repetition frequency (Hz)	50 [a)]
Proton energy (GeV)	10 ± 5
Proton rms bunch length (ns)	2 ± 1
No. of proton bunches	3 or 5 [b)]
Sequential extraction delay (μs)	≥ 17
Pulse duration, liquid-Hg target (μs)	≤ 40
Pulse duration, solid target (ms)	≥ 20

[a)] For a Muon Collider a lower repetition rate, 10–15 Hz, is required.
[b)] For a Muon Collider a single bunch is necessary.

below ~5 GeV (see Fig. 1 below, taken from Ref. 5). The bunch length of 2 nsec is a primary driver. Below a 1 nsec bunch length there is no benefit, but at 3 nsec there is a 10% loss in the muon production intensity. The NF and MC requirements differ in terms of number of bunches per pulse (1 for the MC versus 3-5 for the NF) and maximum repetition rate (15 Hz for the MC and 50 Hz for the NF). Producing a 4 MW proton beam in a single bunch at 15 Hz requires about 10 times the bunch intensity than in 3 bunches at 50Hz. Therefore we adopt the more stringent MC proton beam requirement as the basis of comparison for the mission here, as this will satisfy either mission.

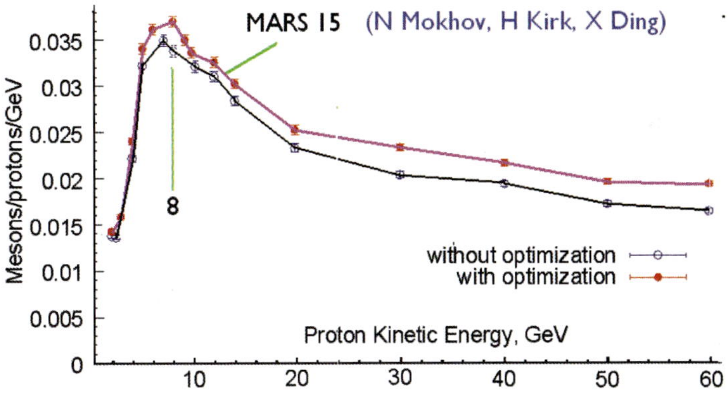

Figure 1. Meson production efficiency vs. proton beam energy (from Ref. 5).

4. Proton Driver Issues

4.1. *General Issues*

Achieving the required MC proton driver parameters is beyond existing accelerator capability and quite difficult, especially with respect to the short bunch structure. Some of the more challenging issues include:

- Keeping uncontrolled beam loss below 1 W/m ($< 3 \times 10^{-7}$) for hands on maintenance.
- Concentrating the collection of beam halo in a well-controlled collimation area (careful beam dynamics and detailed engineering is needed).
- Injection charge exchange stripping, issues such as foil heating, foil lifetime and beam loss from scattering during the multiple foil passages during injection.
- Maintaining a small longitudinal emittance, as required to achieve the short bunch length.

- Collective effects, in particular the large longitudinal phase space density makes longitudinal stability problematic.
- Large longitudinal compression factors required to reach the needed bunch length which necessitate RF gymnastics just before extraction.

4.2. *Potential Solutions*

There are no fully developed NF/MC schemes available, although some promising linac based ideas were presented. In particular, there is not a complete set of beam and machine parameters for the different linac and rings associated with acceleration, accumulation and compression. We note that there is an historical RCS proton driver scheme at FNAL [6], but it does not meet the requirements specified above.

V. Lebedev [7] showed the difficulties in attaining the required MC parameters due to longitudinal microwave instability and tune shift due to transverse space charge. Using the ICD-2 driver alone the beam power is limited to < 1 MW. At 8 GeV, with some intermediate compression 1 MW is possible with a single bunch, and higher powers with multiple bunches compressed (as described below). Going to higher energy tends to alleviate the instability and space charge limitations.

Innovative ideas were introduced which tend to alleviate the compression and charge exchange injection issues. The first is the so-called "trombone/funnel" concept [5, 8] in which multiple 2 nsec bunches are created in a compression ring (each with a fraction of the intensity required by a single bunch 4 MW scheme), the individual bunches are extracted to separate extraction beam-lines of varying length so that they arrive at the target at the same time. Another new concept is a "resonant foil by-pass scheme" [8, 9], which reduces the foil traversal problem in the initial accumulation period for schemes that have very long injection times because of a low linac beam current. This idea involves chopping the CW linac beam to ~ 10% duty factor and incorporates a resonant position bump at the foil in the accumulation ring, synchronized so that the circulating beam passes through the foil during the un-chopped part of injection and is pulled away from the foil during the chopped portion of the injection. However details of this proposed scheme with realistic linac beam distributions and foil heating calculations have not been done.

One complete scheme was presented for a NF application [10], but this feasibility analysis was performed in the context of the CERN accelerator complex, and does not meet the MC requirements.

4.3. *General Comments*

In order to highlight the specific areas needing attention for MC feasibility, the existing proposals should be further pursued. In general the linac based solutions (as opposed to RCS solutions) appear more promising. The relatively low energy (~ 10 GeV) for MC implementations results in long storage times to accumulate the required high intensity beams for the RCS based solutions (at even lower energies). The CW linac implementation (~ 1 mA) appears more difficult than a pulsed linac implementation (~ 50mA), again due to the longer injection and storage times associated with the lower current CW beams. The injection foil and beam loss issues are of particular concern for the CW cases, but the resonant foil by-pass scheme may mitigate this issue.

With respect to the choice of ICD-1 and ICD-2 as a basis for extrapolation to a MC proton driver, the ICD-1 is the preferred option. This case requires the least additional upgrades in an accelerator chain that supports a MC proton driver, and minimizes the physics challenges associated with producing the required extremely short bunch structure. With the ICD-2 option (1 mA CW 2 GeV linac, + RCS) neither the NF nor MC needs can be met. Upgrading the ICD-2 linac to a pulsed higher energy linac (6-8 GeV) may offer some possibility as a proton driver. To this end variable couplers capable of higher peak powers would need to be installed initially to permit reuse of the linac components in the tunnel. Also, if the resonant foil by-pass scheme proves feasible, an energy upgrade of ICD-2 for a MC proton driver may be possible, but this proposal requires more study before adoption.

4.4. *Recommendations*

Judging the feasibility of either ICD-1 ort ICD-2 as the initial accelerating component of a proton driver supporting the MC collider mission is made difficult by the lack of a proposed configuration that meets the requirements, although ICD-1 has more potential. A self consistent set of accelerator parameters for the proton driver acceleration chain supporting this mission should be assembled. These include:

- the charge exchange injection process: including details of the injection scheme, foil temperature rise and lifetime estimates, and beam loss from foil scattering, and analysis of laser stripping charge exchange injection alternative.
- collective effects: such as longitudinal stability and space charge tune shift effects in the compression stage need evaluation.
- bunch rotation and extraction scheme details.

- Finally hardware R&D requirements need to be identified in areas such as:
- Variable power couplers for the linac, which support low current CW operation as well as high instantaneous power in pulsed mode, if the ICD-2 option is pursued.
- RF hardware needs for providing the final bunching and rotation in the compression stage.
- Magnets, as required for the relatively small ring circumferences needed to keep space charge tune shifts low.
- Kickers required for extracting the short bunch lengths.

References

1. S. Holmes, Project X Initial Configuration Document, http://projectx-docdb.fnal.gov/cgi-bin/RetrieveFile?docid=83&extension=pdf
2. Workshop on Applications of High Intensity Proton Accelerators October 19-21, 2009 Fermi National Accelerator Laboratory, Batavia, IL, USA http://conferences.fnal.gov/App-Proton-Accelerator/index.html
3. V. Lebedev, "Accelerator Parameters for Projectt-X: ICD1 &2", Workshop on Applications of High Intensity Proton Accelerators October 19-21, 2009 Fermi National Accelerator Laboratory, Batavia, IL USA, http://indico.fnal.gov/getFile.py/access?contribId=82&sessionId=0&resId=0&materialId=slides&confId=2854
4. M. Zisman, "What's Needed and Why", Workshop on Applications of High Intensity Proton Accelerators October 19-21, 2009 Fermi National Accelerator Laboratory, Batavia, IL, USA, http://indico.fnal.gov/conferenceDisplay.py?confId=2854
5. R. Palmer, "Proton Bunching Options for Muon Colliders", Workshop on Applications of High Intensity Proton Accelerators October 19-21, 2009 Fermi National Accelerator Laboratory, Batavia, IL, USA, http://indico.fnal.gov/getFile.py/access?contribId=15&sessionId=2&resId=0&materialId=2&confId=2854
6. REPORT: Proton Driver Study II, Part 1, Fermilab TM-2169, May 2002, http://www-bd.fnal.gov/pdriver/8GEV/
7. V. Lebedev, "Proton Bunch Compression Strategies", Workshop on Applications of High Intensity Proton Accelerators October 19-21, 2009 Fermi National Accelerator Laboratory, Batavia, IL, USA, http://indico.fnal.gov/getFile.py/access?contribId=63&sessionId=28&resId=0&materialId=0&confId=2854 and http://indico.fnal.gov/getFile.py/access?contribId=63&sessionId=28&resId=1&materialId=slides&confId=2854
8. C. Ankenbrandt, "Accumulation and Bunch Rotation Schemes", Workshop on Applications of High Intensity Proton Accelerators October 19-21,

2009 Fermi National Accelerator Laboratory, Batavia, IL, USA, http://indico.fnal.gov/getFile.py/access?contribId=65&sessionId=28&resId=0&materialId=slides&confId=2854

9. M. Popovic, "CW Linac Options",Workshop on Applications of High Intensity Proton Accelerators October 19-21, 2009 Fermi National Accelerator Laboratory, Batavia, IL, USA, http://indico.fnal.gov/conferenceDisplay.py?confId=2854
10. R. Garoby, "Linac Plus Ring Option", Workshop on Applications of High Intensity Proton Accelerators October 19-21, 2009 Fermi National Accelerator Laboratory, Batavia, IL, USA, http://indico.fnal.gov/conferenceDisplay.py?confId=2854

Working Group 3

SRF Linac Driven Subcritical Systems (ADS)

Conveners

Jean Delayen(Old Dominon University and TJNAF)
Rajendran Raja(Fermilab)
Jean-Pierre Revol(Cern)

PROSPECTS FOR A VERY HIGH POWER CW SRF LINAC*

R. A. RIMMER†

Thomas Jefferson National Accelerator Facility, 12000 Jefferson Avenue, Newport News, VA 23606, USA

Steady development in SRF accelerator technology combined with the success of large scale installations such as CEBAF at Jefferson Laboratory and the SNS Linac at ORNL gives credibility to the concept of very high average power CW machines for light sources or Proton drivers. Such machines would be powerful tools for discovery science in themselves but could also pave the way to reliable cost effective drivers for such applications as neutrino factories, an energy-frontier muon collider, nuclear waste transmutation or accelerator driven subcritical reactors for energy production. In contrast to machines such as ILC that need maximum accelerating gradient, the challenges in these machines are mainly in efficiency, reliability, beam stability, beam loss and of course cost. In this paper the present state of the art is briefly reviewed and options for a multi-GeV, multi-MW CW linac are discussed.

1. Introduction

Large-scale CW SRF installations such as CEBAF [1] and LEP [2] have demonstrated that CW SRF technology can work reliably for demanding applications and facilities with 24/7 user operations, in parameter regimes where no other current technology would be viable. The recent success of the pulsed SNS at ORNL further supports the case, delivering high beam current at high duty factor to meet the milestone of 1 MW beam power for users. Extrapolating this performance to a large CW machine presents the exciting prospect of an unprecedented source of high-energy protons for a wide range of applications. Such ideas are further stimulated by the proposal of a new multi-GeV high intensity proton driver at Fermilab, code named Project X. Initially planned as an 8 GeV pulsed linac based on ILC-like SRF technology, there now appears to be serious consideration of making some or even all of the machine operate CW to provide simultaneous high intensity beam to multiple experiments as well as continuing to support the very high energy users. Such a machine would be the

* This work supported by the U.S. Department of Energy, contract DE-AC05-84-ER40150.

† rarimmer@jlab.org

backbone of the Fermilab program for decades, supporting a world-leading neutrino physics program and other landmark experiments and eventually enabling an energy frontier muon collider to be envisaged on the Fermilab site.

The good news is that many "proton driver optimized" cavities suitable for CW operation exist or are in development e.g. at ANL, Triumf, JLab (Figure 1) and KEK, as well as cryomodule designs that can accommodate them, so designers need not feel constrained by the narrow bounds of the TESLA-like technology

Figure 1. JLab Prototype 750 MHz high-current cavity. This cell shape is optimized for beta=1 high current beams. Higher-order modes are tuned to safe frequencies to minimize power extraction from the beam. Waveguide HOM dampers provide very strong damping of unwanted modes.

SRF cavity performance also continues to advance, not only on the gradient frontier but also in terms of operating efficiency through improved geometry factor and Q_o at typical CW operating gradients. In fact the same electropolishing (EP) technology responsible for gradient advancements for ILC has also been shown to give significant improvements in Q_o at more modest fields.

Experience with operating machines means that the beam-break-up (BBU) limits are quite well understood and mitigations have been developed and demonstrated in the form of HOM damping or detuning (and feedback in the case of multi-pass or re-circulating machines). Average currents in the tens or even hundreds of mA are now quite possible in this regard. The main remaining challenge is halo control to minimize beam loss and activation.

Large 2K cryogenic plants are also becoming more efficient, both in peak capacity and turn down mode, addressing one of the largest cost drivers in operation and construction of such machines.

Even CW RF sources continue to be developed with lower cost, greater efficiency and longer life. The widespread adoption of inductive output tubes (IOTs) by the broadcast industry provides a stable technology base for these devices, which can be adapted to a wide variety of frequencies and power levels. Of course klystrons remain viable and with multi-beam tubes, annular or sheet beams and other developments still in progress they may remain competitive long into the future. Magnetrons are one of the oldest sources of RF power and have good efficiency and low construction cost but being oscillators rather than amplifiers they have not been much favored for large accelerators (although a significant number of medical and industrial machines are powered by them). However advances in digital switching high-voltage power supplies and fast low level RF control do allow the prospect for modulation and phase stabilization of the magnetron output, and if this can be demonstrated to an acceptable level of stability they could become a very favorable option for high power CW machines.

The science case for new machines is very strong and there is encouraging progress in both the technology and the user community interest towards their realization.

2. State of the Art

Large installations such as CEBAF and SNS may be used as benchmarks to assess the state of readiness of a CW machine for applications such as ADS. Although CEBAF is a 1.5 GHz electron accelerator it does operate CW with more than 1 GeV of installed capability from forty cryomodules, Figure 2, (plus two and a quarter in the injector) and has been running for over a decade with high availability for up to three simultaneous user end stations. The average current is modest but because the machine is operated in a 5-pass recirculation mode the maximum energy is 6 GeV (originally 4 GeV) and the beam power can reach 1 MW, the maximum administrative site limit. As CEBAF was first proposed more than 25 years ago the initial gradient specification was very conservative by present standards at 5 MV/m. In fact even during the initial production run many cavities comfortably exceeded this mark. At 4 kW it had what was at the time the worlds largest 2K cryogenic plant. Recently ten of the original cryomodules have been refurbished with a goal of 12.5 MV/m (50 MV per cryomodule), reusing the original cavities but with improved preparation techniques. The machine will shortly be upgraded to 12 GeV, Figure 3, with the addition of ten new high performance cryomodules adding an additional 1 GeV per pass, a fourth hall and an additional 4.5 kW central helium liquefier

(CHL-2). Some parameters for the upgrade cryomodules are listed in Table 1. These along with the demonstrated SNS performance may be useful benchmarks for new projects.

Figure 2. One of the two CEBAF linacs, each comprising 20 original cryomodules assembled in-house, plus $2^1/_4$ in the injector.

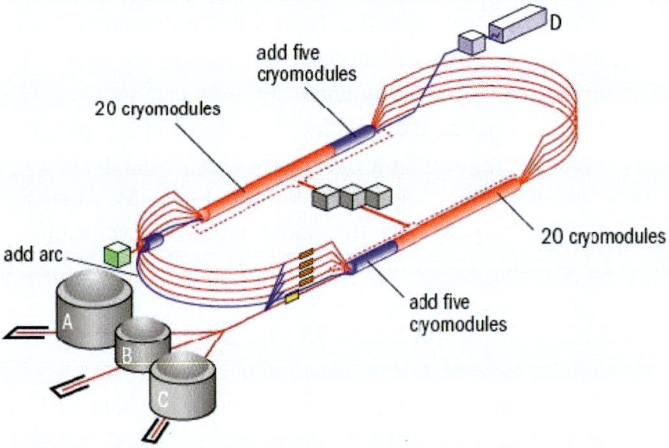

Figure 3. Upgrade of the CEBAF 5-pass CW recirculating linac from 6 to 12 GeV. Five new high performance cryomodules will be installed in each linac along with one extra arc and a new experimental hall.

Figure 4. The Spallation Neutron Source (SNS) superconducting linac at ORNL.

SNS is the most recent large SRF based linac to be completed, Figure 4, and is specifically designed for H^- ions (stripped to protons in the accumulator ring). It also has achieved 1 GeV and 1 MW of beam delivered to users but in pulsed mode with a 5% duty factor (6% design). SNS has 81 independently powered 805 MHz cavities in 23 cryomodules (11 medium beta, 12 high beta), with space reserved for additional modules to upgrade the energy to 1.3 GeV and power to 2-3 MW. The design gradients were 10.1 MV/m for the beta 0.65 cavities and 15.8 MV/m for the beta 0.81 structures with Q_0 of 5E9 at 2.1K. These gradients were achieved on average during cavity qualification and module testing during individual cavity commissioning, however high levels of dark current were observed that resulted in lower ensemble operating limits for the high beta cavities (see section 6). The SNS cavities were also operated at 4K during initial commissioning at low duty factor with no apparent reduction in gradient performance [3].

Overall SNS has been a very successful project having been completed on time and on budget and having met all of the commissioning milestones to date, including delivering 1 MW of beam on target during the 2009 run.

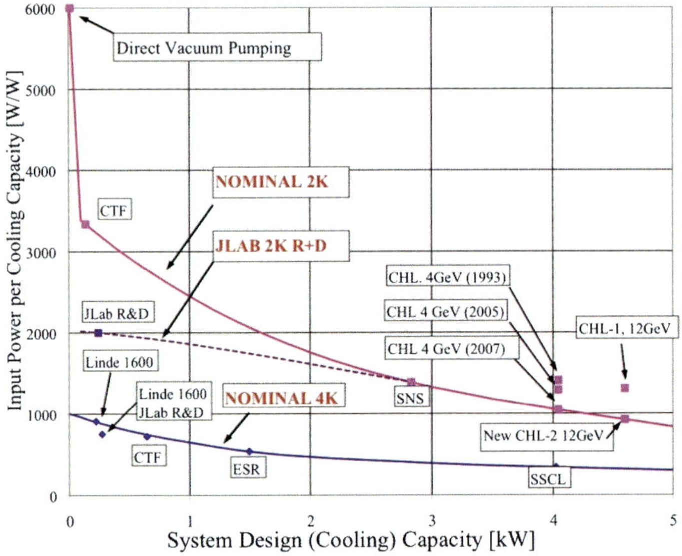

Figure 5. Typical efficiencies of existing and planned 2K and 4K cryogenic plants.

3. 2K Cryogenic Plant Development

Until recently most large 2K cryogenic plants were designed for most efficient operation at their maximum capacity but quickly lost efficiency when turned down. Recent R&D at JLab [4] has resulted in significant efficiency increases (e.g. 20-25% improvement at ~4.5 kW for JLab CHL-2), but also allowing good efficiency at high turn down ratio (as low as 34-40% of maximum output for the full Ganni cycle). This can result in up to a three-fold power reduction when running such plant well below maximum capacity. The JLab floating pressure technology has been applied to a number of existing plants such as SNS, BNL and JLab CHL-1 with consistent improvements in efficiency resulting in very substantial savings in operating costs as well as reduced maintenance. Figure 5 shows some efficiency points for existing or planned machines along with nominal curves for 2K and 4K operation. Note that although the wall-plug power per cryogenic Watt is lower at 4K than 2K this is more than offset by increased SRF losses at the higher temperature for most presently produced SRF cavities, favoring 2K (or lower) temperature operation for CW applications. If cavities could be made with much lower residual resistance than presently achieved then it might become favorable to operate low frequency RF cavities at 4K, greatly simplifying the plant and reducing capital costs. This is a subject of active investigation [e.g. 5].

4. Cost Drivers and Technical Risk

Major cost drivers for any large SRF linac include the cryomodules themselves, RF power and controls, Cryogenic plant and distribution and tunnel and conventional facilities. Some typical parameters and estimated numbers from the JLab 12 GeV upgrade project are shown in table 1 to give a feeling for the magnitude of these costs. As part of any overall project optimization these costs can be traded against each other to find the minimum capital cost, operating cost or combined capital plus some number of years operating. Two such recent studies are enlightening, the Cornell 100 mA ERL design study [6] and the UK new Light Source (NLS) conceptual design, a 1-pass CW linac [7]. Figure 6 shows the breakdown of these cost elements for the Cornell study. Of course these costs depend strongly on assumptions of cavity performance and gradient choice. Costs such as tunnel, cryomodule and controls (e.g. LLRF) typically vary linearly with tunnel length (i.e. inversely proportional to assumed gradient), while cavity RF power and cryogenic costs (which dominate operating costs as well as contributing significant fractions of capital expenditure), vary quadratically with gradient. Beam RF power cost is roughly constant for a given energy and current. The tradeoff of these types of cost tends to produce a rather broad cost minimum with respect to gradient, as shown for example in Figure 7 for the Cornell ERL.

Table 1. Some typical CW cryomodule parameters from the JLab upgrade conceptual design.

Frequency	1.5 GHz
Gradient	15-20 MV/m CW (~10 MV/m real-estate gradient)
Q_o	~ 10^{10} at 20 MV/m has been demonstrated. 8×10^9 assumed.
Module cost	~$2.6M*
RF cost/module Inc. LLRF	~$1.7M*/cryomodule (8x13kW RF stations) at~1mA
Cryo plant 4.5 kW CHL-2	~$30M*/GeV (CHL2) excluding distribution & cold box
Integrated cost/GeV @ 1 MW	~$73M* (excluding tunnel & conventional facilities)

*FY08 Estimate, loaded dollars, no contingency. Actual 12 GeV project costs will be known soon.

Technical risk also plays a significant factor in gradient choice. Given the relatively broad cost optimum there is little or no penalty in choosing a gradient at the low end of the curve, e.g. closer to 15 MV/m than 25 MV/m. This will diminish the risk of degraded machine performance from field emission, reduce coupler power requirements and (in all likelihood) provide operating overhead in case of installed cavities being out of service. However more cavities will mean more systems to maintain, slightly stronger HOM damping requirements, proportionally more emittance growth from wakefields and coupler kicks (if present), etc. All these factors should be taken into consideration in the overall optimization. However it is encouraging that these gradient levels have already been demonstrated in large numbers of cavities in operating facilities. It should be noted that this optimization for CW operation is fundamentally different than for a pulsed energy-frontier machine like ILC, where the highest practically realizable gradient is the cost optimum. The case of the CEBAF upgrade cryomodules is also slightly different in that the tunnel already exists and there are only ten vacant zones available for new cryomodules. This forces the spec to be closer to 20 MV/m, however this performance has been demonstrated and is still conservative compared to machines like ILC and XFEL.

Other technical risks include beam break-up and other instabilities. Although these are rather well understood and modeled there remains some disagreement over the need for HOM damping in high energy but modest current machines. A more significant risk arises from beam loss in the linac that can cause a serious problem of activation. This has already been observed at SNS and although they have been able to operate below the 1 W/m nominal loss for hands-on maintenance they are doing so at considerably lower average current than needed for a very high power CW proton driver. Extensive beam dynamics modeling has not yet identified or reproduced the source of the losses although speculation is that at least some fraction is coming from longitudinal losses out of the RF bucket originating from mismatches upstream in the machine.

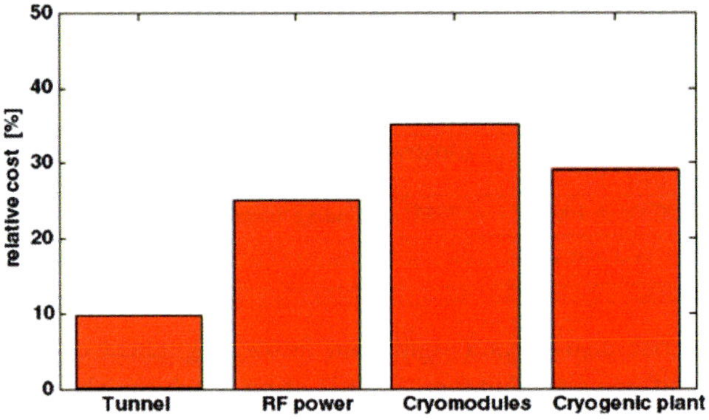

Figure 6. Cost distribution of major systems in the Cornell 100 mA ERL design study.

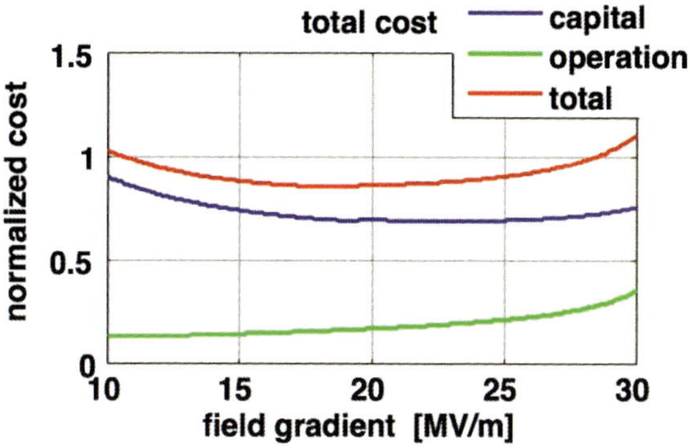

Figure 7. Typical cost variation with gradient for a CW linac (Cornell 100 mA ERL study).

5. Other Parameter Choices

5.1. *Current*

Depending on the application there may be a possible trade-off between current and energy for a given beam power on target, however most new CW projects under consideration seem to be seeking at least 1 GeV and 1 mA, with up to 8 GeV and 30mA being discussed [8], Figure 8. Cavities exist that are designed with sufficient HOM damping for 100 mA or above so this should not be a limitation. Coupler power becomes a factor at high current but this is a matter for careful R&D rather than a hard limit.

5.2. *Frequency*

Exact frequency choices for operating machines are often a result of history such as existence of proven cavity designs, RF sources (e.g. the UHF TV band is very popular), or synchronism with other accelerators. In general BCS theory [7] would predict more efficient operation at lower frequencies with the added benefit of (typically) larger beam apertures and, possibly, fewer cavities for a given active length. However given the typical residual resistances of presently produced SRF cavities the actual benefit in efficiency is somewhat reduced (although the other benefits remain). The downside, at least for elliptical cavities, is the increased size and cost. Handling considerations are also a factor for multi-cell cavities, setting a practical lower limit in the 500-700 MHz range. Spoke cavities and other compact structures are becoming increasingly popular, especially for the low beta end of the linac where they are also more efficient than squashed ellipticals. Beta = 1 spoke cavities are also under development and typically are about half the diameter of an equivalent elliptical cavity at the same frequency, however they also have higher peak surface fields for a given accelerating gradient. Many projects seem to be converging on a combination of quarter-wave or spoke cavities at around 350 MHz for the low beta part of the linac and one or more flavors of elliptical cavities at around 700 MHz for the high-energy end. The transition is typically at about beta= 0.5 or higher. The number of cells or interaction gaps per cavity and the number of different beta values are subjects for detailed beam dynamics optimization.

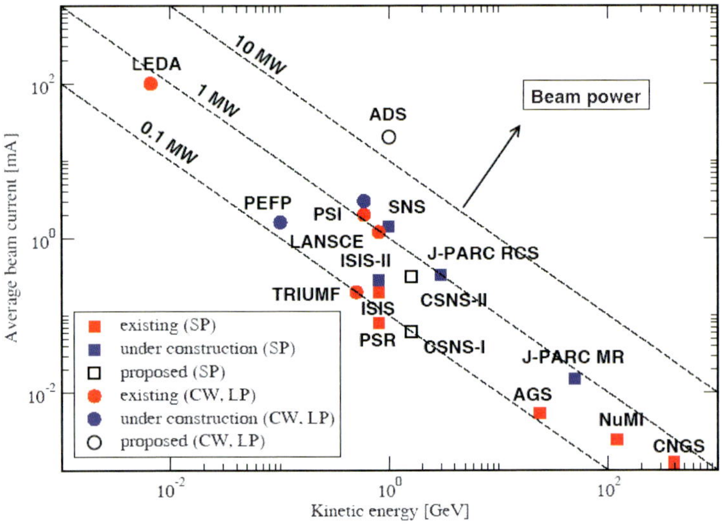

Figure 8. Landscape of existing and planned high power proton sources.

5.3. *Operating Temperature*

For high-gradient CW operation cavities are normally operated in superfluid helium (<2.18K) to benefit from the increased heat transport capability, however both the maximum superfluid heat transport limit and the RF losses are strong functions of temperature in this range. Depending on the achieved residual resistance of the cavities temperatures as low as 1.8K or even 1.6K may be attractive. Figure 9 shows an example of how the BCS surface resistance asymptotes to the residual resistance (assumed to be 10 nΩ) as the temperature falls below about 2K [5]. Clearly in this case there is limited benefit from going much further down in temperature. Any optimization needs to take into account the specific cryogenic system operating characteristics and environment, e.g. long distribution lines and high mass flow may limit the practically attainable minimum pressure in the helium bath. CEBAF was designed for and operates at 2.07K, however the actual dependence of operating efficiency (and therefore cost) on temperature has never been mapped. This would be a worthwhile experiment providing it could be performed without affecting user operations.

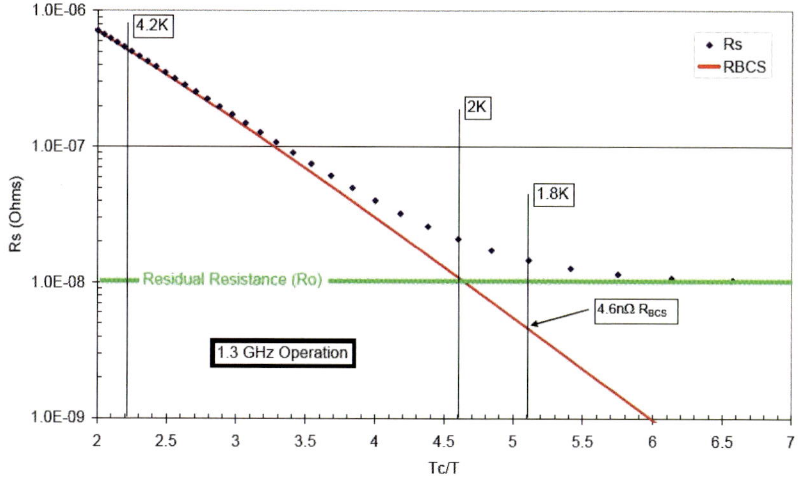

Figure 9. Example of total surface resistance vs. temperate below T_c. Note deviation from BCS resistance to residual resistance at low temperature ($T \ll T_c$).

6. SNS Experience

As mentioned previously the SNS is the most recent example of a large SRF linac and the first designed as a MW class proton source. The decision to use SRF was taken late in the project timeline so there was only a short time available to design and prototype the cavities and cryomodules before launching series production. In light of this the project has been remarkably successful and has met all of its performance milestones to date. However several issues arose during the production and commissioning that are worth taking note of.

6.1. *Initial Cavity Qualification*

Initial cavity qualification was quite problematic with persistent field emission and/or multipacting in many of the cavities. On average it took between two and three processing cycles to get the cavities to qualify in vertical test. Suspicion fell on the effectiveness of the high-pressure rinse system and water quality although investigation at the time and since seems to indicate that probably the water purity was not the issue. The effectiveness of the HPR spray jet at the equator of the 805 MHz cavities has since been measured and found to be significantly less than that for smaller cavities, however it may not be the only factor. Due to the tight production schedule it was difficult to resolve these issues in real time and although many changes were made to the process during the run, such as extended rinse times and different assembly procedures, the net cavity yield was not substantially higher at the end of the project than at the beginning. Another feature of these cavities with BCP process was a strong high-field Q slope even in the absence of field emission. Figures 10 and 11 show some typical curves from medium beta and high beta cavities [10]. This made it harder to meet the Q specification even if the gradient limit was acceptable, especially for the high beta cavities. Many of the qualified cavities still had high levels of field emission at the operating gradient even though they met both the gradient and Q_0 specifications.

Figure 12 shows the residual resistance of the high beta cavities estimated from the measured low-field Q_0 values [11]. The average value is about 8 nΩ but there is a wide spread in values.

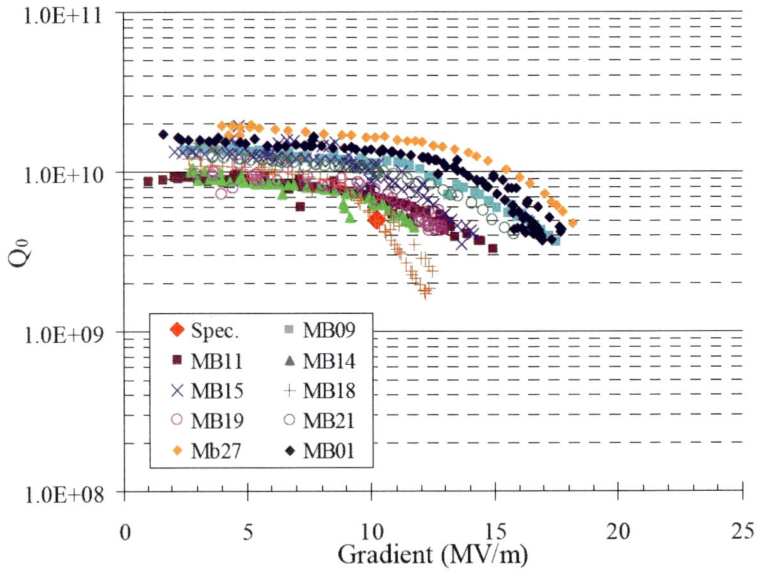

Figure 10. Typical performance of SNS medium beta cavities in vertical test.

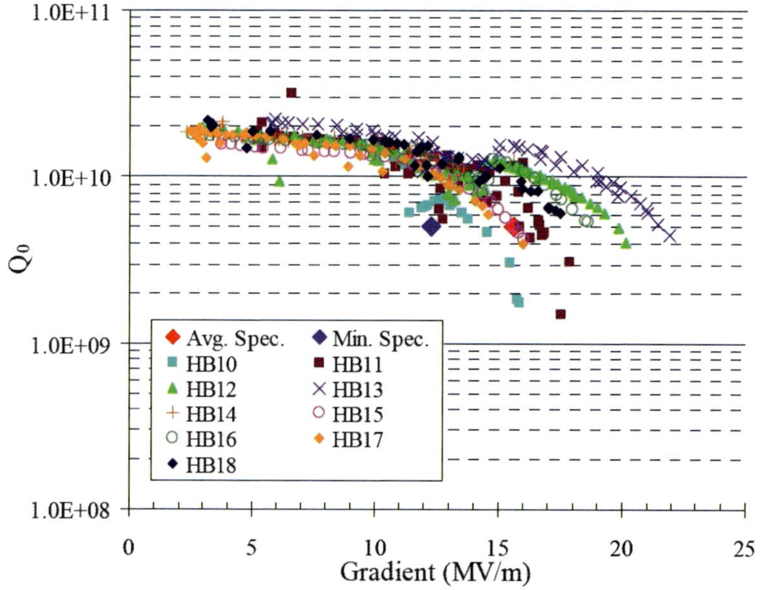

Figure 11. Typical performance of SNS high beta cavities in vertical test.

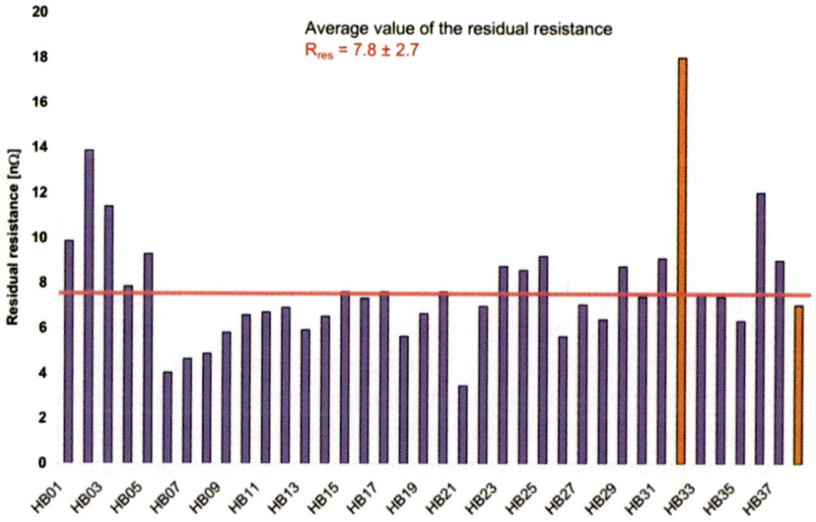

Figure 12. Residual resistance of the SNS high beta cavities estimated from the low-field Q_0.

6.2. *Cavity Performance in Modules*

Early production cryomodules were tested at JLab before shipping to Oak Ridge, affording an opportunity to compare VTA performance with horizontal performance in the cryomodule, Figure 13 [10]. Since the cavities underwent an additional HPR and assembly for the string some scatter in the data is not unexpected, however the average performance in the cryomodule was not degraded compared to VTA, in fact it was slightly better. Thus the cryomodules were delivered to SNS with high expectations of successful operation

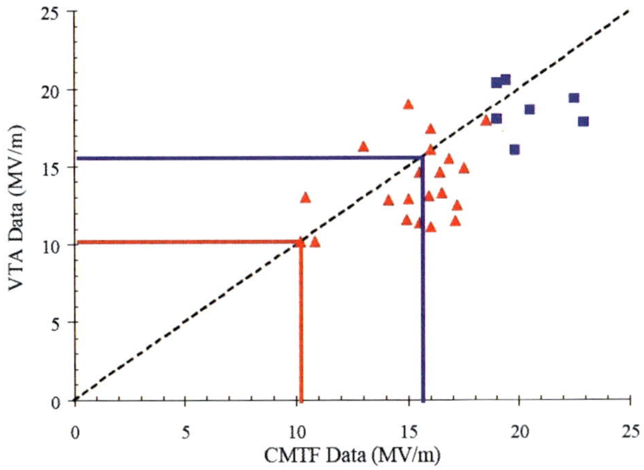

Figure 13. SNS cavity data in vertical test vs cryomodule. Gradient at $Q_0 = 5 \times 10^9$, as measured in the VTA and CMTF, for the medium-β (triangles) and high-β (squares) cavities.

6.3. *Cavity Performance in Initial Ensemble Operation*

Initial commissioning of the modules as an ensemble at Oak Ridge uncovered a number of issues with HOM coupler tuning, power coupler vacuum and some end group thermal limitations from dark current interception that had not been apparent in single cavity limits. As planned many of the cryomodules were not tested at JLab and received their first commissioning at Oak Ridge. As a necessary first step many of the cavity gradients were simply turned down in order to get the linac to run stably. This unfortunately led some people to speculate that there might be a systemic problem in the linac that would prevent SNS from achieving its design energy. Fortunately calmer heads prevailed and after much careful diagnosis and tuning most of the lost gradient has been recovered and the machine is functioning well for users [12]. A few cavities with hardware problems were repaired or removed. The cavities in low beta section of the linac are now operating above the design specification on average. In the high beta section the individual cavity limits now also exceed the specification on average but the cross talk from captured dark current means the ensemble average is still slightly below spec. This is an effect that it was not possible to observe during the prototyping and module testing phases. The net result is that SNS is operating very well and the SRF linac is a very minor contributor to down time. However the operational margin is less than desired. There is an active plan under way to rework some more cavities to provide spare capacity.

6.4. *Prospects for Improvement*

The standard process for SNS cavity qualification employed buffered chemical polishing (BCP) etching for both the bulk material removal as well as final cavity processing. It has been observed that repeated BCP processing can cause the high-field Q slope to become more severe, correlated with visible roughening of the surface. A few cavities were electopolished early in the project with encouraging results but VTA tests at that time were still often dominated by field emission so potential improvements were somewhat masked. Due to schedule and budget pressures the project decided not to continue EP development for the remainder of the production cavities.

Recent success with EP of ILC cavities in the same JLab facilities suggests that in the absence of field emission there is a significant benefit from EP both in gradient reach as well as elimination of the Q slope (after 120°C baking) [13]. This has revived interest in EP as a possible process for the SNS power upgrade project, however the question mark over effectiveness of HPR in these cavities remains.

Very recently JLab has experimented with final light EP on several 7-cell 1.5 GHz cavities that have had bulk BCP, some with repeated cycles of final BCP [14]. As can be seen in Figure 14 the results are very encouraging with only 30 um removal and 120 °C bake for 48 hours the performance improved dramatically and the high field Q slope was virtually eliminated. This further supports the theory that the surface roughness on small length scales is a contributing factor to the Q slope in BCP'd cavities. The residual resistance also appears to be quite low suggesting advantage may be gained by lowering the operating temperature. There is some interest at ORNL in testing this process on spare SNS cavities.

Improvements to the cavity geometry and construction may also be worthwhile. It is known that there is a persistent multipacting barrier in the high beta cell shape and this could be reduced significantly by altering the cell profile at the equator region. Also end group cooling could be improved or higher RRR material could be used to help mitigate the thermal issues from dark current heating. Concerns about possible multipacting in the HOM couplers could be addressed by design changes or possibly these could be eliminated altogether.

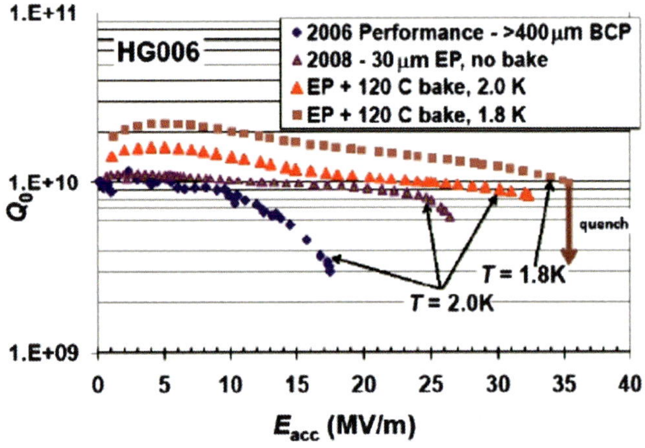

Figure 14. Performance of JLab 7-cell cavity HG006 after nominal BCP (purple triangles), after subsequent heavy BCP (blue diamonds), and after only 30 μm EP + 120°C bake (red triangles). Test at 1.8 K (brown squares) suggests cavity has low residual resistance.

7. Candidate Structures for New High-Power CW Machines

A number of candidate cavity types either exist or are in development in addition to those already proven in existing machines. Half-wave and spoke resonators in a variety of styles are rapidly becoming the default choice for the low beta sections of new machines. For the higher energy sections elliptical cavities are still popular, although high beta and beta=1 spoke cavities are under development. At beta above about 0.5 elliptical cavities tend to have good efficiency and lower peak surface fields than spoke types. At lower betas spoke resonators typically have the advantage. Elliptical cavities tend to have larger apertures, which might be an advantage for minimizing halo interception, but spoke resonators are significantly more compact for a given frequency and tend to have better microphonic performance and higher filling factor in the cryomodule.

7.1. *Low Beta*

An example of a low beta single spoke resonator is the "SSR1" [15], Figure 15 developed for the HINS project. A prototype of this cavity has been built and tested with excellent results.

Figure 15. Low beta single spoke resonator "SSR1".

7.2. *Medium Beta*

Examples of medium beta multi-spoke cavities include the 345 MHz beta 0.5 and beta 0.62 triple spoke resonators developed at ANL [16], Figures 16,17. Originally designed for 4K operation the performance of these cavities at lower temperature is so good that it would clearly be more efficient to operate these cavities CW at 2K or even lower, see Figure 18. Although elliptical designs have been prototyped at beta 0.5 or below, triple spoke resonators seem to be the popular choice. At around beta = 0.6 multi-cell elliptical cavities are a viable option (e.g. SNS low beta = 0.61).

7.3. *High Beta*

At beta approaching unity elliptical cavities are presently the only proven choice, although high beta spoke resonators are being actively developed. The SNS beta = 0.85 cavities have demonstrated performance in the right range for a high current CW proton machine, with potential for further improvement.

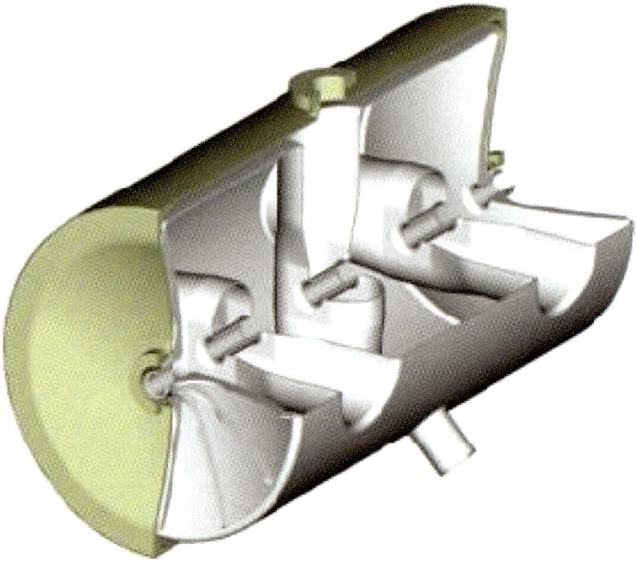

Figure 16. ANL 345 MHz β=0.5 Triple-spoke resonator.

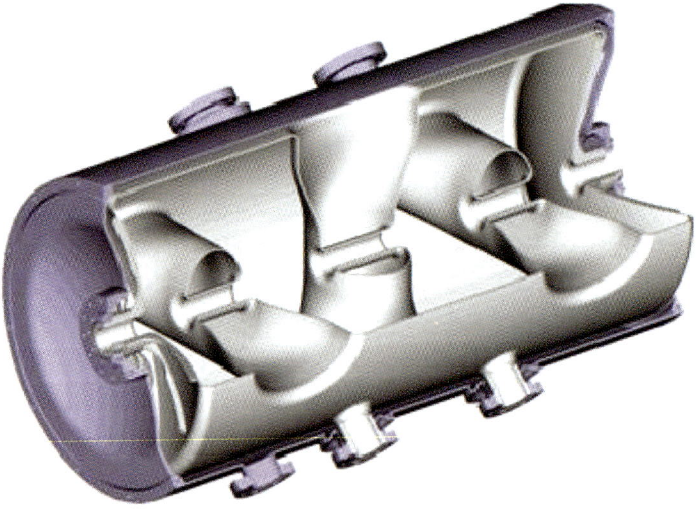

Figure 17. ANL 345 MHz β=0.62 Triple-spoke resonator.

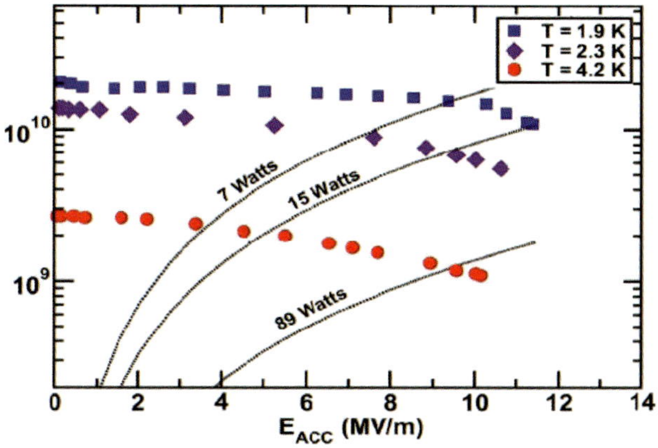

Figure 18. Cryogenic test performance of ANL medium beta triple spoke resonator at various temperatures.

Prototypes such as the 704 MHz BNL electron cooling cavity [17] and 750 MHz JLab high-current cavity [18], although designed for very high current electron beams, could also be candidates. They have the advantage of very large beam apertures and very strong HOM damping (although this is probably overkill for even the highest power proton driver application). Figure 19 shows how a beta < 1 version of the JLab cavity might look and Figure 20 shows a family of such cavities that could share a common end group design. The BNL ECX cavity was qualified at JLab to 20 MV/m with $Q_o \geq$ 1E10 while a single cell beta = 1 version of the JLab high-current cavity shape also showed excellent performance on the first test, Figure 21.

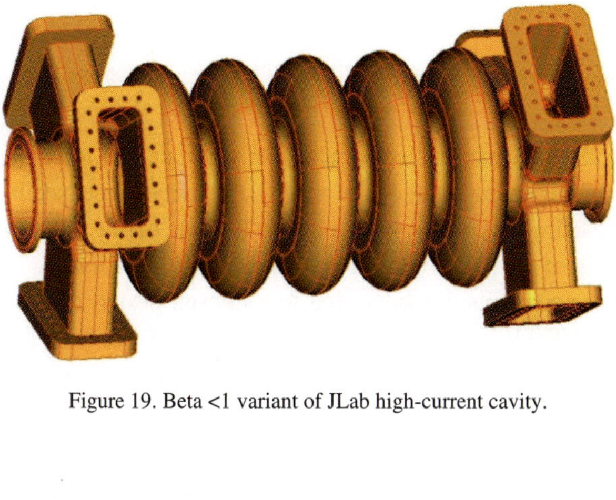

Figure 19. Beta <1 variant of JLab high-current cavity.

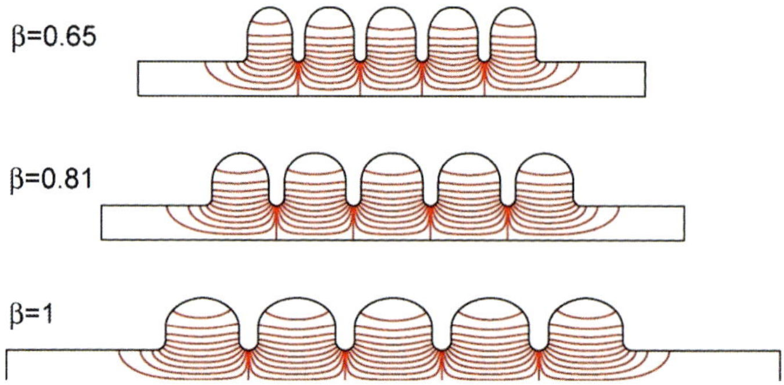

Figure 20. Family of cavities based on the JLab high-current cell shape that could use a common end group and coupler.

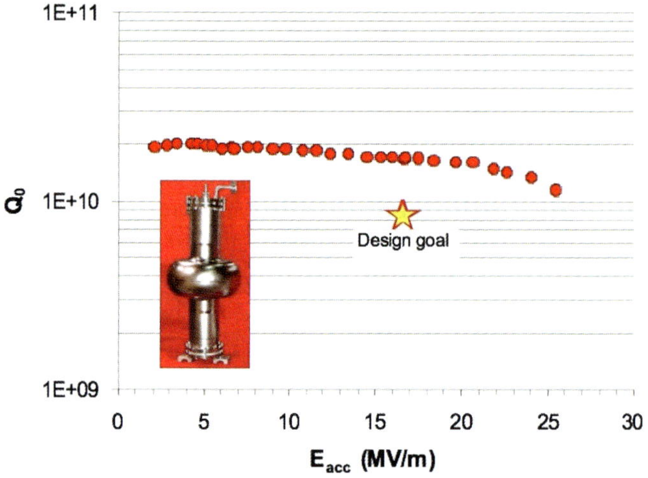

Figure 21. First vertical test result of JLab 750 MHz high-current single cell cavity.

Based on the SNS experience some effort was put into lowering the impact energy in the multipacting zone of the new JLab cavity shape, see Figure 22. No sign of multipacting was observed in the VTA test of the high-current single cell. VTA testing of the 5-cell version is planned in the near future.

Any of these concepts can be scaled to a chosen design frequency although staying in the UHF TV band has advantages in terms of available high power RF sources and proven high-power coupler and RF window designs [19, 20].

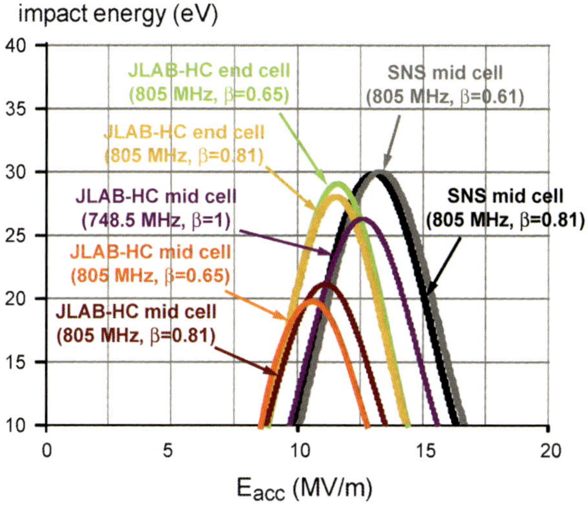

Figure 22. Impact energy of multipacting electrons as a function of gradient in various cell shapes. Flattening the equator region of the cell significantly reduces the impact energy and therefore the likelihood of secondary emission.

8. Cryostat Options

While a lot of attention has been focused on the well-developed TESLA type cryomodule it should not be forgotten that there are other proven designs in existence, many of which are already optimized for CW operation. Examples include SNS, ANL (ATLAS and HINS types), KEK, SACLAY, CERN (LEP and LHC), and CEBAF (original and upgrade types), and more are in development. Depending on the chosen scheme for cryogenic segmentation (e.g CEBAF and SNS are fully segmented with external cryogenic distribution, ILC is minimally segmented with internal distribution), a large internal gas return pipe may not be required or even desirable. The TESLA type module uses the gas return pipe as the cold mass support. Other schemes use a separate support structure or even the vacuum vessel itself for support and alignment of the cold mass, allowing simplified internal helium supply and gas return but requiring external bulk cryogenic distribution. The fully segmented option has great advantages in terms of operability and maintainability of the cryomodule inventory. The primary motivation for internal distribution in the TESLA type module is to minimize static heat load, which is much less of a concern in the CW case where dynamic losses dominate.

Figure 23 shows a concept for a cryomodule containing low beta spoke resonators [15] while Figure 24 shows a concept for the JLab high-current cryomodule based on the CEBAF upgrade/SNS configuration [21]. Figure 25 shows a version of the XFEL type cryomodule. The changes necessary to operate this type of module CW are well understood and have largely been addressed by various projects aimed at CW light source or ERL applications [22]. Should any part of a CW proton linac be chosen to operate at above 1 GHz this could be a viable option.

It is possible to conceive of a common modular cryostat sized to support single and triple spoke resonators e.g. at 352 MHz and one or more families of elliptical resonators at twice that frequency, e.g. 704 MHz using the same support structure and cryogenic distribution throughout and a high degree of commonality in ancillary components such as tuners and couplers. This could provide significant economies of scale and ease of maintenance compared to multiple dissimilar families of cryomodules.

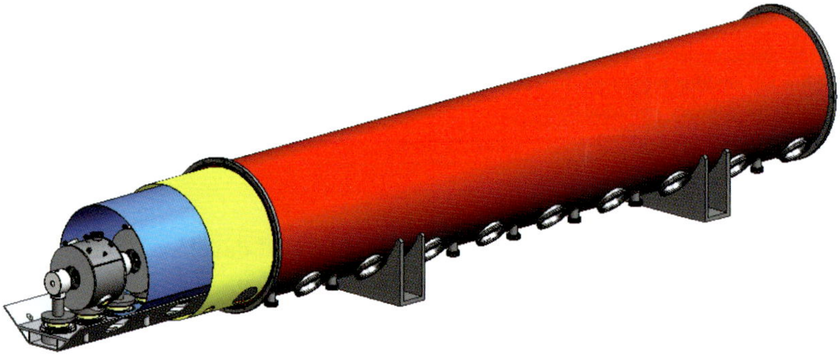

Figure 23. Low beta cryomodule concept for Project X containing spoke resonators.

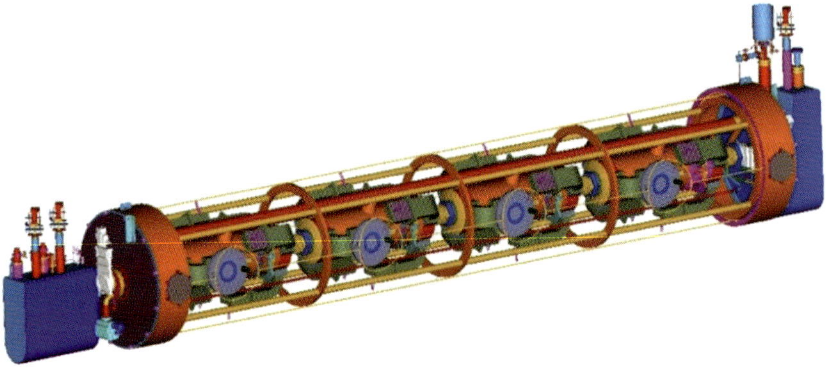

Figure 24. JLab 750 MHz ERL module (based on modified SNS layout).

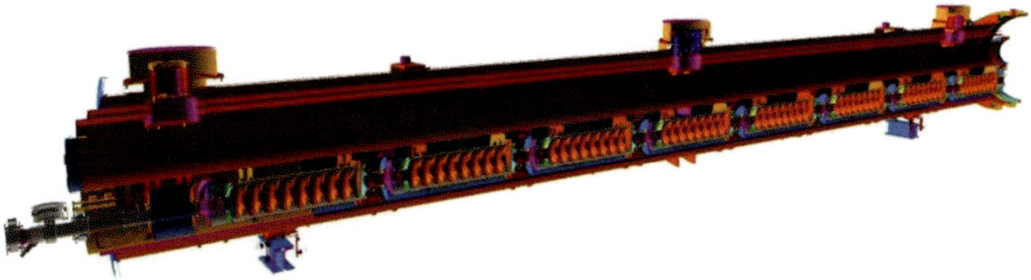

Figure 25. XFEL-type cryomodule modified for CW with larger cryogenic piping (chimney, 2-phase line), higher-power FPC and modified temperature stabilized HOM probes.

9. Topics for Further R&D

A significant cost driver common to all of the CW designs discussed above is the residual RF loss in the superconductor. Understanding the origins of the residual resistance and developing robust protocols to achieve more efficient CW operation would pay large dividends in both capital and operating costs. The wide scatter in observed residual resistance in presently produced cavities indicates that some property of the resultant surface is not being well controlled. This is a complex issue of materials, chemistry and morphology but is fertile ground for significant advancement.

Elimination of field emission is a prerequisite to achieving such benefits but based on SNS experience it remains a challenge in large low frequency cavities.

Microphonics overhead may be a concern for lightly beam loaded cavities but becomes less of a factor as average beam power increases and optimal Qext is reduced. However understanding the sources of excitation and system response may yield improvements in stability and efficiency. Tuning of cavity and support structure stiffness, intelligent design of tuners and application of active feedback may all be viable avenues to address these issues.

A major concern and high priority for further work should be to understand the origins of beam halo and loss mechanisms. Activation of components from beam loss may be a significant performance-limiting phenomenon until this is understood and mitigated.

A demonstration of the technology on a significant scale *with beam* (e.g. a CW version of HINS) is highly desirable, especially in light of the SNS experience where demonstrated individual component performance was not sufficient to guarantee reliable system operation.

BBU analysis and appropriate HOM damping (including real couplers and loads) should be performed in coordination with beam dynamics requirements.

A comfortable safety margin in BBU threshold is advisable but it is possible to be too conservative at the cost of overall system efficiency.

Low trip rate and rapid beam recovery (faster than target thermal time constants) will be very important for some applications such as ADS. Standard reliability engineering techniques such as built-in redundancy and MTBF analysis may go a long way in this regard however safe automated fast beam recovery without operator intervention may take some effort to develop. Experiments at CEBAF have indicated that for many fault conditions this is technically feasible but so far it has not been implemented (JLab policy requires that beam restoration always requires an operator action).

10. Conclusions

Facilities such as CEBAF and SNS operating with MW-class beam power demonstrate that CW SRF is a viable technology for very high power proton drivers. Suitable prototype cavities exist or are in development at multiple institutions for all necessary beta ranges. No "show stoppers" to operating in CW mode at high beam power have been identified. Robust high-power couplers must be used and high overall reliability must be engineered into the machine design. The main challenges may be beam halo, beam loss in the linac, minimizing trip rate and reducing recovery time to well below the thermal time constant of target system components. Cavity operating Q_o (limited by the SRF residual resistance) has a significant leverage on cost. The origins of this residual loss are presently poorly understood but under active investigation. The "cost optimal" design configuration (gradient, frequency, temperature, structure type, etc) depends strongly on assumptions and a full multi-variable optimization using best available world scaling data would be worthwhile.

Acknowledgments

The author would like to thank D. Arenius, G. Ciovati, J. Delayen, S. Henderson, P. Kneisel, M. Liepe, J. Mammosser, F. Marhauser, P. McIntosh, T. Nicol, P. Ostroumov, J. Preble and C. Reece, J. Wei for useful advice and discussion on this subject and for kind permission to reproduce certain of the included Figures. The author would also like to thank the organizers of the AHIPA workshop for providing a unique and stimulating forum to present and discuss these ideas. This paper is based on a presentation of personal opinions by the author. These opinions are not necessarily shared by Jefferson Science Associates or the DOE, but they should be.

Editors' Note

This paper exceeded the set limits for a parallel session paper by a large margin. Because of the pressure to publish on time and also due to the fact that several speakers did not submit their contributions, we have decided to publish the paper as is, rather than ask for a shortened version. We regret any impression of unevenness this may have created.

References

1. "The Continuous Electron Beam Accelerator Facility", H. A. Grunder, J. J. Bisognano, W. I. Diamond, B. K. Hartline C. W. Leemann, J. Mougey, R. M. Sundelin, R. C. York, proceedings of PAC 87.
2. "Performance of the LEP2 SRF System", D. Boussard, proceedings of PAC97.
3. "Thermal Stabilities and Optimal Operating Parameters for the Oak Ridge Spallation Neutron Source Superconducting Linear Accelerator", Sang-ho Kim, Isidoro Campisi, PRST-AB 10, 032001 (2007).
4. Dana Arenius, private communication (presented at the 2008 JLab internal Science and Technology review).
5. "High Field Q-slope and the Baking Effect", Gianluigi Ciovati, Proc 14th International Conference on RF Superconductivity (SRF2009) Sept. 20 – 25, 2009, Berlin, Germany.
6. "SRF System Optimization Process", M. Liepe, Proc. ERL09, Cornell University, Ithaca, NY, USA, June 8-12, 2009.
7. "New Light Source (NLS) Project: Science Case And Outline Facility Design", J. Marangos, et. al, 7/09, http://www.newlightsource.org.
8. "China Spallation Neutron Source Design", J. Wei et. al., Proc. APAC 2007, Raja Ramanna Centre for Advanced Technology (RRCAT), Indore, India.
9. "Theory of Superconductivity", John Bardeen, L.N. Cooper, J.R. Schrieffer, (Illinois U., Urbana). Jul 1957. Published in Phys. Rev. 108: 1175-1204, 1957.
10. "Performance Overview Of The Production Superconducting RF Cavities for the Spallation Neutron Source Linac", J. Delayen, J. Mammosser, and J. P. Ozelis, Proc. PAC05, Knoxville, TN.
11. Gigi Ciovati, private communication to be published?
12. "The Spallation Neutron Source Linac: Performance and Operational Experience", Stuart Henderson, presentation at the 2009 Linear Collider Workshop of the Americas, LCWA 2009, September 29, 2009. http://panda.unm.edu/LCWA09

13. "Overview of High Gradient SRF R&D for ILC Cavities at Jefferson Lab", Rongli Geng, JLAB-ACC-09-1086, Proc. 14th International Conference on RF Superconductivity (SRF2009), Sept. 20 – 25, 2009, Berlin, Germany.
14. "Improving Gradient And Q Performance of BCP Etched Multi-Cell Cavities by Applying a Light EP", R.L. Geng et. al., Proceedings of SRF2009, Berlin, Germany, 2009.
15. "325 MHz Cryomodule Design Status and Plans", Thomas H. Nicol, presentation at the Project X collaboration meeting, FNAL, September 11-12, 2009.
16. "Triple-Spoke Cavities for a CW Proton Linac", P.N. Ostroumov, presentation at the Project X collaboration meeting, FNAL, September 11-12, 2009.
17. "Some Aspects Of 704 MHz Superconducting RF Cavities", R. Calaga, Proceedings of SRF2009, Berlin, Germany, 2009.
18. "A Family of L-Band SRF Cavities for High Power Proton Driver Applications", R.A. Rimmer, F. Marhauser, Proceedings of PAC09, Vancouver, BC, Canada 2009.
19. "Fundamental Power Coupler Development for the SNS Superconducting Linac Cavities", M. Stirbet et. al., Proceedings of LINAC2002, Gyeongju, Korea, 2002.
20. "A High-Power L-Band RF Window", R.A. Rimmer et. al., Proc. PAC 2001, Chicago., LBNL-47968, LAUR 01-2574.
21. "The Jlab Ampere-Class Cryomodule Conceptual Design", R.A. Rimmer et. al., Proceedings of EPAC2006, Edinburgh, Scotland. 2006.
22. "Assembly Preparations for the International ERL Cryomodule at Daresbury Laboratory", P. A. McIntosh et. al., Proceedings of SRF2009, Berlin, Germany 2009.

INDIAN ACCELERATOR PROGRAM FOR ADS APPLICATIONS

V. C. SAHNI

Bhabha Atomic Research Centre (BARC), Mumbai-400 085, India

PITAMBER SINGH

BARC, Mumbai-400 085 & RRCAT, Indore-452 013, India

We describe accelerator R&D being pursued in Indian (mainly DAE) labs for Accelerator Driven Sub-critical system (ADS) applications. Currently design and prototype development of most subsystems is on, that should pave the way for a construction of full system in future.

1. Introduction

Currently the world produces nearly 18 Trillion Units (1kWH) of electricity per year that is about 40% of humanity's total energy needs. About 60 % of this is produced using fossil fuels that results in ~10 Giga-tons of carbon dioxide being added to the environment causing undesirable consequences. Clearly, from amongst non-carbon based sources, nuclear energy is a good option to meet the base load needs, if we can find an acceptable solution for nuclear waste problem. In this context, main limitation of Uranium based reactors -in the form of long lived minor actinides (MA) & long-lived fission products (LLFP)- can be traced to poor neutron yield of the fission process and a way to tackle it can be to integrate the fission process to neutron rich spallation process. This can be done using a high energy proton beam falling on a high Z (like Pb, W, U, Th) material surrounded by a sub-critical reactor. (Except for a tail of high-energy neutrons, spectrum of spallation neutrons is similar to that of fission neutrons.) These neutrons can then multiply in the sub-critical core through finite fission chain reactions and deliver an overall energy gain. ADS systems have the capability to incinerate MA and transmute LLFP to help reduce radiotoxic waste and also aid in the use of Thorium as an alternate nuclear fuel. The last fact is especially important for India, offering a way for an early start of large-scale thorium usage in our nuclear power program. But R&D needs to be pursued as a priority if we want to advance in this area. An ADS system[1] would typically require a proton beam of ~1 GeV, ~10 mA and a sub-critical core based on fast neutrons or thermal neutrons or a combination thereof. All facets of this program are being studied in India; but here we will only focus on the accelerator aspects.

2. Accelerator R&D Related to Indian ADS Program

Some years ago Indian DAE embarked upon the design and development of an ADS linac[2]. Visualizing a plan for 1 GeV, 30 mA machine (Fig 1), detailed physics design was initiated, keeping in view that beam loss in the accelerator must be minimum allowing a hands-on-maintenance of the entire linac. The linac involves a 50 keV ion source, 4 vanes 3 MeV Radio Frequency Quadrupole (RFQ), Drift Tube Linac (DTL) up to 40 MeV, Cavity Coupled DTL (CCDTL) up to 100 MeV and elliptical Superconducting (SC) 5 cell cavities to get to 1 GeV beam energy[3]. Initially we planned SC sections with gradient of 5 MV/m; but with improved SC technology ~25-30 MV/m gradients seemed possible. Still, in our revised design 15 MV/m gradient was used as an optimum choice between accelerating gradient and cost to operate the machine, since this linac must work in Continuous Wave (CW) mode requiring large amount of RF power and cryogenics. The increased gradient reduced the length of 1 GeV linac from ~880m to 400m. The SC cavities -designed to perform over the given velocity range- are identified in our design by so called the geometric velocity, β_G. Our design exploits the large velocity acceptance of superconducting cavities. Transverse and longitudinal phase advances per unit length are maintained constant at all transitions between the structures to provide a current independent match into the next structure. While operating frequency for RFQ and DTL was chosen as 352.21 MHz, that for CCDTL and SC linac was chosen as 704.42 MHz. Simulation results revealed that over full (400m) accelerator length, beam transmission is 98%, with 2% loss (confined to RFQ during the beam bunching) posing no serious radiation problem as energies involved are low. Linac's scheme is given in Fig.1, and design parameters in Table 1, while variation of beam size & aperture with energy are shown in Fig.2.

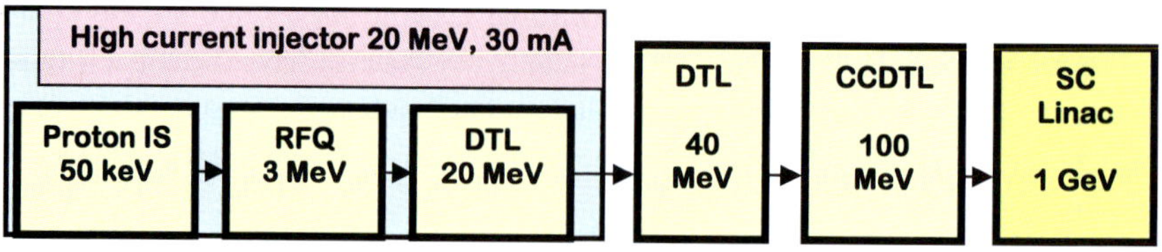

Figure 1. Schematic arrangement of the 1 GeV linac designed for ADS application.

Table 1. Design parameters of various parts of the 1 GeV linac.

Parameters		DTL	CCDTL	SC
Input Energy (MeV)		3.0	40.12	100.24
Output Energy (MeV)		40.12	100.24	1016.5
Frequency (MHz)		352.21	704.42	704.42
Synch. Phase (deg)		-30	-30	-30/ -35
Accelerating field gradient (MV/m)		2.5	1.37	15.0
Total Power (MW)		3.0	6.4	26.9
Total Length (m)		22.7	69.6	306.2
Normalised rms transverse	In x	0.0233	0.0232	0.0351
emittance (π cm-mrad)	In y	0.0242	0.0236	0.0313
Longitudinal emittance (deg-MeV)		0.1150	0.2373	0.2549

Most challenging part of a CW accelerator is the low-energy injector, usually in the 10-20 MeV range, where space-charge effects dominate. Hence a 3 phase program (with target energies of 20 MeV, 100 MeV and 1 GeV) was thought of and work was started at BARC on a low energy (20 MeV) high intensity (30 mA) proton accelerator (LEHIPA) to validate the design of front-end injector. In what follows we describe the details of LEHIPA[4] (Fig. 3).

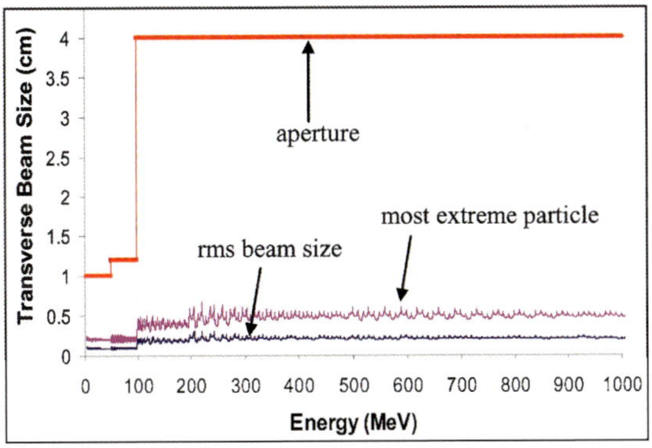

Figure 2. Beam size and aperture variation as a function of energy.

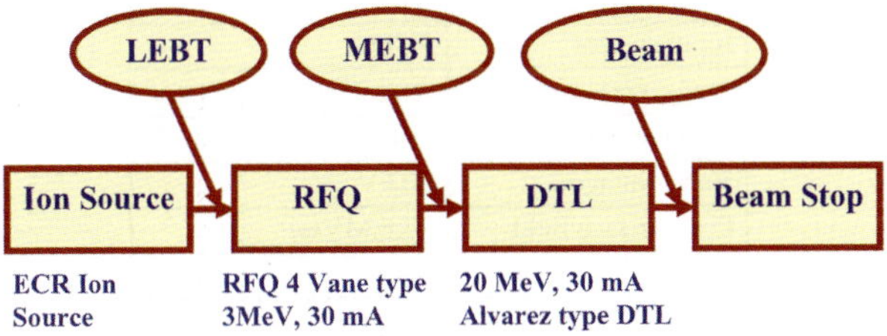

Figure 3. Layout of LEHIPA.

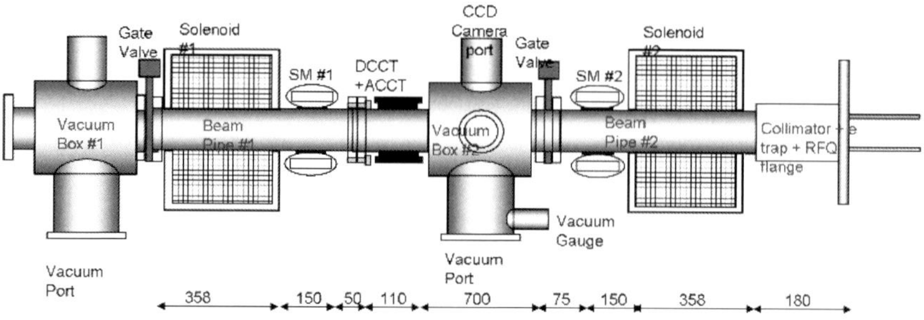

Figure 4. Layout of the LEBT for LEHIPA.

Proton beam extracted from the ECR ion source is injected into RFQ through a Low Energy Beam Transport System (LEBT). LEBT (Fig. 4) contains two solenoids meant to match the beam from the ion source to the RFQ and includes useful beam diagnostics like ACCT, DCCT and beam position monitors. Space charge expansion of a high current beam is one of the major processes responsible for increase in beam size and emittance. To facilitate transport of high current beam and to reduce emittance growth due to space charge forces, space charge compensation of the beam is done in LEBT. LEBT's solenoids have been designed, fabricated and also tested at RRCAT.

RFQ design has been done using the equi-partitioning scheme, where longitudinal and transverse beam temperatures have to be equal. By this technique the emittance growth was observed to be less than 2%. In our design the vane voltage has been kept constant, keeping the peak surface field less than 1.8 times the Kilpatrick limit. Table 2 shows the RFQ parameters. Full RFQ (4 m length) will be built in 4 segments, which will be coupled with coupling

Table 2. Parameters of the RFQ.

Parameters	**Value**
Frequency	352.21 MHz
Input energy	50 keV
Output energy	3 MeV
Input current	30 mA
Transverse emittance	0.02/0.0204 π cm-mrad
Synchronous phase	-30^0
Vane voltage	80 kV
Peak surface field	32.8 MV/m
Length	4 m
Total RF power	500 kW
Transmission	98 %

cells. The total RF power requirement is 500 kW that includes 88.5 kW of beam power. The cavity optimization of the RFQ was done using SUPERFISH. In order to include 3D features, like, end cells, coupling cells, tuners, vacuum and RF ports, MAFIA and CST Microwave Studio were used.

3. Status of the Ongoing Activities and Plans for Future

The LEHIPA project involves handling of very large RF power and fabrication of complex structures. To learn such technologies we first decided to build a 400keV RFQ based machine that will replace an existing dc accelerator (of a 14 MeV neutron source coupled to a sub-critical assembly). This machine consists of a 50keV/1mA deuteron ion source, a LEBT and a RFQ. After design of RFQ we made a prototype in ETP Cu (Fig.5) and another 600mm RFQ has been made in OFHC copper and tests on it are in progress. The RF group of Vacuum Physics Instrumentation Division of BARC has built many subsystems for this RFQ based 400 keV machine. Foremost amongst these is a tetrode based RF source at 352 MHz, which is currently undergoing elaborate tests. A LEBT line along with 1mA ion source and 2 solenoids for focusing the beam are also being assembled for this program. An indigenous power coupler development has also been initiated. Two 50 kW CW coaxial couplers are under fabrication at BARC. The design incorporates disc type alumina windows with double barrier for better mechanical stability and reliability. A shorted stub provides RF matching along with providing support for central conductor and cooling channels. A prototype RF coupler fabricated by our colleagues at BARC is shown in Fig.6a. (A 250 kW iris type coupler, with tapered transition to WR 2300 half height wave-guide, is also being designed. Its thermal loss estimation and mechanical tolerance level requirements are being worked out.)

Figure 5. The 1.02 m long RFQ made by Centre for Design & Manufacturing, BARC for 400 keV machine.

(a) (b)

Figure 6. (a) Prototype coupler fabricated at BARC and (b) prototype 1.2m long DTL under fabrication.

In parallel with 400 keV machine work, design of DTL is going on at BARC to hike beam energy in LEHIPA from 3 to 20MeV. The 12.8m long DTL will be built in 4 tanks. Its power requirement is 1.2 MW. Based on this design, a 1.2m long prototype DTL has been fabricated (Fig. 6b). Frequency measurements on it have been done and found to be 349.5 MHz. For focusing the beam we will use permanent magnet based quadrupoles which will be placed inside the drift tubes. Design of the quadrupoles has been done and a prototype has been built.

To reach an energy of ~1 GeV one must also develop Niobium based Superconducting Radio Frequency (SCRF) cavities. Infrastructure is being built (mostly at RRCAT) for multi cell cavity fabrication, chemical processing and cleaning, materials characterization etc. Efforts have also begun towards design of cavity test set up which will be later followed up with facilities to test the cryo-modules. In this regard, while R&D to make and characterize β=0.47 SCRF cavities is in progress at BARC, RRCAT is rapidly making parallel progress. In partnership with Fermilab (FNAL) RRCAT scientists and engineers have learnt and started to contribute to the development of 1.3 GHz (or sub-harmonic) niobium SCRF cavities, needed for Project-X. With assistance provided by IUAC, New Delhi they have made the first Indian single cell solid niobium

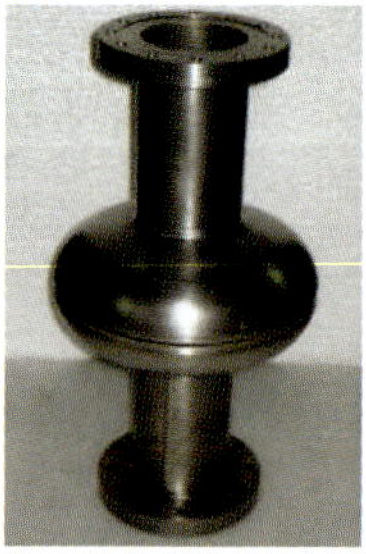

Figure 7. First solid Nb cavity fabricated by a team from IUAC & RRCAT.

cavity (see Fig. 7). They have also been closely working with FNAL scientists in doing value engineering on many present system designs evolved elsewhere. Some of the items on which work has been done are: (a) A new way to make the elliptical cavity half cells to ensure better parallelism between planes passing through the equator and iris (b) A more economical way to make cavity end groups, wherein number of e-beam welding operations is greatly reduced to save many kilo dollars (c) A new way of mounting the cavities in a cryomodule that not only relaxes the stringent specifications on helium return gas pipe (thus reducing the cost) but also simplifies the procedure of aligning the cavities in the cryomodule (d) An exhaustive analysis of beta=0.81 cavity that has brought out new options and (e) SCRF infrastructure design and construction. RRCAT has also built and delivered to FNAL tools to make half cells for the elliptical cavities. These tools are being used by TRIUMF in Canada to make cavities. In addition BARC, RRCAT, VECC, IUAC and FNAL teams are closely working on many facets of the R&D related to Project-X. Overall, a strong collaboration between Fermilab and Indian labs has emerged beneficial to both the sides. (To learn more details of how this partnership is progressing, please see the site http://iifc.fnal.gov/.) BARC is developing other technologies relevant for spallation target design and reactor configurations that have been reported at ADS meetings organized by IAEA from time to time. Interested reader can look at them for details.

Acknowledgement

Work reported here is a joint effort with many colleagues at RRCAT & BARC whom we wish to thank for their excellent contributions.

References

1. F. Carminati, R. Klapisch, J.P. Revol, J.A. Rubio and C. Rubbia CERN/AT/93-47 (ET); C. Rubbia et al; CERN/AT/95-44 (ET); CERN/LHC/96-01 (EET); CERN/AT/95-53 (ET); CERN/LHC/97-01 (EET).
2. S.S. Kapoor, "Roadmap for development of Acccelerator Driven Sub-Critical Reactor Systems", BARC/2001/R/004.
3. Rajni Pande, Shweta Roy, T. Basak, S.V.L.S. Rao and P. Singh, Beam Dynamics for the 1 GeV Proton Linac for ADS, Proc. Indian Particle Accelerator Conference (InPAC-2006), BARC, Mumbai, Nov. 1-4, 2006, page 207.
4. P. Singh, S.V.L.S. Rao, Rajni Pande, T. Basak, Shweta Roy, M. Aslam, P. Jain, S.C.L. Srivastava, Rajesh Kumar, P.K. Nema, S. Kailas and V.C. Sahni, Accelerator Development in India for ADS Programme, Pramana – J. Phys. **68**, 33 (2007).

ION ACCELERATOR ACTIVITIES AT VECC (PARTICULARLY, OPERATING AT LOW TEMPERATURE)

RAKESH KUMAR BHANDARI†

Variable Energy Cyclotron Centre, 1/AF, Bidhannagar
Kolkata, West Bengal, 700064, India

VECC has developed excellent infrastructure for advanced research and development in the field of accelerator technology particularly in the superconducting magnet technology. The main activities are focused on development and operation of the K130 cyclotron, K500 superconducting cyclotron and the ECR ion sources. We have recently commissioned the superconducting cyclotron with internal ion beam and subsequently beam extraction program has been taken up. This large superconducting magnet (1.42 m pole diameter) operates at 4.8T average magnetic field. This cyclotron will be used to accelerate a variety of heavy ion beams for basic nuclear physics and allied experiments. In this paper, details of the design, construction and operation experience of various systems of this superconducting cyclotron will be presented.

1.1. *Introduction*

VECC has one of the premier accelerator facilities of India, the Variable Energy Cyclotron (VEC) facility, which was developed in nineteen seventies based on Berkeley 88" cyclotron design (K=130). Initially it delivered light ion (proton, deuteron and alpha) beams using an internal hot cathode PIG type ion source. Since 1998, it started delivering heavy ion beams using an indigenously developed 6.4 GHz ECR ion source. Later, another 14.5 GHz source was coupled to it for heavier mass gaseous as well as metallic ions. This cyclotron has served a large number of research institutions and universities all over the country for research in basic nuclear physics, condensed matter physics, radiation damage, radio chemistry, biophysics and many other related fields. At present, this cyclotron is being refurbished to improve the overall performance and to use it as the primary source for the upcoming Radioactive Ion Beam (RIB) facility.

In late nineties, a superconducting cyclotron project was conceived based on the design of Texas K500 superconducting cyclotron [1]. This cyclotron will deliver light ion beams with maximum energy of 80 MeV/nucleon and heavy

† bhandari@veccal.ernet.in

ion beams with typical energy of 5-10 MeV/nucleon. The low energy beam extracted from ECR ion source is transported about 25 meters and axially injected into the cyclotron. It is then accelerated with the help of three accelerating electrodes (called dees), which can develop peak voltage of 100 kV in frequency range 9-27 MHz. On 25th August 2009, it has successfully accelerated internal ion beams (Ne^{+3}) up to extraction radius. The beam extraction process is in progress and very soon beam will be available in experimental caves.

Figure 1. (a) The K=130 room temperature cyclotron, (b) The K=520 superconducting cyclotron.

VECC is setting up a 30 MeV, high intensity (500 μA) proton cyclotron facility in a new campus, about 15 km away from the main campus. It will be able to produce two simultaneous beams with different energies and intensities from two opposite ports, which will increases its utilization time and flexibility. There will be three beam lines dedicated for production of PET[*] and SPECT[†] radioisotope and fully automated radiochemistry system for the remote production of radiopharmaceuticals from the irradiated target. Besides, two beam lines will be available for material science research with high power (15 kW) proton beam. The other major research and developmental activities includes, design and development of high current proton ECR ion source, 10 MeV proton cyclotron for accelerator driven sub-critical system and development of Radioactive Ion Beam facility.

The superconducting cyclotron will provide opportunities to probe the frontier areas in nuclear physics around the Fermi energy domain. Several experimental facilities are being developed at VECC as a part of the utilization

[*] Positron Emission Tomography

[†] Single Photon Emission Computed Tomography

program. These include a general-purpose scattering chamber, 4π charged particle detector array, 4π neutron multiplicity detector, TOF neutron detector array and high-energy gamma detector array. The three segment cylindrical scattering chamber is one of the largest in the country, providing the unique opportunity to mount various kinds of detector systems in different geometry and to carry out TOF measurements involving large flight paths. It has already been installed in the beam hall-1 of the superconducting cyclotron. In this paper, we discuss about the development of Kolkata superconducting cyclotron.

1.2. *Construction and Commissioning of Superconducting Cyclotron*

Construction

The main magnet of the superconducting cyclotron is designed to operate at about 800A current in each of its two coils, named as α and β coil, producing a maximum of about 5.8 T magnetic field. The main magnet iron structure, weighing about 80 tonnes, was machined out from large forged blocks. The Nb-Ti superconducting coil was wound over the stainless steel bobbin and placed inside the main magnet cryostat.

Figure 2. (a) Lower half of magnet iron, (b) The superconducting coils wound on the SS bobbin.

The iron along with the coil and cryostat had been installed in 2004. The liquid He plant, cryogenic delivery system, power supplies and the control systems were successfully commissioned. The magnet was cooled down to liquid He temperature and energised first time in the early 2005. In the subsequent months, the magnet was operated satisfactorily for almost one year at 4.5 K temperature, while extensive magnetic field measurements were done for correction of the manufacturing defects and assembly errors in the iron structure, centering of the main coil, calculation of operational settings, calculation of extraction trajectory etc. After completion of magnetic field measurement activity in April 2006, the coil was warmed up to facilitate assembly of the RF system and other major systems of the machine.

The main magnet consists of the specially contoured pole tips, iron pole caps, return path rings and the cryostat. The pole diameter of the magnet is 1.42 meters and the gap varies from 6.35 cm in hills to 91.4 cm in valleys.

The cryostat, a fully confined annular chamber, consists of 3 major subassemblies – the liquid helium chamber (bobbin), liquid nitrogen cooled radiation shield and vacuum chamber (coil tank). The bobbin is an annular housing made of stainless steel (SS316L). The small (α) and large (β) coils are wounded on its inner wall and the outer walls are welded to enclose the chamber. The liquid helium volume within the coil is about 300 litres and at above is about 40 litres. Seven level sensors read the helium level within and above the coils. The bobbin is surrounded by another annular chamber called coil tank. This chamber works as the vacuum enclosure for the bobbin. It is made of low carbon steel (AISI 1020). In between bobbin and coil tank, the copper (CDA 110) radiation shield is placed. The bobbin, weighing about 7 ton, is supported inside the coil tank by three horizontal and six vertical support links made of glass epoxy material (scotch ply). There are three ports at the top of the cryostat for connecting the current leads, helium lines and safety devices to the helium chamber. There are also three ports at the bottom for liquid nitrogen supply, vacuum pump connection and over-pressure safety flange connection. There are 20 numbers of radial penetrations in the median plane of cryostat to access the beam space from outside. Various beam extraction and diagnostic devices like magnetic channels, deflectors, beam probe etc. pass through these penetrations to guide the beam in to the desired path and finally extract from cyclotron through exit port (figure 3).

Figure 3. (a) The bobbin wrapped up by multiplayer insulation is being put in to the annular iron jacket called coil tank, (b) Twenty radial penetrations in the median plane are brazed to the cryostat to access the beam space from outside for installing various beam extraction and diagnostic devices.

The inner walls of bobbin and inner & outer walls of the coil tank were made of ring rolled forgings to avoid longitudinal joints. They were rough machined and welded with the end flanges by TIG process. Overall welding length is about 200 m. Each assembly was finally machined to achieve dimensions within specified tolerances. The bobbin assembly was cryo-shocked and subsequently vacuum-leak tested before it was taken for coil winding. After the winding was over, the outer walls were welded to make it a closed chamber for the liquid helium. The chamber was helium-leak checked and afterwards insulated with multi-layer insulations. It was then covered with the copper radiation shield, insulated again with more multi-layer insulation and placed inside the vacuum chamber supported with the vertical links. Finally, all the links, ports, connections and median- plane-penetrations were welded to complete the assembly work. The whole assembly was vacuum-leak tested successfully. Then the cryostat was assembled with the magnet iron in the cyclotron building. The alignment of the cryostat with the magnet iron was ensured for correct positioning of the cryostat.

Commissioning

The moisture level in the liquid helium chamber was reduced below 10 ppm, before starting the cool down process, by repetitive evacuation and purging with pure helium gas and heating the coil with 5A current. Then the temperature was brought down by sending cold helium gas from the plant to the cryostat. Initially, the liquid nitrogen flow to the radiation shield was not available. Still the liquid helium plant was capable enough to cool the coil to 4.5K and to immerse it with liquid helium (fig 4a). Subsequently, the liquid nitrogen flow to the shield was provided. Some leaks were detected on the current lead bellows during cooling down and those were fixed.

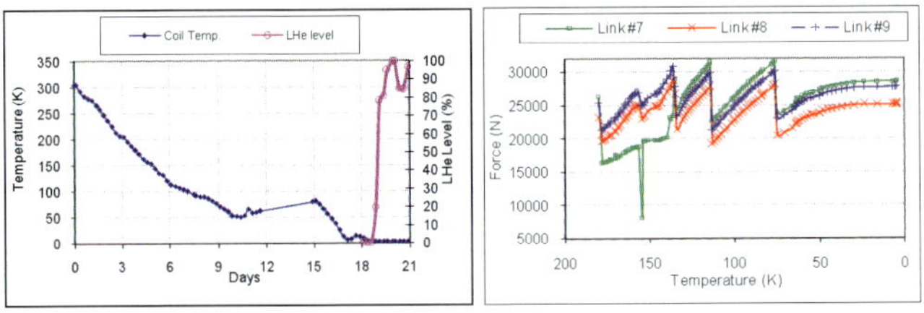

Figure 4. (a) Coil temperature and liquid He level vs. time during the cool down, (b) Radial support link force adjustment during cool down of cryostat.

During the cool down, the radial support link forces were adjusted to keep the forces within safe limits. At 154.4 K temperature, the safety nut of the link #7 happened to share the entire load instead of the strain gauge stud due to a human error. This is indicated as a sharp dip in fig. 4b. Liquid He boil-off measurement was carried out to find out the cryostat liquid He evaporation rate (< 40 litres/hr).

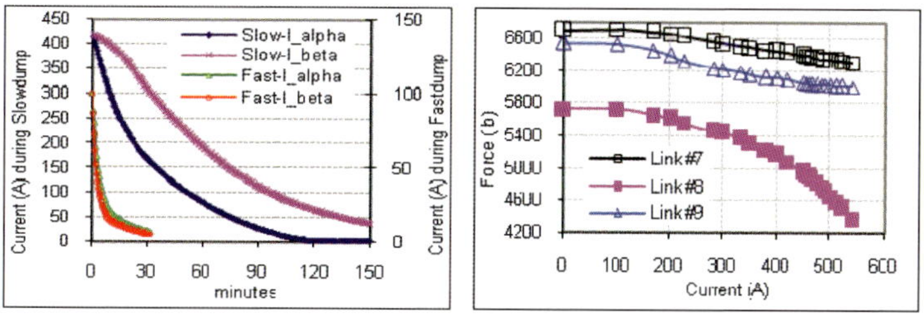

Figure 5. (a) Slow-dump and fast-dump decay curves, (b) radial support link force adjustment during cool down of cryostat.

After stabilizing the required liquid helium level in the cryostat for a couple of weeks, the magnet excitation was initiated. All the safety interlock systems were tested by forcing the coil current decay through slow and fast dump resistors (fig. 5a). Several iterations were required for coil the centering process before reaching high excitations. The coil experiences position and current dependent forces, which were observed as change in the radial support link force. The coil needs to be moved to a position where the coil axis matches with the magnetic axis of symmetry. At this position, the forces on the radial support links decrease monotonically with current due to expansion of the coil in response to the magnetic hoop stresses as shown in fig. 5b.

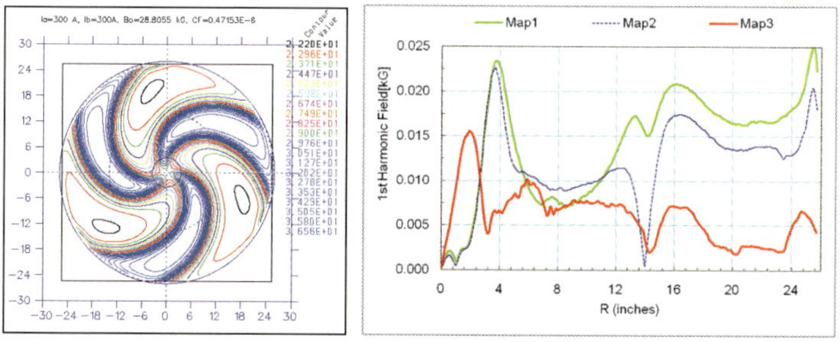

Figure 6. (a) Isogauss contour plot of measured field, (b) Radial distribution of first harmonic field.

Detailed magnetic field mapping has been done at different main coil excitations. The field has been measured with a search coil and NMR (for calibration) set up, with linear and angular positioning accuracy of 10 μm and 0.5 arc-seconds, respectively. The required field uniformity to lock the NMR signal was found at the centre of the magnet and near the RF holes in the valley region. Fig. 6a shows the iso-gauss plot of the measured field. The radial distributions of 1st harmonic component at different stages of corrections are shown in fig. 6b as Map1, Map2 and Map3. The 1st harmonic amplitude and phase distribution were carefully investigated to determine the size and position of the iron shims. One of the observations was the movement of the pole-tips when magnet was excited. Faulty bolts were detected and replaced. Finally, the 1st harmonic component was minimized to within 7 gauss near the extraction radius, which can be adequately handled during the beam extraction process with the help of the harmonic trim coils.

1.3. *Other systems*

There are 13 epoxy potted trim coils wound on each pole tip as shown in fig. 7a. There are 78 trim-coil feed-thrus and several vacuum fittings on the back of the pole caps, those were thoroughly leak tested, after installing RF liners on the pole tip sectors wound with trim coils. Figure 7b shows the median plane liners and dees installed in position over the pole tips and trim-coils.

Figure 7. (a) Trim coils are wound around the spiral poletips, (b) copper liner covers the trim coils and three spiral shaped dees are placed in the valley region.

The radio-frequency system of superconducting cyclotron consists of three $\lambda/2$ RF resonator cavities (fig. 8a) powered individually by 80 kW radio-frequency amplifiers via three hydraulically driven coupling capacitor through 3-1/8 inch rigid coaxial transmission line with 50Ω characteristic impedance. The fine frequency tuning (±0.3%) of the cavity is achieved by a hydraulically driven trimmer capacitor.

Main beam chamber vacuum (~8 x 10^{-7}mbar) is achieved by combination of scroll pump and turbo-molecular pump, which is further reduced to 2x10^{-7}mbar pressure with the liquid nitrogen cooled cryo-panels.

The low energy beam from a 14.5 GHz ECR ion source (fig. 8b) is injected axially in to the cyclotron and bent in the median plane with a spiral inflector. The beam extraction system of superconducting cyclotron consists of two electrostatic deflectors, eight passive magnetic channels and one active magnetic channel. Figure 9a shows the experimental beam line layout. At present 5 m external beam line (fig. 9b) has been installed for beam extraction.

Figure 8. (a) Upper resonator cavity installed on upper pole, (b) ECR Ion source and the injection beam line.

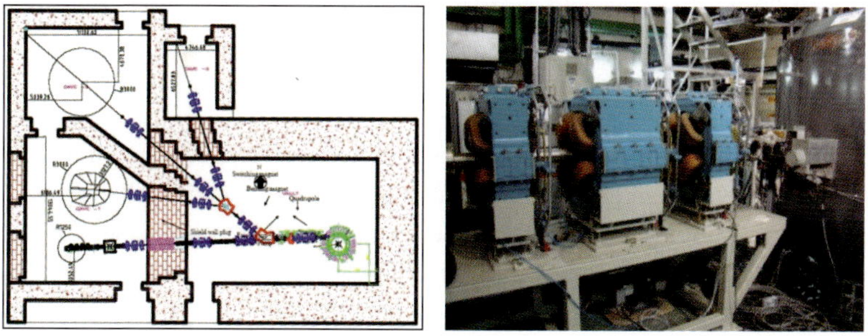

Figure 9. (a) Layout of experimental beam lines, (b) External beam line.

Conclusion

The VECC K500 superconducting cyclotron has been commissioned in August 2009 with internal beam accelerated up to extraction radius. The extracted beam is expected to be available for experiments in the middle of next year.

Acknowledgments

Contribution of several dedicated scientists, engineers and administrative staffs of the Centre are gratefully acknowledged.

References

1. R. K. Bhandari, Int. Conf. on Cyclotrons and their Applications, *2007, p51.*

CHINESE EFFORTS IN HIGH INTENSITY PROTON ACCELERATORS

SHINIAN FU†
,JIUQING WANG, SHOUXIAN FANG
Institute of High Energy Physics, Beijing 100049, China

High intensity proton accelerators have two important applications in China in recent years: Accelerator Driven Sub-critical System for nuclear waste transmutation and China Spallation Neutron Source. This paper focuses on the R&D activities of the key technology of high intensity proton accelerators in China, including the normal conducting and superconducting accelerator technologies. Chinese efforts in ADS basic research will also be outlined.

1. Introduction

High intensity proton accelerators have important applications in China. It focused on the Accelerator Driven Subcritical System and spallation neutron source. ADS is a long-term program and China now is still in the status of basic research and key technology development. China Spallation Neutron Source (CSNS) project [1] has been approved by the Chinese government and is expected to start construction in 2010 and completed in 2016.

China, as a developing country with a great population and relatively less energy resources, is rapidly developing nuclear energy. The nuclear electricity is foreseen to reach 75GWe, and meanwhile 30GWe power will be under development in 2020, according to the recently revised plan of the Chinese government in April 2009. To develop nuclear power in such a large scale, long-lived radioactive nuclear wastes have to be safely disposed to reduce the impact on the environment and to eliminate public fear of nuclear power. Right now only a small amount of spent fuels from nuclear power plants has been accumulated in China. But the situation will become very serious in the future. The accumulated waste is estimated to be more than 10k tons in 2020, and it will be doubled in 2030. ADS has been recognized as the best option for the nuclear

† Work partially supported by the National Key Basic Study Program of China.

waste disposal and some basic study programs on ADS have been supported by the Chinese government since 1999 [2]. High intensity proton accelerator is one of the major subjects in these programs.

CSNS accelerator consists of an H^- linac of 81MeV and a rapid cycling synchrotron of 1.6GeV with an output beam power of 100kW in the first phase. In the future upgrade plan to 500kW beam power, the linac energy is expected to increase to 250MeV with superconducting cavities.

2. R&D Activities on Proton Linac

A high current front end consisting of an ECR ion source, LEBT and an RFQ accelerator of 3.5MeV has been built in the ADS program, as shown in Figure 1.

ADS demands an extremely high reliability of the driver accelerator and it is identified that the injector is a major source of the beam trip. So, great efforts were made on the reliability and stability of the ion source operation. A high reliability of 99.9% is achieved during an 120 hours continuous operation. The output beam current is around 70mA in normal operation, and now is pushing to 120mA in test.

Figure 1. The ECR ion source (left) and the RFQ accelerator (right).

The RFQ is designed for CW operation, but as the first step it is operated at 7% duty factor with a beam kicker in the LEBT. This is a four-vane type structure with two coupled resonators and dipole stabilizer rods on the middle coupling plate and the two end-plates, similar to LEDA RFQ [3]. The major design parameters are listed in Table 1. A beam current of 46mA was obtained from the RFQ at 7% duty factor with a beam transmission rate of 93%. The beam energy spectrum is measured in a good agreement with the simulation result from PARMTEQM code [4], as plotted in Figure 2. The RF power source is kindly provided by CERN with capability up to CW operation. A digital RF control system based on FPGA was developed at IHEP and added to the RF system. With this new feedback system the operation stability became better in the case of long pulse and heavy beam loading. The RF amplitude and phase stability reached $\pm 1\%$ and $\pm 1^0$ respectively.

Table 1. RFQ major design parameters.

Input Energy (MeV)	0.075
Output Energy (MeV)	3.5
Peak Current (mA)	50
Duty Factor	6% to 100%
RF frequency (MHz)	352
Maximum Es (MV/m)	33
Total RF Power (kW)	590
Total Length (m)	4.75

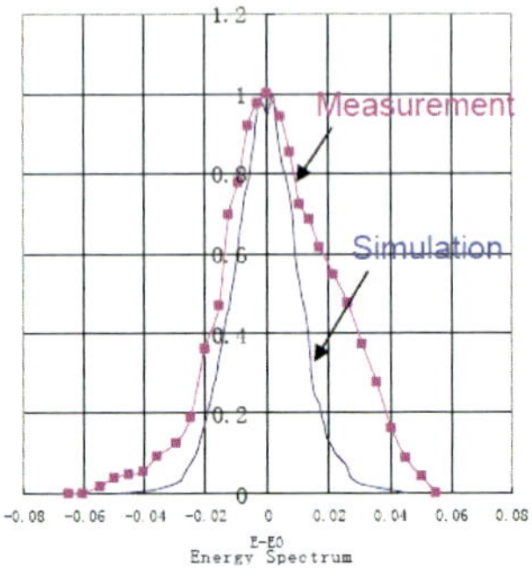

Figure 2. Energy spectrum of the RFQ output beam.

In recent we are raising the duty factor to 15%. A beam line with a periodic focusing channel and beam diagnostics are building for the experimental study of the beam halo formation.

The technology of medium-β superconducting RF cavity has been studied for the medium energy section of 1GeV proton linac for ADS. Three superconducting single ellipsoid cells of 1.3GHz at β =0.45 were manufactured with assistance from KEK. It is the scaled-down prototype of 700MHz cell, as shown in Figure 3. It reaches the surface field E_{sp}=42.4MV/m.The measured Q_0 versus E_{sp} is plotted in Figure 3. With the successful experience of 1.3GHz cell, 700MHz cell is now under development.

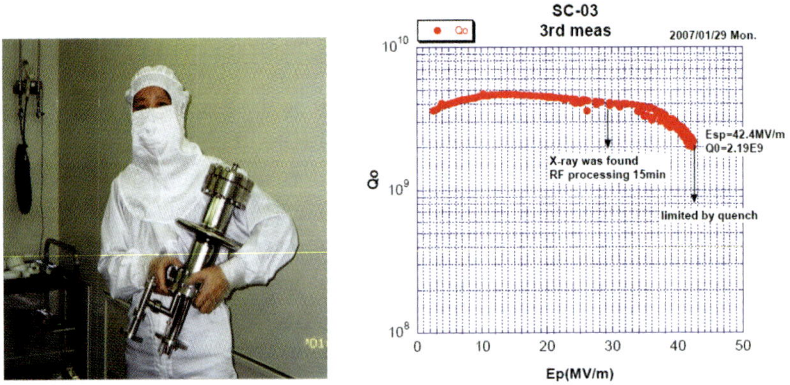

Figure 3. 1.3GHz single cell (left) at β =0.45 and its measured Q_0 versus Esp (right).

For the development of superconducting RF technology at IHEP, an RF superconducting laboratory of 250m^2 has been set up for the RF cell processing and measurement. A cryostat for 1.3GHz cavity has been built for vertical measurement at the working temperature of 1.5~4.2K. The lab does not only used for proton linac R&D, but also serve for the cavity development of ILC and photo cathode gun.

Spoke cavity is an option for the low-β section of the ADS linac. Cavity design study on the RF property and mechanical structure is in progress. In recent, a new budget from Chinese Academy of Sciences was allocated to IHEP for the superconducting spoke cavity development.

3. Status of ADS Study

In reactor physics study, a series of neutron multiplication experimental study has been carried out. The VENUS facility with coupled fast and thermal regions has been constructed as the basic experimental platform for neutronics study in subcritical blanket. It's a zero power sub-critical neutron multiplying assembly driven by external neutrons produced by a pulsed neutron generator. The theoretical, experimental and simulation study on nuclear data, material properties and nuclear fuel circulation related to ADS is carrying on to provide the database for ADS system analysis. A moderate style multi-purpose verification system is under consideration. In the near future, we plan to build a MW swimming pool light water sub-critical reactor driven by a 150MeV accelerator.

4. Summary

High intensity proton accelerator has two major applications in China: ADS and CSNS. Some key technologies in high intensity proton accelerator for ADS are under development. ADS basic study has gained support in two phases from the Ministry of Science and Technology, China. CAS presented an ADS new proposal to the government in recent to speed up ADS from the present basic study to small-scale system construction.

References

1. S. N. Fu, H. F. Ouyang, et al, *Proceedings of LINAC 08*, 21(2008)
2. Xia Haihong, Zhao Zhixiang, et al, *Frontiers of Energy and Power Engineering in China* 1(2), 135 (2007).
3. L. M. Loung et al, *Proceedings of LINAC 2000*, 583 (1998).
4. K. R. Crandall, R. H. Stokes and T. P. Wangler, *Proceedings of the LINAC conference*, Montauk, NY, 1979.

ADSR ACTIVITY IN THE UK

R. J. BARLOW
School of Physics and Astronomy, The University of Manchester, Manchester M13 9PL, UK
and the Cockcroft Institute, Daresbury Science and Innovation Campus, Warrington WA4 4AD
E-mail: Roger.Barlow@manchester.ac.uk

A review is given of present and planned involvement in ADSR research in the UK, and the activities of the ThorEA association.

Keywords: ADSR; High Current Proton; Thorium; FFAG

1. Introduction

The UK currently emits approximately 10 tons of CO_2 equivalent per person per year, broadly in line with the European average. [1] The government has set ambitious targets, including the cutting of CO_2 emissions by 80% by 2050 from the 1990 baseline [2], but there is no firm plan as to how such goals will be met. There is some scope for development of wave and tidal power and for offshore wind farms, less for onshore wind and solar power, however it appears virtually impossible to meet such targets without the use of nuclear power.

Although the UK currently obtains about 20% of its electricity from nuclear power, the reactors are old and are all but one due to be out of service by 2023. Steps are being taken to replace and even expand these plants, though progress is slow as there still considerable public opposition.

The Thorium ADSR (accelerator-driven subcritical reactor) system is well placed as a match between the pragmatic need and the distrustful public: it is manifestly safe, as switching off the accelerator switches off the reactor, it produces no (or almost no) long lived waste, and it is highly proliferation resistant. Although there has been little UK activity in this area until recently, this is changing very rapidly and the number of groups and people interested in, and working on, the topic has grown greatly in recent years.

2. ThorEA

ThorEA (Thorium Research for the Energy Amplifier) is a not-for-profit organisation and learned society. It aims to promote thorium-fuelled energy amplifier systems as a safe, sustainable and publicly-acceptable form of nuclear power. Its goal is the construction of a thorium-fuelled ADSR in the United Kingdom. It provides a framework within which individuals and organisations can co-operate in pursuit of this.

There are over 80 individuals on the mailing list, and some 40 names appear (with permission) as members on the organisation website. They cover a very broad range of disciplines: Accelerator Scientists, Particle Physicists, Nuclear Physicists, Nuclear Engineers, and Economists. They also come from many different institutes. The UK Accelerator Institutes, Cockcroft and John Adams, are well represented, as are the Rutherford Appleton and Daresbury Laboratories. There are members of 10 different universities, and several from industry. Although most are UK based there are several from continental Europe, the US and elsewhere that use this as a way of keeping in touch, and they are very welcome – we are aware that we are behind in many areas. Anyone who wishes to join should send an email to the author.

It maintains a website `http://www.thorea.org` as a central point of information and contact, which includes details of the organisation, a news feed of relevant stories, links to other sites, and details of meetings. These open and informal workshops take place 3-4 times a year, and normally get around 30 participants and about 10 talks, covering all aspects of ADSR systems from the accelerator to the economics.

It also acts as a forum for partnerships to respond to funding opportunities. There have been several small successful bids: further ones are in progress. It seeks out and exploits opportunities to publicise Thorium and ADSRs, through the web, in the press, by public talks and private lobbying.

3. Design Choices

In order to focus ideas one has to propose a particular system. In doing so we are aware that choices may need to be reconsidered as circumstances change and more knowledge is obtained.

- We propose a Thorium fuelled fast reactor. Uranium ADSRs do not solve the waste or proliferation problems, and thermal Thorium reactors do not produce the fast neutrons needed for transmutation.
- We emphasize energy production, with transmutation as an extra benefit, as we think this will have a stronger political appeal (in the UK situation) than a system purely for waste disposal.

- We propose a 1 GW Thermal power station, large enough to make an appreciable contribution to the electricity generation capacity.
- We propose a production reactor as the first system, rather than a prototype, which would be expensive with no return on investment.
- We operate at a criticality k of 0.985. We believe that this is low enough to be safe.
- For this we require an accelerator (or multiple accelerators) delivering $\approx 30\,mA$ at 1 GeV This would be a 3 stage process, with a conventional cyclotron to 35 MeV, a ring accelerating protons to 400 MeV, and another for the final acceleration to 1 GeV.
- We use lead as target, coolant and moderator, and operate at a high temperature for high Carnot efficiency. We have benchmarked different simulation codes (MCNPX, FLUKA, GEANT4 and models within these) for the spallation process, and shown how target optimisation can increase the neutron yield. [3].
- Fuel must be recycled. ^{233}U must be reclaimed not only as fuel but also as it will be a proliferation hazard after a few hundred years, when the ^{232}U has decayed. There is also a long term waste problem from the ^{233}U decays. The design of fuel elements to survive in high temperature lead is a challenge, but should be possible.

The FFAG is our preferred accelerator [3]. Its DC magnets are cheaper and more reliable than those of a synchorotron. The acceleration rate is limited by the RF rather than the magnets, and we envisage acceleration up to 1 GeV in around 1000 turns, which would take 1 ms for a 50 m ring. The beam energy of 1 GeV is very hard to attain using a cyclotron, and 10 mA is very hard to attain using a synchrotron. A Linac can achieve both, but its capital cost is inevitably large due to its length. We are currently engaged in construction of EMMA, first nsFFAG, at Daresbury, which will start operating in 2010 [4] and hopefully this low energy electron accelerator will show the viability a high current proton machine.

At 1 KHz, if an FFAG is run with a ‘synchrotron style’ duty cycle, in which individual pulses are injected, accelerated, and extracted, then a 10 mA current requires $\approx 6\,10^{13}$ protons/bunch. This is well over typical space charge limits, which are of order 10^{13}. A large harmonic factor (several bunches per turn) can only help slightly. However if it can be run ‘cyclotron style’ with a 100% duty cycle the bunch is only $\approx 6\,10^{10}$ particles which presents no space charge challenge, however the sweeping of the RF frequency (in itself an unsolved problem) limits the duty cycle.

4. Reliability: The Next Accelerator Frontier

The accelerator for an ADSR must attain levels of reliability far beyond those of present accelerators. Requirements stated in the literature vary between 3 and 1000 trips per year, with varying definitions of a 'trip'. We urgently need more investigation to harden up the actual requirements. Thermal stresses in the window and target are directly beam-related, thermal stresses in the core and its components are moderated by the time constant of the reactor and cooling, and there are also limits from economics: a 1 GW plant cannot just drop out of the Grid with impunity.

Reliability can be achieved by redundancy, robustness, under-rating and planned maintenance. These have a cost penalty, and the design must include them only where necessary. Ion Sources are known to be fallible and two or more should be used. The DC magnets will probably be sufficiently reliable. RF components will fail, and the design must cope with this: this probably rules out clever designs such as the Separated-Orbit Cyclotron, Recirculating Linear Accelerators, and Harmonic number jumping, as these require the bunch to have a specified energy at a specified point in the cycle. There are also mundane components such as vacuum and power. Multiple accelerators may be needed to provide adequate reliability.

5. Conclusions

ADSR involvement in the UK is increasing rapidly. We have a wide range of specialists working together across discipline boundaries. Activities are increasing in many areas, and those interested are welcome to browse the ThorEA website and help us towards the goal of establishing Thorium powered ADSRs as the sustainable carbon-free energy source of the future.

References

1. D. Mackay, *Sustainable Energy – without the hot air* (UIT Cambridge Ltd, 2009).
2. Climate Change Act 2008 UK Government Office of Public Sector Information, (2008).
3. C. Bungau, R. Barlow, R. Cywinski and S. Tygier, Accelerator driven systems for energy production and waste transmutation Proc. European Particle Accelerator Conference, Genoa, (2008).
4. R. Barlow, J. Pozimski, K. Peach, N. Bliss, N. Marks, H. Owen, M. Poole and T. R. Edgecock, The CONFORM project: construction of a nonscaling FFAG and its applications Proc. Particle Accelerator Conference, Albuquerque, (2007).

ADS DEVELOPMENT IN JAPAN

KENJI KIKUCHI

Frontier Research Center for Applied Atomic Sciences, Ibaraki University, Shirakata 162-1, Tokai-mura, Ibaraki-ken, 319-1106, Japan

Accelerator driven nuclear transmutation system has been pursued to have a clue to the solution of high-level radioactive waste management. The concept consists of super conducting linac, sub-critical reactor and the beam window. Reference model is set up to 800MW thermal power by using 1.5GeV proton beams with considerations multi-factors such as core criticality. Materials damage is simulated by high-energy particle transport codes and so on. Recent achievement on irradiation materials experiment is stated and the differences are pointed out if core burn-up is considered or not. Heat balance in tank-type ADS indicates the temperature conditions of steam generator, the beam widow and cladding materials. Lead-bismuth eutectics demonstration has been conducted. Corrosion depth rate was shown by experiments.

1. Specification of ADS in JAPAN

Accelerator driven nuclear transmutation system aims at transmuting Minor Actinides such as Am, Cm and Np with having long-life decay periods into nuclei with short–life ones or stable nuclei at higher efficiency. Fig. 1 indicates the conceptual model proposed in JAPAN. Super conducting linac accelerator (SC) makes high intensity and high power protons beams. The beams are bent and injected to the liquid lead-bismuth target at the inner core region of sub-critical fast reactor through the beam window, which is the interface between accelerator and reactor. High intensity proton beam will produce neutrons through spallation process and increase neutrons more through fission process.

The energy of proton is to be 1 - 1.5 GeV from points of views of neutron production economy by spallation reaction. It is decided to be 1.5 GeV. Accelerated proton will produce about thirty neutrons by the spallation process of target nuclei. [1] Nishihara described the core power by the Eq. (1)

$$P_c = \frac{C_t(E_p,k,S)N_p(E_p)}{\upsilon^{fiss}}\frac{\phi^*(E_p,k,S)}{\rho(k)}E^{fiss} \tag{1}$$

where P_c (W) is the core power, $C_t(E_p,k,S)$ (A) is the beam current, E_p is an energy of proton (GeV), S is an shape of the beam, $N_p(E_p)$ is the number of the spallation neutron, ν^{fiss} (= 3.0) is the number of fission neutron by fission, $\phi^*(E_p,k,S)$ is spallation source effectiveness that is the ratio of multiplication of spallation source to fission source in minor actinide (MA) core, $\rho(k)$ is reactivity defined as $\rho(k)=1/k-1$, k is core criticality, and E^{fiss}(=202.6 x 10^6) (eV) is a power generated by fission. [2]

A choice of target materials was considered by neutron yield, system simplification with reactor coolant, and heat removal engineering. Spallation target favors to high atomic number substance. Except for non-actinide materials, Ta, W and Pb are candidates. Engineering consideration fixed firstly to use common materials for both spallation and reactor coolant, secondly to use good physical property with having a low melting temperature at 124.5°C and a high boiling temperature at 1670°C, thirdly to use good chemical property without aggressive reaction with water, and fourthly to use liquid material that is free from radiation damage. Our choice is to use lead-bismuth eutectics (LBE), which has a chemical concentration of lead 45wt% and Bi 55wt%.

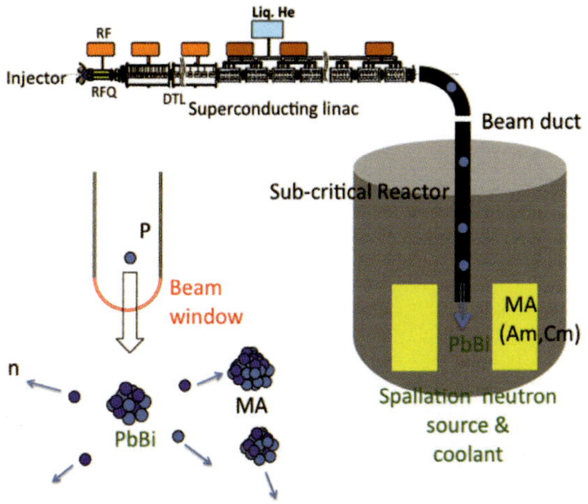

Figure 1. The concept of accelerator driven nuclear transmutation system.

The reference specification of ADS [3] is 800 MW of thermal power and the burn-up period is 600 effective full-power days. The amount of MA transmutation is 250 kg per year. The core fuel is (MA, Pu)-nitride. Inert matrix in the fuel is zirconium-nitride (ZrN). Nitrogen with 15N enriched (99%) is to be used for both (MA, Pu)-nitride and ZrN to reduce the production of 14C from

14N(n, p) reaction. The proton accelerator for ADS has high intensity power about 20MW with good economy and high reliability. High reliability should be emphasized for realizing ADS. In order to satisfy the conditions SC accelerator is a promising choice. A conceptual design study was to use a cryomodule with 972MHz radio frequency for protons accelerated from 100 MeV to 1.5 GeV. Preparatory study about superconducting twin niobium-made cavities showed that the maximum surface electric field (MSEF) was 42 and 40 MV/m at 4.2K, and 37 and 35 MV/m at 2.1K, respectively. The design study with using 30 MV/m of MSEF shows that the estimated total length is 472m by using 89 cryomodules.

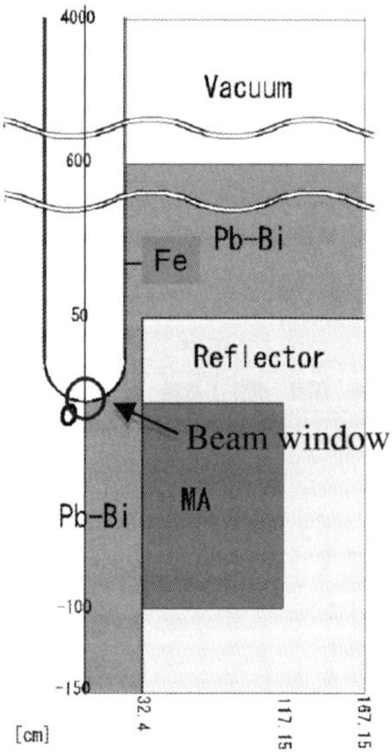

Figure 2. Damage calculation model.

2. Irradiation Material Damage of Beam Window

Fig.2 indicates material damage calculation model of the beam window. Digits means size in cm scale. The beam window locates in a depth of 600 cm and a thickness of shielding Pb-Bi is 150 cm against 1.5 GeV proton beam. For the evaluation, flux of the proton and neutron is calculated by the PHITS [4] (particle and heavy ion transport code system) and TWODANT [5] (two-dimensional diffusion-accelerated neutral-particle transport code system). The heat cross section is calculated by PHITS for whole energy region of proton and neutron. Displacement by proton and neutron above 150 MeV and production of hydrogen and helium by high-energy particle above 20 MeV are also calculated by PHITS. DPA cross-section below 150 MeV is calculated by processing LA-150 by NJOY code [6]. Cross sections in a library of DCHAIN-SP [7] are available for production of hydrogen and helium by neutron below 20 MeV. Table 1 shows heat deposition, displacement per atom (DPA), and light –cluster production (H, D, T, 3He, and 4He) at the beam window material, which consists of JPCA, 14Cr-16Ni-2.34Mo-1.54Mn-0.22Ti-Fe-Balanced wt%, with 2mm in thickness. Each of values is separated by the contribution to incident proton, produced proton, spallation neutron (>=10MeV), spallation neutron (<10MeV), and total values during the first 300 FPDs in ADS. Incident proton deposits 229 W/cc. It is evaluated that the incident proton generates over 60% to the total heat. Irradiation damage DPA is caused by the incident proton to 4.2 DPA and the ratio to the total value is 7.6%. Neutron below 10 MeV contributes to 47 DPA, 85%. This is attributed to the fuel burn-up. A large number of hydrogen is produced in comparison with other light clusters. The incident proton and the secondary or higher order particles give 50% and 30% to the total production, respectively. 4He and 3He productions contributed by the incident proton are 86% (709appm) and almost 100% (130appm), respectively. The ratio of 4He/3He due to the incident proton is 5.5 and the ratio due to all proton and neutron is 6.4.

Table 1 Beam window material damage in the first 300 FPDs.

Effects/Particles	Incident Proton	Produced proton	Neutron(>10MeV)	Neutron (<10MeV)	Total
Heat (W/cm3)	229	75	7.2	63	375
DPA (300 FPDs)	4.2	0.31	3.6	47	55
H (appm,300 FPDs)	3119	1831	725	503	6179
D (appm,300 FPDs)	727	1.8	7.2	0.082	736
T (appm,300 FPDs)	163	0.27	0.72	0.054	164
3He (appm,300 FPDs)	130	0.2	0.28	3.90E-06	130
4He (appm,300 FPDs)	709	5.5	45	65	825

3. Translation of Material Damage Parameter in Spallation

The SINQ target irradiation program (STIP) initiated in 1996 and conducted under collaboration with many laboratories at Paul Scherrer Institute. Irradiating space was provided in the neutron source, which was bombarded by 580MeV proton beam. Through a post irradiation examination of materials mechanical properties of austenitic stainless steel, JPCA, and ferritic-martesitic(FM) steel, F82H (8Cr-2W-Fe-Balance wt%), were investigated. They are candidates for the beam window material. Material data irradiated up to 20DPA are stored in the material data sheet. [8]

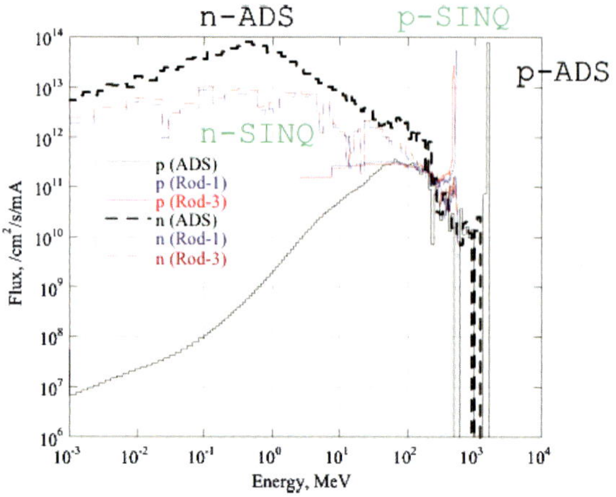

Figure 3. Comparison of proton and neutron.

Consideration will be paid to the difference of incident proton energy for applying STIP data to ADS. Translation of material data among different irradiation fields will be issue. Figure 3 shows a comparison of proton and neutron fluxes at ~~in~~ the beam window of ADS with rod-1 and -3 in SINQ target3. [9] The incident proton flux ($/cm^2/s/mA$) at ADS shows almost the same level with that of SINQ. The difference is the absolute value of incident proton energy. The neutron flux at ADS is higher than spallation target. As the reason is already mentioned, the higher neutron flux is due to fuel burn-up.

Materials property of DPA base evaluation at fission and spallation fields is reported to be different. A ductility brittle transition temperature (DBTT) of the different FM steels increases more or less linearly with dose after irradiation in spallation environment up to 20DPA. But the trend of DBTT in the neutron irradiation deviates from that of spallation cases. The reason is likely attributed to helium effects. [10] The helium production rate depends on the incident proton energy. TEM observation of irradiated samples showed small bubbles in the materials. Helium gas remained in the irradiated samples below 1000°C. [11]

4. Materials Compatibility with LBE and Heat Balance in the System

The knowledge for a use of materials such as ferritic-martensitic and austenitic stainless steels in LBE is to keep an oxide layer on the surface of base metal by controlling an oxygen concentration. Thermo-dynamics consideration indicates that an adequate oxygen concentration in LBE exits between, for example, 10^{-6} to 10^{-4}wt% in the temperature region of 400 – 700°C. Gromov et al. showed that too little oxygen solution in LBE will dissolve iron oxide and excess oxygen solution in LBE will make lead oxide. [12] Alternative is to cooperate with anti-corrosion elements like Al, which has a good function to guard base metal. [13]

For the ADS a primary mission is to transmute MA efficiently but not to generate electricity efficiently so that LBE working temperature can be set up in the low temperature regime comparative with power reactors. This point is an advantage to suppress the material corrosion in flowing LBE. In order to remove heat in the subcritical reactor a tube-in-shell type steam-generator (SG) is to be applied. It is desirable to downsize the structure by keeping LBE temperature over 300°C. The lower temperature will make a heat transfer coefficient of water inefficient. Under the condition of both LBE velocity less than 2m/s and 375W/cc of heat deposition at the beam window, the beam window temperature was designed less than 520°C. Then Maximum temperature at fuel claddings is evaluated to be about less than 600°C, while LBE temperature is <505°C. [3]

In the last ten years the study of LBE engineering has been demonstrated in material corrosion[14], flow velocity profile measurement in the target model[15] and heat transfer experiments at the beam window model[16]. A choice of candidate materials for ADS beam windows was JPCA and F82H irradiated at SINQ, which had the large database of neutron irradiation. Candidates for fuel clad are HCM-12A(10Cr-2W-SiMnMoCuNiV-Fe Balance,wt%) and high Cr-ODS steels, which have been developed in the field related to GEN-IV project.

The oxidation kinetics of the steel in the flowing LBE was proposed by the next equation [17]:

$$D = (K_p t)^{0.5} - \frac{2}{3} K_r t \qquad (2)$$

The equation means that the corrosion mechanism is explained by both an oxide scale formation and a scale removal, where D is corrosion depth, K_p is a parabolic oxide growth rate constant, K_r is a scale removal rate constant and t is time. On the contrary, for a log time range evaluation of corrosion depth in flowing LBE, a mechanism of mass loss due to continuous scale removal or discrete cracking will be important. We observed cracks in the oxide layer parallel to the material surface of both JPCA and F82H at 500°C. The next simple equation was proposed for design use where D: μm, t: h. [18]

$$D = 0.01t \qquad (3)$$

This suggests that 5000hrs operation will make 50μm corrosion depth.

Furthermore it should be mentioned that the beam window takes quasi-static pressure in the submerged condition into LBE filled core. Irradiation hardening changes the buckling mode and the critical buckling stress.[19]

5. Summary

Development of ADS in Japan is overviewed. Reference concept is a tank-type 800MW thermal subcritical reactor. Intensified proton beam with 1.5GeV energy is accelerated by super-conducting linac. Beam power will be 15-30MW, depending on fuel burn-up. Irradiation data of the beam window materials were obtained by STIP program and comprehensive understanding needs considerations the effect of fuel burn-up. LBE techniques are demonstrated on corrosion, velocity measurement and heat transfer performance. Buckling analyses optimized the beam window shape against quasi-static pressure and irradiation effect.

References

1. N. Watanabe, JAERI-Review 2000-031 (2000) 26.
2. K Nishihara et al., Journal of Nuclear Materials 377 (2008) 298–306.
3. K. Tsujimoto, H. Oigawa, K. Kikuchi, Y. Kurata, M. Mizumoto, T. Sasa, S. Saito, K. Nishihara, M. Umeno, and H. Takei, Nuclear Technology vol. 161Mar. (2008).
4. H. Iwasa, K. Niita, T. Nakamura, J. Nucl. Sci. Technol. 39 (2002) 1142.
5. R.E. Alcouffe et al., LA-10049-M, 1990.
6. M. Harada et al., J. Nucl. Mater. 343 (2005) 197.
7. T. Kai et al., JAERI-Data/Code 2001-016, 2001.
8. S. Saito, K. Kikuchi, D. Hamaguchi, K. Usami, A. Ishikawa, Y. Nishino, S. Endo, M. Kawai, Y. Dai, J. of Nucl. Mater, doi:10.1016/j.jnucmat.2009.10.009.
9. Y.Dai and G.S. Bauer, J. Nucl. Mater. 296 (2001) 43-53.
10. Y.Dai, W. Wagner, J. Nucl. Mater. 389 (2009) 288-296.
11. B.M. Oliver, Y.Dai, R.A.Causey, presented at IWSMT7, (2005).
12. B. F. Gromov, Yu I. Orlov, O. N. Martynov, K. D. Ivanov, V. A. Gulevsky, Liquid Metal Systems, Edited by H. U. Borgstedt and G. Frees, Plenum Press, New York, (1995) 339-343.
13. G. Muller, G. Schumacher, F. Zimmermann, J. of Nucl. Mater., 278 (2000) 85-95.
14. K. Kikuchi, S. Saito, D. Hamaguchi, M. Tezuka, J. of Nucl. Mater., doi:10.1016/j.jnucmat.2009.10.018.
15. K. Kikuchi, Y. Takeda, H. Obayashi, M. Tezuka, H. Sato, J. of Nucl. Mater., 356, (2006), 273-279.

16. K. Hayashi1, M. Ono, K. Kikuchi, N. Tokunaga, T. Kitano, H. Oigawa, Transactions of the Atomic Energy Society of Japan, Vol. 7, No. 1, (2008), 44-57.
17. J. Zhang, N. Li, Oxid. Met. 63 (516) (2005) 353.
18. K. Kikuchi, K. Kamata, M. Ono, T. Kitano, K. Hayashi, H. Oigawa, J. of Nucl. Mater., 377 (2008) 232–242.
19. T. Sugawara, K. Kikuchi, K. Nishihara, H. Oigawa, J. of Nucl. Mater., doi:10.1016/j.jnucmat.2009.10.040.

PROJECT-X, SRF, AND VERY LARGE POWER STATIONS*

CHARLES M. ANKENBRANDT[1], ROLLAND P. JOHNSON[1†], MILORAD POPOVIC[2]

[1]*Muons, Inc., 552 N. Batavia Ave., Batavia, IL, 60510, USA*

[2]*Fermilab, P.O. Box 500., Batavia, IL, 60510, USA*

We seek to develop accelerator-driven subcritical (ADS) nuclear power stations operating at more than 5 to 10 GW in an inherently safe region below criticality, generating no greenhouse gases, producing minimal nuclear waste and no byproducts that are useful to rogue nations or terrorists, incinerating waste from conventional nuclear reactors, and efficiently using abundant thorium fuel that does not need enrichment. First, the feasibility of the accelerator technology must be demonstrated. Fermilab is developing concepts for Project X, which would use a superconducting RF (SRF) linear proton accelerator to provide beams for particle physics at the intensity and energy frontiers. We propose to extend this linac design to serve as a prototype for a practical accelerator that can drive several ADS reactors at once and also provide beams for reactor development.

1.1. *Overview*

A commercial GW-scale ADS power plant requires a proton accelerator with a beam power of at least 10 MW. Recent accelerator developments promise to make even more powerful accelerators feasible. There is a new opportunity to explore the relevant concepts in concert with another project, thereby achieving considerable synergies and cost savings. Namely, Fermilab is developing concepts for Project X, which would use a superconducting RF (SRF) linear accelerator that could deliver megawatts of beam power to provide beams for particle physics at the intensity and energy frontiers. One concept calls for an 8-GeV pulsed SRF linac; another concept is for a CW linac with a lower initial energy of about 2 GeV. One of the steps in proceeding through the Department of Energy's critical decision process from CD0 to CD1 is to look at alternative designs. In that spirit, Muons, Inc., Fermi National Accelerator Laboratory (Fermilab, High Energy Physics), Thomas Jefferson National Accelerator Facility (JLab, Nuclear Physics), and the Oak Ridge National Laboratory Spallation Neutron Source (SNS Basic Energy Sciences) have proposed to examine alternative designs for Project X that would be consistent with the needs

---

* Supported in part by Fermilab subcontract to Muons, Inc. under DOE DE-AC02-07CH11359

† rol@muonsinc.com

of ADS and ATW. For example, the use of continuous-wave (CW) RF may enable production of tens of MW of beam power, considerably more than what is required for the intermediate-term HEP program at Fermilab, at a modest incremental cost relative to the baseline Project-X. The linac could serve as a prototype of a device that could drive several ADS reactors at one location, an approach which will become increasingly attractive with the development of the national power grid using low-loss transmission lines based on new superconductors.

The first major milestone of the project discussed here is to produce an enhanced or alternative design for Project X that includes ADS and ATW development needs. The planning, component development, construction, and operation of the machine will be the first step toward a practical accelerator for ADS and ATW based on SRF. Once constructed, the proton beam would allow tests and development of reactor components. Combining the goals of the High Energy, Nuclear Physics, and Basic Energy Sciences communities of DOE with the national energy and environment goals will lead to many desirable outcomes including lower costs, better technology, faster implementation, and the synergies that come from talented people working together to solve critical national and global problems.

Figure 1 shows the present and planned accelerator parameters that can be compared to the potential of a CW Project-X.

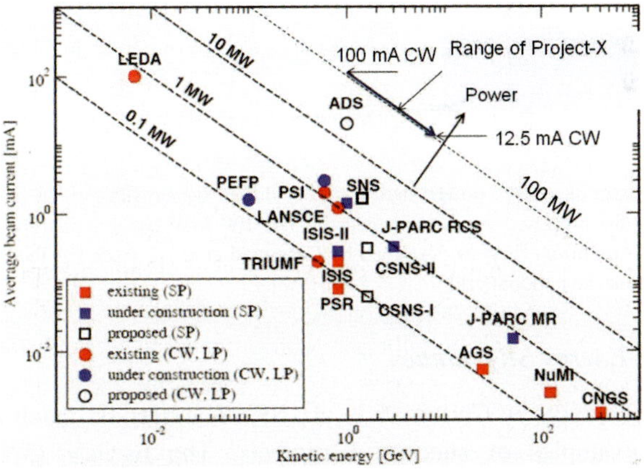

Figure 1: Present and planned high-intensity proton accelerators. Original data compiled by J. Wei and S. Henderson. The present power record is held by the ORNL SNS. The range of parameters that could be explored by Project-X is indicated on the 100 MW line.

In ADS schemes, spallation neutrons are produced by a 10 MW beam of protons on a high Z target. The fast neutrons (1-10 MeV) interact with Thorium

232 (fertile nucleus) to convert it to Protactinium which in turn decays into Uranium 233 (Fissile nucleus). (Similarly for U 238, one can make Plutonium 239 which is fissile). Additional neutrons induce fission to produce power.

It has been shown that neutron production from a proton beam increases almost linearly with proton energy for energies above 1 GeV. Consequently, it is reasonable to consider beam power as the relevant variable such that a lower beam current accelerated to higher energy can provide the needed beam power. Or, as we propose here, a large current at higher energy can supply several ADS reactors in parallel.

Essential advantages of using a higher-power higher-energy machine to drive several ADS/ATW reactors simultaneously compared to one accelerator for each reactor include better efficiency and lower cost. By creating most of the beam power with higher-gradient, more-efficient SRF cavities operating where the proton velocity is close to the speed of light (beta=1), capital and operating costs are reduced.

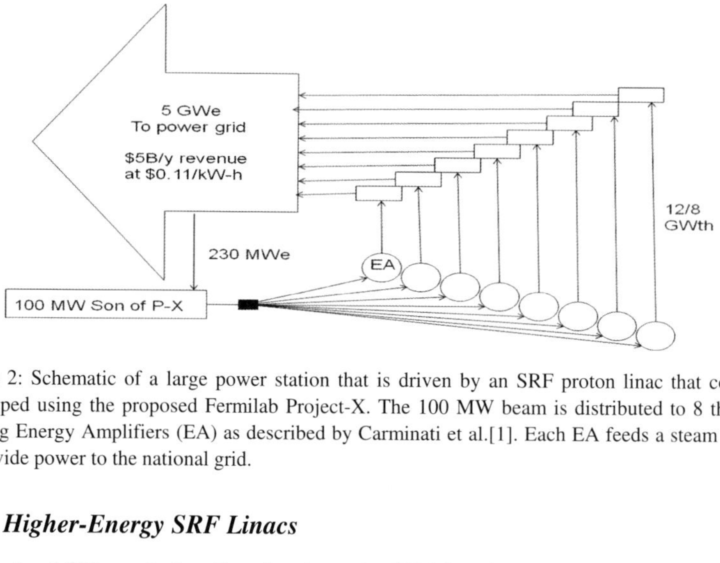

Figure 2: Schematic of a large power station that is driven by an SRF proton linac that could be developed using the proposed Fermilab Project-X. The 100 MW beam is distributed to 8 thorium-burning Energy Amplifiers (EA) as described by Carminati et al.[1]. Each EA feeds a steam turbine to provide power to the national grid.

1.2. *Higher-Energy SRF Linacs*

Since the 1993 study by Carminati et al., SRF has become much more mature, with many examples of successful projects. The 6 GeV CW Continuous Electron Beam Accelerator Facility (CEBAF) at JLab has demonstrated reliable SRF operation, while advances in cavity construction and processing have shown higher gradients and quality factors that will lead to lower construction and operating costs for future machines. The 1 GeV SRF linac at the Spallation Neutron Source at ORNL, while operating in 60 Hz pulsed mode, is being used to explore many of the issues relevant to reliable operation and control of losses at high beam power that will be essential for ADS applications. A proton beam power near the MW-level has already been achieved at SNS, thereby demonstrating the feasibility of one of the key technologies required for ADS. Free Electron Lasers and synchrotron light sources that are based on CW SRF are likewise becoming commonplace.

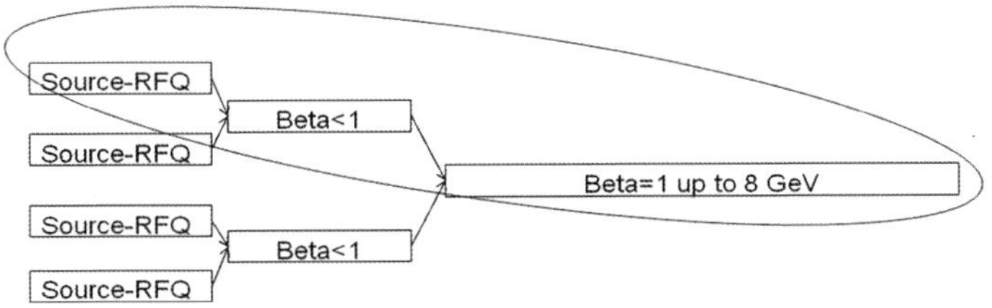

Figure 3: Schematic of an accelerator with sufficient redundancy to serve as a practical driver for the power station described in figure 2. The components enclosed by the ellipse represent Project-X.

The special additional requirement for ADS uses, and an important reason to have an ADS prototype, is that the accelerator must be extremely reliable. This requirement is motivated not so much by the desire for steady power output but by the concern that reactor components might be damaged by sudden changes in power level. We will propose to demonstrate this reliability by invoking a combination of component selection and redundancy, where figure 3 indicates how Project-X can be used for this development. For example, instead of fanning out power from one klystron to many RF cavities, we can use individual power sources for each cavity. A power source failure in this latter case can be compensated by adjusting the synchronous phase of the other cavities in the linac.

The original motivation for this proposal to operate Project X as a CW machine was to make sure that there would be sufficient beam power to satisfy future needs that might not yet be known. Considering that the present Booster synchrotron will have operated more than 40 years by the time it is replaced by Project-X, it seems also likely that there will be intensity needs that are yet to be imagined in the lifetime of Project-X. Developing beams for ADS reactors may be the first example of an otherwise unanticipated reason to have a generous supply of several-GeV protons at Fermilab. Potential high energy physics beneficiaries of such a proton source include a muon colliders and neutrino factories.

Reference

[1] F. Carminati, R. Klapisch, J.P.Revol, Ch. Roche, J.A,Rubio, and C. Rubbia, , CERN-AT-93-47. http://ppd.fnal.gov/experiments/e907/raja/energy_amplifier/convert_at-93-047.pdf

Working Group 4

SRF Linac Driven Subcritical Core

Conveners

Yousry Gohar(Argonne)
Alex Stanculescu(IAEA)

POWER PRODUCTION AND ADS

RAJENDRAN RAJA

Fermilab, Batavia, Illinois 60510

We describe the power production process in Accelerator Driven Sub-critical systems employing Thorium-232 and Uranium-238 as fuel and examine the demands on the power of the accelerator required.

1. Introduction

We can transmute fertile nuclei such as Thorium-232 and Uranium-238 into fissile nuclei such as Uranium-233 and Plutonium-239 by bombarding them with neutrons. The reactions in question for Thorium are

$$^{232}Th + n \Rightarrow {}^{233}Th + \gamma \xrightarrow{(22m)} \Rightarrow {}^{233}Pa + \beta^- \xrightarrow{(27d)} \Rightarrow {}^{233}U + \beta^-$$

The neutron is absorbed by ^{232}Th which converts into ^{233}Th which decays in 22 minutes (half-life) by the emission of a photon to protoactinium ^{233}Pa which beta decays in 27 days to produce the long-lived isotope of Uranium, ^{233}U.

^{233}U undergoes fission by means of the thermal neutrons in the system to produce power. The neutron flux must not be such that significant amount of ^{233}Pa is broken up before it converts into ^{233}U. This places an upper limit of $\sim 10^{14}$ neutrons/cm^2/s for the neutron flux in the system[1]. This corresponds to a proton accelerator of 10MW power operating at ~ 1-2 GeV beam energy producing spallation neutrons on a high Z material such as lead or lead-bismuth eutectic, which are then used with the breeding process.

In the case of ^{238}U, the following reactions take place under neutron bombardment.

$$U^{238} + n \rightarrow U^{239} + \gamma \xrightarrow{(23m)} \Rightarrow Np^{239} + \beta^- \xrightarrow{(2.3d)} \Rightarrow Pu^{239} + \beta^-$$

The transmutation reactions occur with an order of magnitude more efficiency with fast neutrons (energy >1000eV) where as the fission of Uranium-233 and

Plutonium-239 proceeds more efficiently by thermal neutrons (<1 eV). Conventional nuclear reactors attain criticality by moderation of the fission neutron spectrum (which peaks at 1MeV) slowing them down to thermal energies. The average energy released per fission is ~200MeV.

Fast reactors and the ADS approach both utilize fast neutrons (1kilo eV or higher); fast reactors by means of a Plutonium oxide/U235 unmoderated core which supplies fission neutrons to the surrounding fuel and the ADS by means of a spallation target on which an intense proton beam (10 MW, 1 GeV) is targeted.

2. ADS Power Production

A 10mA 1 GeV proton beam has 10MW of power. The gain G of an ADS (or energy amplifier) is defined [2] as the ratio of the thermal power output/power of the incoming beam. The neutron multiplication factor **k** is the average number of neutrons available for fission as a result of each neutron induced fission. If N_0 is the initial number of neutrons, after **n** fission generations, the total number of neutrons will be $N_0/(1-k)$, as n tends to infinity. The gain G, can be expressed as

$$G = \frac{G_0}{1-k} = \frac{2G_0}{2-\eta(1-L)}$$

Where η is the (spectrum averaged) number of fission neutrons produced by a neutron absorbed in a fissile isotope, and L is the sum of fractional losses of neutrons absorbed or lost in a variety of ways (captures, diffused outside volume and so on). G_0 is a gain proportionality constant, typically 2.4-2.5.

$$k = \frac{\eta}{2}(1-L)$$

To achieve criticality (k=1) neutron losses have to be brought down to be below $L_{crit}=(1-2/\eta)$. Since L>0, in order to achieve criticality η has to be greater than 2, one neutron being used for fission and the other for breeding. Rubbia et al [2] assume a value of k=0.98, for which they obtain a gain factor of G=120. A 10MW beam will then produce 1.2GW of thermal power. Assuming a conversion efficiency of 0.45, they then expect to produce 540MW of electrical power[3].

Reactor type	Temper ature	Carnot Eff.	Actual
	Deg K		
PWR	648	0.409	0.330
EA	973	0.606	0.489
Coal with water below critical point			0.36-0.40

Table 1. Table of operating temperatures, Carnot efficiencies and actual efficiencies for a PWR, ADS and a coal powered plant. ADS and PWR expected to operate with an output temperature of 110 degrees C.

3. Thermal Efficiencies

Table 1 contains the table of operating temperatures for a pressurized water reactor (PWR) and an ADS. An ADS runs at a higher operating temperature than a PWR. The Carnot efficiencies for both are calculated. The actual obtained efficiency for a PWR is 0.33. Assuming the same ratio of Carnot/actual efficiencies for an ADS, we obtain an ADS efficiency of 0.49, which is close to that assumed in [2]. Note that a coal power plant (with water below critical point) has typical efficiencies of 0.36-0.40. The ADS with a lead bismuth target can in principle be designed to operate at even higher temperatures yielding greater thermal efficiencies.

4. Comparison to the NEA/OECD Report Numbers

Figure 1 shows plots of the accelerator current (in mA) vs the proton energy (in MeV) for various power outputs of the ADS as calculated in the joint NEA/OECD comparative study of fast reactors and ADS [4]. Curves for constant beam power are also shown. It can be seen that as the proton beam energy increases to 1 GeV, the ADS becomes more efficient at producing power. However, a 5MW beam at 1 GeV produces slightly under 100MW electrical power. So a 10MW 1 GeV beam will produce slightly below 200MW electrical power according to this study. The k value used for this study is 0.97. Scaling up the power by the ratio of the two (1-k) numbers produces a Rubbia equivalent power of ~280MW electrical power at a k value of 0.98. This is nearly a factor of 2 below the Rubbia number.

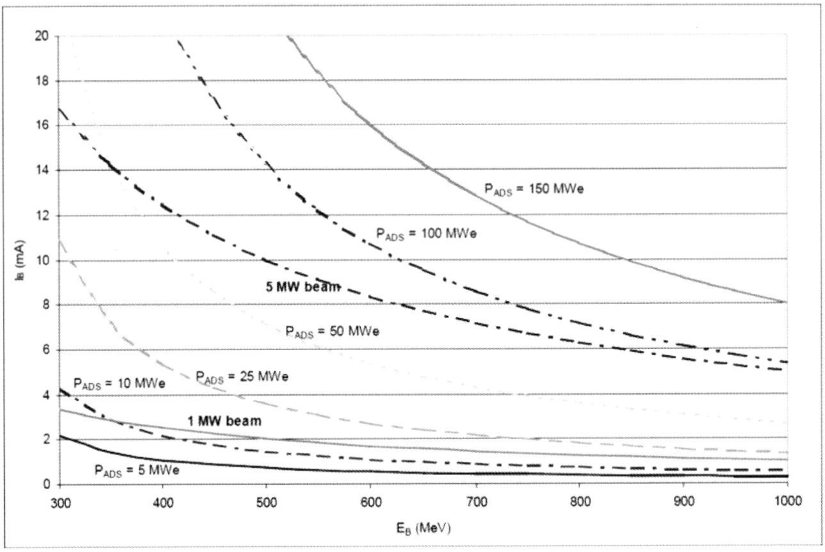

Figure 1 Plots of Accelerator current (mA) vs beam energy (MeV) for various ADS electrical power outputs. A k value of 0.97 was assumed. Curves for constant beam power (1MW and 5MW) are also shown. NEA/ OECD study[4].

5. First Principles Calculation

We assume that a 1 GeV proton on a lead target produces 30 spallation neutrons all of which interact with the fuel. Half of these neutrons are involved in the fission process and the other half in breeding. Each fission yields on average 0.2 GeV of energy. Then a 10mA 1 GeV proton beam (10MW) power will produce 30MW of fission which gets amplified by the k factor of 0.98 to 1500MW of thermal power. With an efficiency of 0.45, this will yield 675MW of electricity. This is closer to the Rubbia number than the NEA/OECD figure. It is difficult to pin down the causes of the discrepancy without detailed investigation into the assumptions made by both studies.

6. Conclusions

A 10-20 MW proton accelerator operating in a beam energy range 1-2 GeV should suffice for ADS systems that operate in the power production mode. The demonstration of such a high intensity proton accelerator would provide a great deal of impetus to the ADS approach that would provide yet another tool in our arsenal to combat global warming, transform nuclear waste into a usable commodity providing energy for thousands of years to come.

References

1. An energy amplifier for a cleaner and inexhaustible nuclear energy production driven by a particle accelerator, F.Carminati et al, CERN/AT/93-47(ET)
2. Conceptual design of a fast neutron operated high power energy amplifier, C. Rubbia et al, CERN/AT/95-44(ET)
3. J-P Revol, these proceedings.
4. ADS and fast reactors in advanced fuel cycles, A comparative Study, NEA,/OECD

EXPERIMENTAL NEUTRON SOURCE FACILITY BASED ON ACCELERATOR DRIVEN SYSTEM

YOUSRY GOHAR

Argonne National Laboratory, Argonne, Illinois 60439

An experimental neutron source facility has been developed for producing medical isotopes, training young nuclear professionals, providing capability for performing reactor physics, material research, and basic science experiments. It uses a driven subcritical assembly with an electron accelerator. The neutrons driving the subcritical assembly were generated from the electron interactions with a target assembly. Tungsten or uranium target material is used for the neutron production through photonuclear reactions. The neutron source intensity, spectrum, and spatial distribution have been studied to maximize the neutron yield and satisfy different engineering requirements. The subcritical assembly is designed to obtain the highest possible neutron flux intensity with a subcriticality of 0.98. Low enrichment uranium is used for the fuel material because it enhances the neutron source performance. Safety, reliability, and environmental considerations are included in the facility conceptual design. Horizontal neutron channels are incorporated for performing basic research including cold neutron source. This paper describes the conceptual design and summarizes some of the related analyses.

1. Introduction

Argonne National Laboratory (ANL) of USA has developed a neutron source facility [1 to 11]. The facility has an accelerator driven subcritical assembly with low enriched uranium (LEU) fuel. An electron accelerator will be used for generating neutrons through photonuclear reactions with high mass number material, tungsten or natural uranium, for driving the subcritical assembly. The facility will be utilized for producing medical isotopes, basic research, and training young specialists.

Several studies have been performed to develop the facility conceptual design and to investigate the main design choices and the design parameters for satisfying the facility objectives. For the electron beam target assembly, the spatial energy deposition in the target materials, and the neutron source intensity, spectrum, and spatial distribution have been studied as a function of the electron beam parameters, target materials, and target configurations. The main objectives are to maximize the neutron production from the available electron beam power and to satisfy the different engineering design requirements. Conceptual target designs have been developed based on these studies and the engineering analyses were performed including heat transfer, thermal hydraulics, thermal stresses, and material requirements. The target geometrical configuration has been designed to maximize the neutron utilization in the subcritical assembly and to match the geometry of the fuel assemblies. The neutron flux distribution of the subcritical assembly design has been examined as a function of the uranium fuel enrichment, the uranium density, the reflector material selection, the reflector thickness, and the target material for k_{eff} of ~0.98. The facility design is configured to maximize its utilization by using simple and efficient procedures for its operation and maintenance; to emphasize safety, reliability, and environmental considerations; and to allow future upgrades and new functions. The medical isotope production capability of the facility has been analyzed to define the irradiation locations and the sample sizes for more than fifty radioactive isotopes. The current facility conceptual design, key results from the design analyses, and its capability for medical isotope production are summarized.

2. Neutron Source Main Components

The neutron source facility consists of several components, which are integrated for steady state operation. The main components are the target assembly, the subcritical assembly, the biological shield, and the radial neutron ports. Auxiliary components are also designed for supporting the operation including cooling loops, target and fuel replacement machine, removable biological shield, and cold neutron source. The main components are presented with their key performance parameters.

2.1. *Target Assembly*

Several studies have been carried out to investigate the target design choices and the accelerator beam parameters for a satisfactory design and an acceptable operating performance. The main focus is to maximize the neutron production for the available beam power. The target design has been configured to operate satisfactorily with electron energy in the range of 100 to 200 MeV without changing the target operating parameters. The MCNPX [12] is utilized to determine the neutron source intensity, the neutron spectrum, the spatial neutron generation, and the spatial energy deposition in the target assembly as a function of the beam parameters, the target materials, and the target design details. The Computational Fluid Dynamics (CFD) software package STAR-CD [13] is used for the thermal-hydraulics analyses. The coolant velocity profiles and the spatial temperature distributions in the target assembly have been studied using the spatial energy deposition distributions obtained from the MCNPX analyses. The NASTRAN [14] structure analysis computer code has been used to calculate the thermal stresses in the target materials using the spatial temperature distributions from the CFD analyses. These analyses have been iterated to satisfy the temperature and the thermal stress limits for a satisfactory operation. Conceptual target designs have been developed based on the results of these studies and the engineering practices including nuclear physics, heat transfer, thermal hydraulics, structure, fabrication, and material requirements. Samples from the target studies are summarized in this section.

The electron beam generates x-rays with a continuous energy spectrum (Bremsstrahlung radiation) from the interactions with the target materials. These x-rays are absorbed in a variety of photonuclear reactions in the target materials and neutrons are produced from these reactions. Target materials with high atomic number are required to maximize the neutron yield. In addition, high melting point, high thermal conductivity, chemical inertness, high radiation damage resistance, and low neutron absorption cross section are the desirable properties for the target materials. The physics analyses of the possible target materials show that uranium, tungsten, lead, and tantalum produce the highest number of neutrons per electron. The physical properties of the uranium and tungsten materials, their neutron yields, and the operating experience from different accelerator facilities around the world lead to their selection for generating neutrons. Tungsten has the highest melting point of all metals (~3695 K). Uranium target material produces the highest neutron yield per electron due to its photo-neutron reactions and the additional neutrons from the photo- and the neutron-fission reactions. The operating lifetime of the uranium target is shorter relative to the tungsten because of the extra swelling caused by fission gasses.

In order to characterize each target material, several performance parameters were analyzed. These parameters include the total neutron yield, total energy deposition, neutron spectrum, neutron and energy deposition spatial distributions, and required target length. Each of these parameters has a particular role for defining the target performance and design. A high neutron yield enhances the neutron source intensity, which defines the neutron flux level and the total power of the subcritical assembly. The target energy deposition influences the design of the target coolant system and the neutron yield. The neutron spectrum affects the system's effectiveness for performing material characterizations and producing medical isotopes. The spatial neutron distribution of the target determines the target position inside the subcritical assembly and the neutron source utilization. All these parameters were studied to define the target performance characteristics and the design configuration. The calculated neutron yield for the tungsten target is 0.0584 neutrons per electron (n/e with 200 MeV Energy) generating 1.82×10^{14} neutrons per second (n/s) from the 100 KW beam. For the uranium target, the neutron yield is 0.1067 n/e and the corresponding neutron intensity is 3.33×10^{14} n/s. The spatial energy deposition normalized to 2-KW/cm^2 power density on the beam window is plotted in Fig. 1 for two different geometrical configurations. These two configurations are resulted from parametric and optimization studies for the target performance.

Heat transfer and thermal-hydraulics parametric studies have been performed to help defining the target mechanical configuration, the size of the water coolant channels, and the temperature distribution in the target materials. The 100-KW electron beam with ~2-KW/cm^2 uniform beam power density, the 7-m/s water coolant velocity inside the target manifold, and the 4-atm coolant pressure

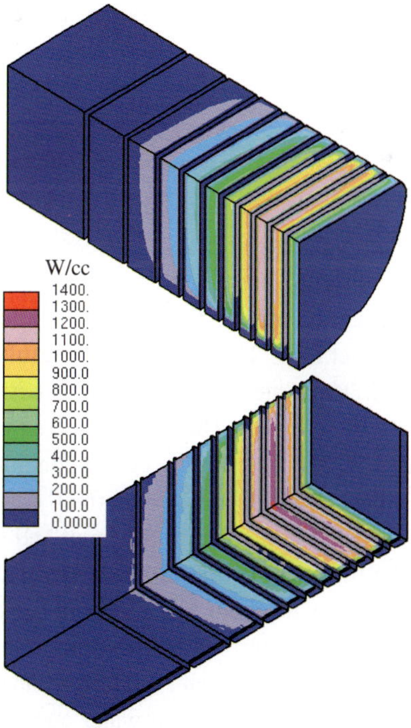

Fig. 1. Power density distributions (W/cc) in U-target disks with circular electron beam (top) and square electron beam (bottom). Only ¼ target segment is shown for 200-MeV electrons.

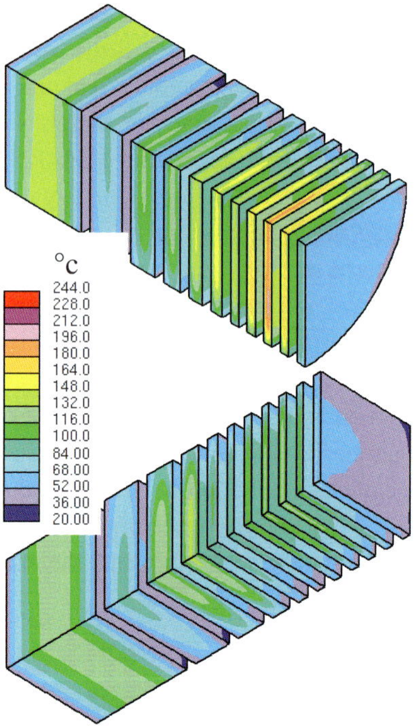

Fig. 2. Temperature distributions (°C) in Al-clad U-target disks with circular electron beam (top) and square electron beam (bottom). Only ¼ target segment is shown. The results correspond to Fig. 1.

are used. Higher coolant velocity is also considered to enhance the target performance. The target material has a cylindrical or square geometry and its axis coincides with the beam tube axis. The flow direction of the water coolant is perpendicular to the target axis. This arrangement results in a stack of disks forming the target design. The water coolant channels between the target disks have a constant thickness of 1.75 mm. The average increase in the water coolant temperature is less than 5 °C for the current coolant conditions. Each target disk is cooled from both sides to minimize the thermal deformation. The water coolant channels are connected in parallel to the input and the output manifolds. The analyses defined the thickness of the different target disks for limiting the maximum surface temperature to about 80 °C, for the electron energy in the range of 100 to 200 MeV. This surface temperature was selected to provide about 65 °C margin away from the water boiling temperature. The uranium disks have a 0.7-mm aluminum clad to avoid water coolant contamination with fission products.

The results from the MCNPX studies using three-dimensional geometrical models were used for STAR-CD thermal hydraulics analyses. The three-dimensional geometrical models account for the target design details. The STAR-CD calculations were performed to define the thickness of each target disk according to the above design criteria. In these analyses, the water velocity field was calculated for each coolant channel and the spatial temperature profile was calculated for each target disk and each coolant channel using a single geometrical model for the target assembly and the three-dimensional MCNPX energy deposition results. The three-dimensional temperature profile in the uranium material is shown in Figure 2 for the-100 KW uniform beam using 200-MeV electrons. These temperatures satisfy the adopted temperature design criteria.

The three-dimensional model and the corresponding temperature profiles of the uranium and the tungsten targets were imported into NASTRAN computer code for thermal stress analyses. The intensity and distribution of the thermal stresses were evaluated for the normal operating conditions. In the target design, the disks are allowed to expand in the radial and axial directions to reduce the operating stresses. The thermal stress intensity distribution during the normal operation was evaluated. Although thermal stresses are secondary stresses, their maximum values are limited to a fraction of the material yield stress. This is a very conservative approach leaving a large design margin to account for radiation damage and thermal-cycling effects. These effects are important to define the target operating lifetime. The tungsten target results show the peak thermal stress is less than 100 MPa.

2.2. *Subcritical Assembly*

Parametric studies were performed to maximize the neutron flux of the subcritical assembly with a subcriticality level of 0.98. Different fuel designs, uranium enrichments, and reflector materials have been considered. The target assembly is placed at the subcritical assembly center. LEU with < 20% uranium-235 and HEU with 90% uranium-235 were analyzed with two uranium densities of 1.0 and 2.7 g/cm^3. The analyses focused on the possibility of utilizing LEU fuel instead of HEU fuel without penalizing the neutron source facility performance. Several three-dimensional models of the subcritical assembly including the detailed target design were developed to define the subcriticality level and the neutron flux distribution. The target and the fuel assemblies are explicitly modeled without any geometrical approximation or material homogenization to get an accurate prediction of its performance. The

MCNPX computer code with continuous energy data libraries and S(α,β) thermal data was used for the analyses. Three reflector materials, beryllium, water, and carbon were considered. Parametric analyses were performed to assess the number of fuel assemblies required to get k_{eff} of ~0.98 for different combinations of the reflector and target materials, fuel densities, and fuel enrichments. The corresponding neutron fluxes were also calculated.

The results show that the use of uranium target requires the smallest number of fuel assemblies to achieve the desirable k_{eff} value. This small number of assemblies is due to the extra neutron multiplication produced by the natural uranium target material. The subcritical configurations with water reflector require a larger number of fuel assemblies as compared to the configurations with beryllium and carbon reflectors, because of the neutron absorption in the water. The very small number of fuel assemblies limits the subcritical assembly flexibility to study different geometrical configurations. In addition, the fuel assembly arrangements with beryllium reflector and uranium target are significantly asymmetric. The fuel design of the Kiev research reactor with low enriched uranium (LEU) produced higher neutron flux relative to other fuel designs.

In order to compare the performance of the subcritical assembly for irradiation experiments, with the LEU and the HEU fuels, the average neutron flux in the radiation channels around the target was calculated. The results show that no significant difference between the subcritical assemblies with HEU and LEU fuels was observed. However, the HEU subcritical assembly has an extremely small number of fuel assemblies, which limits its utilization for reactor physics experiments, generates difficulties for reactivity measurements, and calibration procedures. Consequently, it is desirable to use the LEU fuel instead of the HEU fuel to avoid such difficulties.

The calculated neutron spectra show that about half of the neutrons in the irradiation channels have energies below 100 keV and the other half are in the energy range of 100 keV to 20 MeV. The neutron fraction with energy above 20 MeV is very small. The use of beryllium or carbon reflectors produces higher neutron flux in the irradiation channels relative to the water. The use of uranium instead of tungsten as a target material doubles the neutron flux intensity because of the higher neutron yield as shown from the target analyses. Figure 3 shows the subcritical assembly configuration with uranium target, LEU fuel, and carbon reflector and its X-Y and R-Z total neutron flux distribution are shown in figs. 4 and 5, respectively, for the 100 KW electron beam. The corresponding X-Y power deposition distribution is shown in Fig. 6.

Fig. 3. Subcritical assembly configuration indicating four irradiation locations, the first is next to the target.

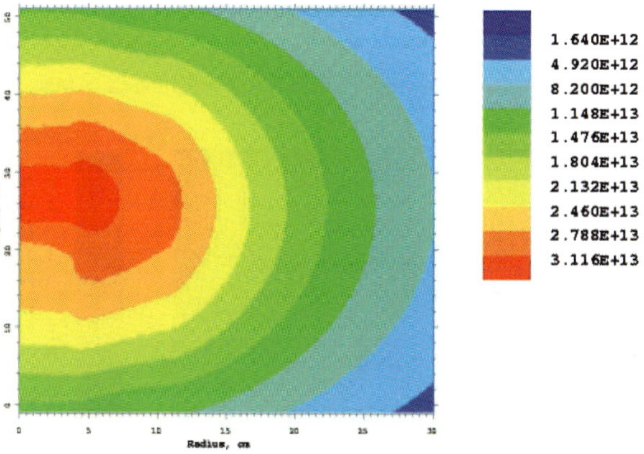

Fig. 4. Subcritical assembly total neutron flux distribution (R-Z Map, $n/cm^2.s$).

2.3. *Biological Shield*

Biological shielding analyses have been performed to define the shield thickness for permitting personal access around the subcritical assembly during operation. The shielding design criterion limits the biological dose to < 2.5 mrem/h. Such dose value allows the worker to have 40-hour working week without exceeding the allowable international exposure limit. In the design analyses, this value is reduced by a factor of 5 to account for uncertainties in the nuclear data, the calculational method, and the modeling details. Elaborate three-dimensional models have been developed for performing the shielding analyses with MCNPX starting with the electron beam. In the analyses, the biological shield of the subcritical assembly is divided into radial and top sections. The top section includes the top cover of the subcritical assembly and the biological shield of the electron beam.

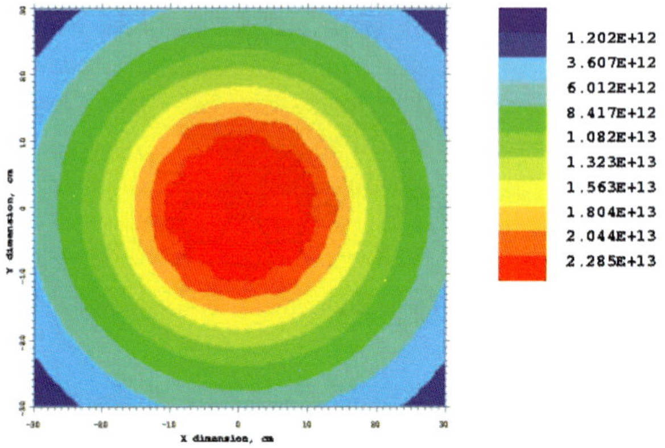

Fig. 5. Subcritical assembly total neutron flux distribution (X-Y Map, n/cm^2.s).

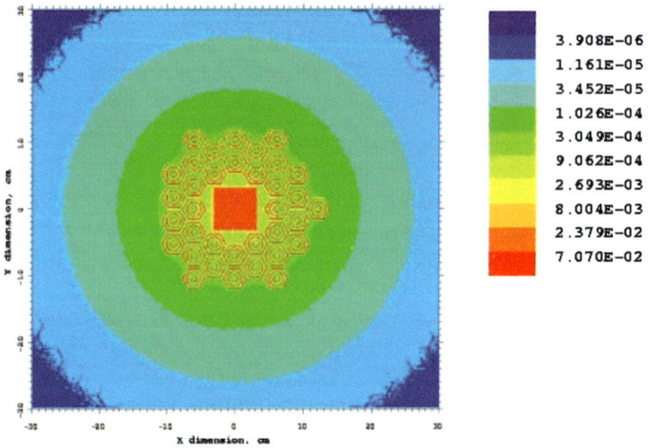

Fig. 6. Subcritical assembly energy deposition distribution (X-Y Map, KW/cm^3).

The parametric shielding results showed that the use of a steel shield zone in front of the heavy concrete has a small impact on the biological dose and the shield thickness. The heavy concrete has a density of 4.8 g/cm^3 and it is used for shielding the facility. In the radial direction, a shielding thickness of 180 cm is required to reduce the total (neutron and photon) dose to < 0.5 mrem/h. On the top of the subcritical assembly, the water coolant in the subcritical assembly pool acts as a shielding material, which reduces the required heavy concrete shield thickness. However, the radiation streaming from the beam tube complicates the shielding design in this area. In the vertical direction along the subcritical assembly axis, where the beam tube is located, the required heavy concrete shield thickness is 140 cm. This thickness is tapered as the radial distance from the beam tube increases to reach a minimum of 60 cm at the subcritical assembly boundary

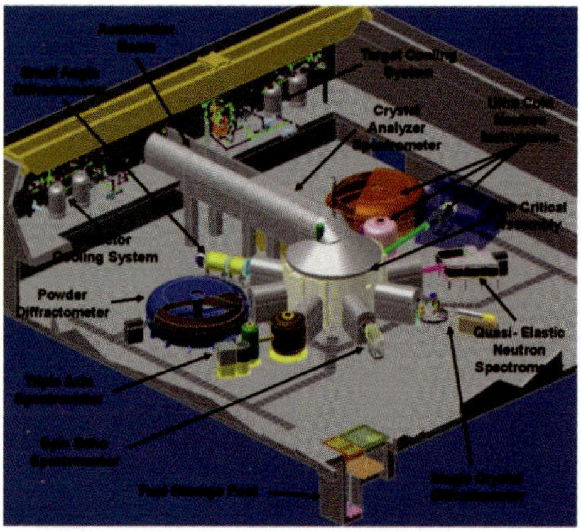

Fig. 7. Facility conceptual design overview.

3. Neutron Source Facility Conceptual Design

The subcritical assembly facility uses proven techniques and practices for its design, operation, and maintenance to enhance its utilization. The main facility components are the target assembly, the subcritical assembly, the biological shield, and the auxiliary supporting systems. Figure 7 shows an overview of the facility and an example of possible experiments using the radial neutron beams.

4. Medical Isotope Production

One of the main functions of the facility is the production of medical isotopes. A comprehensive study [7] has been carried out to determine the production rates, the irradiation locations, and optimize the sample sizes for more than fifty radioactive isotopes. Four locations were considered as shown in Fig. 3. MCNPX models have been developed to perform the study and the obtained neutron spectra of the four locations are shown in Fig. 8. The study has shown the importance of the self-shielding effect for determining the production rate, the optimum location, and the sample size. The study showed the excellent capability of the facility for performing this function.

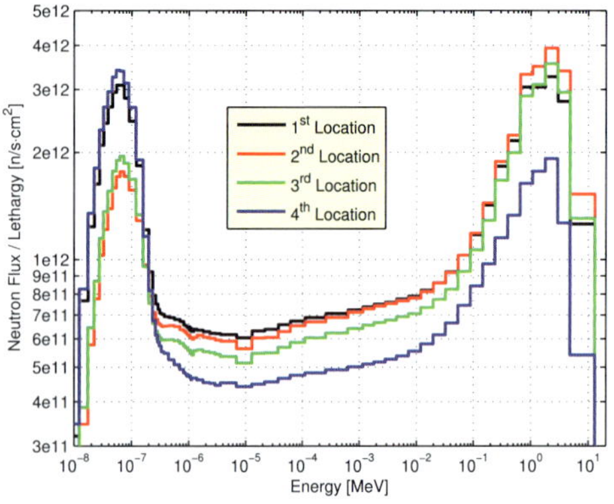

Fig. 8. Neutron spectra of the four-irradiation locations.

Conclusions

A conceptual design of an accelerator driven subcritical assembly facility has been successfully developed using an electron accelerator. The facility is designed with the low enriched uranium fuel design of the Kiev research reactor. The developed concept satisfies the facility objectives and it has flexibility for future upgrades and new functions. It has excellent capabilities for producing medical isotopes, performing basic research using its radial neutron beams, performing physics studies, and training young nuclear scientists.

Acknowledgments

Argonne National Laboratory's work was supported by the U.S. Department of Energy, Office of Global Nuclear Material Threat Reduction (NA213).

References

1. Y. Gohar, J. Bailey, H. Belch, D. Naberezhnev, P. Strons, and I. Bolshinsky, "Accelerator-Driven Sub-critical Assembly: Concept Development And Analyses," Proceeding of The RERTR-2004 International Meeting on Reduced Enrichment for Research and Test Reactors, Vienna, Austria, November 7-12, 2004.
2. Yousry Gohar, Igor Bolshinsky, Dmitry Naberezhnev, Jose Duo, Henry Belch, James Bailey, "Accelerator Driven Sub-Critical Facility: Conceptual Design Development," paper presented and submitted to AccApp05, Venice, Italy, August 29 – September 1, 2005.
3. D. G. Naberezhnev, Y. Gohar, J. Bailey, and H. Belch, "Physics Analyses of Accelerator Driven Sub-Critical Assembly," paper presented and submitted to AccApp05, Venice, Italy, August 29 – September 1, 2005.

4. Yousry Gohar, Igor Bolshinsky, Dmitry Naberezhnev, Jose Duo, Henry Belch, James Bailey, "Accelerator Driven Sub-Critical Facility: Conceptual Design Development," Nuclear Instruments and Methods in Physics Research A, Volume 562, June 2006, Page 870-874.
5. D. G. Naberezhnev, Y. Gohar, J. Bailey, and H. Belch, "Physics Analyses of Accelerator Driven Sub-Critical Assembly," Nuclear Instruments and Methods in Physics Research A, Volume 562, June 2006, Page 841-844.
6. Yousry Gohar, Igor Bolshinsky, Henry Belch, Dmitry Naberezhnev, Tanju Sofu, Alberto Talamo, and Zhaopeng Zhong, "Accelerator Driven Subcritical Facility Design Concept," the Eighth International Topical Meeting on Nuclear Applications and Utilization of Accelerators (AccApp07), Pocatello, Idaho, 30 to August 2, 2007.
7. Alberto Talamo, Yousry Gohar, and Dmitry Naberezhnev, "MCNPX Analysis Of The Medical Isotope Production In The Kharkov Electron Driven Subcritical Facility," the Eighth International Topical Meeting on Nuclear Applications and Utilization of Accelerators (AccApp07), Pocatello, Idaho, 30 to August 2, 2007.
8. Tanju Sofu, Yousry Gohar, Henry Belch, Zhaopeng Zhong, and Dmitry Naberezhnev, "Neutronic, Thermal-Hydraulic, And Structural Analyses Of The Electron Target For An Accelerator-Driven Subcritical System," the Eighth International Topical Meeting on Nuclear Applications and Utilization of Accelerators (AccApp07), Pocatello, Idaho, 30 to August 2, 2007.
9. Yousry Gohar, Igor Bolshinsky, Henry Belch, Tanju Sofu, Alberto Talamo, and Zhaopeng Zhong," Design and Analyses of KIPT Electron Accelerator Driven Subcritical Assembly Facility Concept," the XX International Workshop on Charged Particle Accelerators, the Accelerator Driven Subcritical Assemblies Workshop, Alushta, Ukraine, September 9-15, 2007.
10. Henry Belch, Yousry Gohar, Tanju Sofu, and Zhaopeng Zhong, "Conceptual Design Configuration for Electron Accelerator Driven Subcritical Assembly," the XX International Workshop on Charged Particle Accelerators, the Accelerator Driven Subcritical Assemblies Workshop, Alushta, Ukraine, September 9-15, 2007.
11. T. Sofu, Y. Gohar, H. Belch, D. Naberezhnevthe, "Multi-Physics Analysis of an Electron Target for the KIPT Subcritical Assembly," XX International Workshop on Charged Particle Accelerators, the Accelerator Driven Subcritical Assemblies Workshop, Alushta, Ukraine, September 9-15, 2007.
12. MCNPX Version 2.5.0 User's Manual (LA-CP-05-0369) Los Alamos National Laboratory, (http://mcnpx.lanl.gov/) April 2005.
13. STAR-CD Version 3.26 User's Manual, (http://www.cd-adapco.com/products/STAR-CD/), CD-adapco, Melville, NY.
14. MSC NASTRAN and MARC User's Manual, (http://www.mscsoftware.com/), MSC Software Corporation, Santa Ana, CA.

TRANSMUTATION MISSION

WON SIK YANG

Argonne National Laboratory, Nuclear Engineering Division
Argonne, Illinois 60439, USA

This paper presents a brief summary of the objectives of transmutation mission and relevant fuel cycle strategies and transmutation systems, including a short comparison of the main characteristics of fast reactor and accelerator-driven systems for actinide transmutation.

1. Introduction

Nuclear energy is a significant contributor to U.S. and international electricity production, about 20% in the U.S. and over 16% in the world. Given the recent concern over carbon gas emissions, there may be significant growth in nuclear utilization worldwide.

In the U.S., a once-through fuel cycle has been employed to date. As a result, large quantities of spent fuels are stored at reactor sites, but final waste disposal is not secured. With the expected nuclear expansion, this is not a sustainable approach. Thus, advanced fuel cycles to transmute the hazardous transuranics (TRU) of spent nuclear fuel are being explored.

The primary goal of the transmutation mission is to remove TRU elements from the waste stream and to fission them in advanced nuclear systems such as fast reactors and accelerator-driven systems, producing about one MW-day of energy for every gram.

The transmutation of TRU would reduce the long-term environment burden of nuclear energy through significant reduction of released dose and radiotoxicity and efficient utilization of permanent disposal space. It would also enhance overall nuclear fuel cycle proliferation resistance by avoiding disposal of weapons usable materials and enhance energy security by extracting energy recoverable in spent fuel, avoiding uranium resource limitations [1].

2. Transmutation Strategies

The currently considered advanced fuel cycles can be divided into two groups

[1]: limited recycle in which spent fuel is recycled a few times in current light water reactors (LWRs) and then directly disposed, and continuous recycle where spent fuel is recycled continuously and only fission products are disposed. The continuous recycle can be further divided into single and double tier systems [2]. In the single-tier system, the TRU from LWR spent fuel is fed into fast systems while the TRU from fast system spent fuel is recycled continuously. In the double-tier system, the plutonium or TRU of LWR spent fuel is recycled in LWRs limited times before fed into fast systems.

2.1. *Radiotoxicity Reduction*

Figure 1 shows the impact of the fuel cycle strategies on long-term radiotoxicity [1]. For the conventional once-through strategy, the spent fuel radiotoxicity remains above the level of the original source material (uranium ore) for about 300,000 years. The limited recycle approach produces little improvement over the once-through case due to only partial consumption of the transuranics.

The continuous recycle is required to meet the environmental goals through significant reduction of radiotoxicity. It can significantly improve the basic nature of nuclear waste disposal through the reduction of thermal load and isolation period.

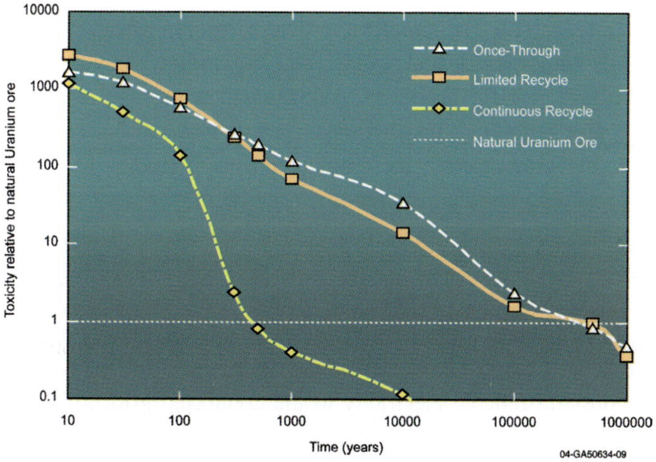

Figure 1. Radiotoxicity for once-through, limited recycle, and continuous recycle.

2.2. *Spectral Effect on Transmutation Physics*

Table 1 shows the fission-to-absorption ratios of major actinides in typical pressurized water reactor (PWR) and sodium-cooled fast reactor (SFR) systems.

As can be seen, fissile isotopes are likely to fission in both thermal and fast spectrum systems, with higher fission fraction in fast system.

However, fertile isotopes have very small fission probabilities in thermal system, and thus they are transmuted into higher actinides instead of being consumed by fission. In fast system, the fission probabilities of actinides increase significantly (up to 50%), and more excess neutrons are produced. As a result, transuranics are consumed efficiently with less generation of higher actinides.

Table 1. Actinides fission-to-absorption ratios in thermal (PWR) and fast (SFR) systems.

	PWR	SFR		PWR	SFR
U235	0.821	0.807	Am241	0.011	0.233
U238	0.099	0.194	Am242m	0.812	0.870
Np237	0.014	0.296	Am243	0.008	0.334
Pu238	0.072	0.711	Cm242	0.212	0.579
Pu239	0.637	0.865	Cm243	0.847	0.943
Pu240	0.003	0.576	Cm244	0.054	0.479
Pu241	0.747	0.873	Cm245	0.865	0.914
Pu242	0.015	0.550			

3. Comparison of Fast Reactor and Accelerator Driven System

Either fast reactor (FR) or accelerator driven system (ADS) can effectively destroy TRU with closed fuel cycle (i.e., continuous recycle of TRU). In reducing the legacy TRU inventory, ADS has an advantage over FR since its TRU conversion ratio (i.e., production to destruction ratio) can be reduced to zero by utilizing uranium-free fuel. However, recent studies [3,4] showed that the TRU consumption rate reaches about 80% of the maximum theoretical value of uranium-free fuel when TRU conversion ratio is in the range of 0.25 to 0.35 and that fast reactors can achieve this low conversion ratio without compromising passive safety behavior of metallic-fuel FR.

On the other hand, the electricity generation cost of ADS is higher than that of FR approaches by 10 to 25%, depending on the level of first-tier burning [2]. In order to provide reliable electricity to the connected grid, the reactivity loss due to TRU fuel burning must be compensated by increasing the external source strength through increased accelerator power and the reliability of proton beam should be enhanced significantly. In addition, FR and ADS have different level of maturity and technical risk. Fast reactors are closer to deployment, while fuel cycle research and development are required for both transmutation scenarios.

From these aspects, FR is ideal as a single-tier system since significant electricity can be generated while consuming TRU. ADS is best suited to the second tier of a double-tier system since the number of the second tier systems can be reduced by targeting deep burnup in the first tier, preferably using commercial reactors.

It is also noted that the future nuclear energy scenario is a key consideration for deploying ADS or FR. Growing scenarios favor FR with a moderate support ratio, while contraction scenarios favor ADS (with uranium-free fuel) that destroys TRU more quickly.

4. Concluding Remarks

In summary, either FR or ADS system with closed fuel cycle can be used for TRU transmutation mission. However, the preference for either system highly depends on the purpose and strategy of TRU transmutation such as nuclear growth or end, TRU stabilization or burn-down, etc.

Dynamic scenario studies need to be performed to quantify deployment approaches and impacts to nuclear park, since instantaneous support ratios of scenarios calculated by dynamic and equilibrium simulations differ because of time lag impacts.

Uranium-free and/or minor-actinide-only fuels are attractive for dedicated transmutation systems but would need a significant development time. Material performance degradation in ADS due to high-energy proton and neutron irradiation also needs to be considered. Innovative design solution may be required to provide inherent beam control and/or shutdown, since subcritical operating state greatly reduces the effectiveness of inherent passive safety reactivity feedbacks.

References

1. "Advanced Fuel Cycle Initiative: Objectives, Approach, and Technology Summary," Report to Congress, U.S. Department of Energy (May 2005).
2. "Accelerator-driven Systems (ADS) and Fast Reactors (FR) in Advanced Nuclear Fuel Cycles: A Comparative Study," NEA3109, OECD/NEA (2002).
3. W. S. Yang, "Trends in Transmutation Performance and Safety Parameters Versus Conversion Ratio of Sodium-Cooled Recycle Reactors," *Proc. of 10th Information Exchange Meeting on Actinide and Fission Product Partitioning and Transmutation*, Mito, Japan, October 6-10, 2008.
4. J. E. Cahalan, M. A. Smith, R. N. Hill, and F. E. Dunn, "Physics and Safety Studies of Low Conversion Ratio Sodium Cooled Fast Reactor," *Proc. of PHYSOR 2004: The Physics of Fuel Cycles and Advanced Nuclear Systems: Global Developments*, Chicago, Illinois, April 25-29, 2004.

SAFETY PERFORMANCE AND ISSUES

JAMES E. CAHALAN
Argonne National Laboratory, Nuclear Engineering Division
Argonne, Illinois 60439, USA

This paper identifies the basic safety functions in nuclear reactor design, and the safety design issues for accelerator-driven subcritical reactors.

1. Nuclear Reactor Safety Basics

1.1. *Basic Safety Objectives*

The first objective of nuclear reactor safety is to protect public health by reducing the risk from releases of radioactivity to acceptable levels. This is accomplished by providing a design that complies with established regulatory requirements, and also provides additional margins beyond the requirements.

The second safety objective is to protect the plant worker health and safety by providing a safe working environment that reduces the risk of injury and complies with the regulatory requirements for worker protection.

The next safety objective is to build, operate, and decommission the plant in a manner that protects environmental quality.

Finally, the plant must be designed, equipped, operated, and maintained to preserve investor equity and produce electricity reliably.

The design basis for radiation protection consists of a) barriers to contain radioactive materials and prevent human exposure or release to the environment, b) distance to provide spatial margins to reduce the intensity of radiation exposure, and c) time to isolate radioactive material until it has decayed to a stable or less harmful state.

2. Safety Design and Defense-In-Depth

2.1. *Safety in Design*

By design, the plant and all its systems are configured and constructed in a manner that assures safe, stable, and reliable operation while preserving protection mechanisms. Engineered safety systems are provided to *prevent* conditions that can defeat the basic radiation protection mechanisms, and to *mitigate* the consequences of equipment failure or inappropriate operator actions. Inherent protection margins can be provided by selection of materials and arrangement of components. The design principle of *defense-in-depth* is applied for important safety-related functions: containment, reactor shutdown, and residual heat removal

2.2. *Defense-In-Depth*

Plant functions that are important to preservation of safety protection mechanisms are designed according to the *defense-in-depth* principle, which provides multiple layers of safety assurance. At the first level, the aim is to provide a conservative design with large safety margins that can be constructed and operated normally without challenges to safety limits. The next level provides additional design features that protect against a single, unlikely fault (~once in the plant lifetime) resulting in limited damage that requires minor repair. The third level provides additional design features that protect against a single, extremely unlikely fault (not expected in the plant lifetime) that may cause extensive damage requiring major repairs. These foreseen events are within the plant safety design basis, and the most demanding events are usually identified as *Design Basis Accidents (DBAs).* DBAs are analyzed to demonstrate compliance with regulatory requirements, and the analysis results are documented in safety reports to verify safety margins for licensing.

For key safety functions, multiple, diverse, and independent structures, systems, or components are provided, each one of which is capable of achieving the defined safety function. Redundancy, diversity, and independence assure that all safety function can not be lost due to a single failure, either internal (equipment failure, operator action) or external (earthquake, fire, flood). *Safety grade* systems, components, and structures are designed and maintained to criteria that assure their reliable operation, with special attention to quality assurance and provisions for inspection, testing, and repair.

Containment of radioactive material is assured by multiple physical barriers: the fuel cladding, the primary coolant system boundary, and the containment structure.

Reactor shutdown is assured by multiple reactor control and protection systems: a primary system for startup, shutdown, reactor power changes, and power distribution management, and a secondary shutdown system, always available for activation.

Residual heat removal is assured by multiple heat transport paths and systems: the normal heat removal system (steam generator, condenser), and dedicated emergency shutdown heat removal systems.

3. Accelerator-Driven Systems (ADS) Safety Design Issues

3.1. *ADS Requirements*

Specific safety design requirements for ADS systems will be documented during licensing of the facility, much in the way that the General Design Criteria (GDC) for Light Water Reactors (LWRs) are now determined (See Code of Federal Regulations, Title 10, Part 50, Appendix A). Although the case for subcritical reactor safety is often stated in terms of decoupling the reactivity and power and the degree of subcriticality, the total safety case depends on many factors that include performance of containment, cooling, and control systems.

Residual decay heat removal requirements and design features for a subcritical system will be similar to a critical system. Reactor cooling after the driving beam has been shut down will require both normal and safety-grade emergency shutdown cooling systems, just as are required for a critical reactor.

However, special requirements and design features will be needed in ADS for containment and for reactor control and protection.

3.2. *ADS Containment*

For ADS, the driving beam must penetrate two of the three barriers normally in place to contain radioactive material, the containment building and the primary coolant system. Design decisions are necessary to determine the interface configurations between the beam tunnel and the containment building boundary, and between the beam and the primary coolant system. The design questions are: Does the beam tunnel become part of the containment building? Does the target become integral with the reactor coolant system? Are the interfaces able to be inspected, tested, and repaired? The containment function must be maintained and account for the beam penetrations.

3.3. *ADS Reactor Control and Protection*

For ADS, the reactor control and shutdown functions focus on beam management and cooling systems operation. The normal reactor instrumentation used for critical reactor reactivity control and protection will be used for ADS beam management. Mechanisms are needed for controlled reactor start-up and shutdown, as well as for emergency shut down.

The accelerator/reactor control interface is a two-way portal. A fault on the accelerator side (beam interruption) must be communicated to the reactor to allow normal shutdown. A fault on the reactor side must be communicated back to the accelerator to command beam shutdown (or diversion) to permit graceful termination of power production. The control system must be able to terminate power deposition in the target and the reactor, to accomplish programmed reduction of reactor coolant flow, and to avoid thermal shocks in the reactor system.

Past studies have highlighted the issue of frequent beam interruptions causing excessive thermal fatigue in reactor structures. Reactors are normally designed for one emergency shutdown per year (total of a few dozen or less in the life of the plant). The ADS reactor must be designed to accommodate unavoidable beam interruptions.

Current licensing regulations require an inherent (passive) mechanism separate from the active systems for protection against external events that can suddenly increase reactor power (e.g. seismic motions, inlet temperature decrease, etc.). In a critical system, prompt negative reactivity feedback is provided by the fuel Doppler effect. In a subcritical system, depending on the level of sub-criticality, the reactivity and power are uncoupled, and the normal reactivity feedbacks may not be effective.

3.4. *Beyond-Design Basis Reactor Safety Considerations*

The level of initial ADS subcriticality is a design parameter. As the degree of subcritcality is reduced (operation closer to critical), reactor multiplication increases and greater reactor power is achieved for a given beam power. For fast spectrum subcritical reactors, the issue of beyond-design basis safety (multiple fault initiators, severe accident prevention and consequence mitigation) is as relevant as for critical reactors. Fast reactors (critical and subcritical) normally operate with multiple critical masses to promote efficient neutronic performance (transmutation). Fuel reconfigurations in severe accidents with fuel melting and compaction can result in reactivity additions that exceed the initial subcriticality.

4. Conclusion

Safety-related design criteria and features for nuclear reactors are well defined. The regulation process for commercial nuclear power reactors is mature and the top-level safety requirements are known. ADS subcritical reactors will be subject to the same top-level criteria and requirements as critical reactors. Design solutions will be needed to address containment, reactor protection and control, and decay heat removal requirements. The accelerator/reactor interface and safety issues can be resolved within the normal engineering design process, subject to the usual performance and cost trade-off assessments.

Acknowledgement

The submitted manuscript has been created by the U Chicago Argonne, LLC, Operator of Argonne National Laboratory under Contract No. DE-AC02-06CH11357 with the U.S. Department of Energy. The U.S. Government retains for itself, and others acting on its behalf, a paid-up, nonexclusive, irrevocable worldwide license in said article to reproduce, prepare derivative works, distribute copies to the public, and perform publicly and display publicly, by or on behalf of the Government.

SPALLATION TARGET DESIGN FOR ACCELERATOR-DRIVEN SYSTEMS

YOUSRY GOHAR
Argonne National Laboratory
Argonne, Illinois 60439

A design methodology for the lead-bismuth eutectic (LBE) spallation target has been developed and applied. This methodology includes the target interface with the subcritical assembly and the different engineering aspects of the target design, physics, heat-transfer, hydraulics, structural, radiological, and safety analyses. Several design constrains were defined and utilized for the target design process to satisfy different engineering requirements and to minimize the time and the cost of the design development. Target interface requirements with the subcritical assembly were defined based on performance parameters and material damage issues to enhance the lifetime of the target structure. Different structural materials were considered to define the most promising candidate based on the current database including radiation effects.

1. Introduction

The developed target design has a coaxial geometrical configuration to minimize the target footprint and it is installed vertically along the subcritical assembly axis. LBE is the target material and the target coolant with ferritic steel (HT-9 alloy) structural material. The proton beam has 8.33-mA current uniformly distributed and 8.14-cm beam radius resulting in a current density of 40 $\mu A/cm^2$. The beam power is 5 MW and the proton energy is 600 MeV. The beam tube has 10-cm radius to accommodate the halo current. A hemi-spherical geometry is used for the target window, which is connected to the beam tube. The beam tube is enclosed inside two coaxial tubes to provide inlet and outlet manifolds for the LBE coolant. The inlet and the outlet coolant manifolds and the proton beam are entered from the top above the subcritical assembly. The paper describes the design criteria, engineering constraints, and the developed target design.

Design methodology and spallation target design have been developed to generate the required neutron source for the subcritical assembly of the accelerator driven system. A design methodology was developed to guide the target design process and to satisfy the target design requirements. Different engineering constraints were also developed to insure satisfactory target performance based on the current experience and the existing database. The physics analyses were performed using the Monte Carlo code MCNPX [1] to account for the geometrical details, the spallation process, and the production and the transport of the spallation particles and the generated photons. The thermal hydraulic analyses were performed to define the velocity distribution and the flow stability of the lead-bismuth eutectic and the temperature distribution in the target structure and the target coolant. The analyses utilized the commercially available Computational Fluid Dynamics (CFD) software package STAR-CD [2] to account for the intrinsic feature of the target design. Structural analyses were performed in conjunction with the thermal hydraulics to check the design compliance with the stress and buckling design criteria developed for the Accelerator Production of Tritium project [3] and the International Thermonuclear Experimental Reactor [4] for irradiated structural materials. The results are used to select the shape and thickness of the beam window to maximize the engineering margins. Radiological analyses were performed to define the spallation products. These products define the radiological toxicity and the decay heat source from the lead-bismuth target material as a function of the time after shutdown. The design analyses utilize the decay heat source to check the design performance during normal and abnormal conditions with respect to the maximum allowed temperature for the structural material. Also, the dose rate from the gamma rays of the LBE spallation products was calculated to define the required input for calculating the appropriate time and the shielding requirements for maintaining the target system. All these analyses were iterated to achieve a target design that satisfies the design constraints and the design requirements.

The spallation target design is based on a coaxial geometrical configuration to satisfy the SCM configuration requirements for minimizing the space requirements and to maximize the SCM utilization of the spallation neutrons. The target is installed vertically along the subcritical assembly axis. Lead-Bismuth Eutectic (LBE) is the target material and the target coolant. Ferritic steel (HT-9) alloy is the selected structural material for the target based on the current database and the design analyses. Austenitic steel (Type 316 stainless steel) is the second choice. A uniform proton beam is employed to perform the spallation process. The beam power is 5 MW and the proton energy is 600 MeV. The inlet and the outlet coolant manifolds and the proton beam are entered from the top above the subcritical assembly. The LBE flow cross-section area is maintained at a constant value along the axial direction to maintain a constant average velocity, which improves the target hydraulic design. The geometrical configuration has been carefully designed to insure flow stability and adequate cooling for the beam window and the structure material. Target design objectives were defined for the design process. Several design constraints are defined and used in the target design process to satisfy different engineering requirements, to minimize the design development time and cost, to insure a satisfactory operating performance, and to maximize the operating lifetime of the target structural material.

2. Design Requirements, Engineering Constraints, and Design Criteria

The main objective of the target design is to generate the required neutron source to drive a subcritical assembly. The neutrons are generated from the spallation process driven by the 600-MeV proton beam. The beam has a total power of 5 MW and it has a uniform spatial distribution over the beam cross-section area. The subcritical assembly design requires a small target diameter to simplify the fuel and the target replacement procedures, to reduce the neutron losses in the beam direction, to decrease the shield volume, and to lower the required number of the SCM fuel assemblies for a specific power level. However, the structural material and the heat transfer considerations require a large beam diameter to reduce the energy deposition and the irradiation damage densities in the beam window. A 40-$\mu A/cm^2$ current density was selected as a compromise to satisfy the engineering requirements for the window design and to extend its operating life without a significant impact on the subcritical assembly design. The other main objectives for the target design are to protect the subcritical assembly from the high-energy protons and neutrons, to contain the spallation products, to help achieving the availability goal of the facility, and to reduce the shut down time for the target replacement during normal and abnormal conditions. Also, the target has to generate a uniform neutron source along the beam axis as much as possible to minimize the subcritical assembly axial power peaking.

Several design constrains are imposed on the target design process to satisfy different engineering requirements and to minimize the design development time and cost. Existing structural materials, HT-9 alloy is the selected structural material for the target design. LBE is used as a target material and coolant to simplify the design. The surface temperature of the structural material in contact with the LBE is limited to less than 550 °C to reduce erosion and corrosion concerns. This temperature limit assumes that the coolant chemistry is controlled to maintain an oxide layer on the structural material surface for corrosion protection. The stress analysis of the irradiated structural materials limits the maximum temperature to less than 550 °C for HT-9. The average coolant velocity is limited to ~2 m/s based on the current database to avoid erosion and corrosion concerns. The coolant pressure is minimized to avoid high primary stresses in the structural material. The selected coolant inlet temperature is 200 ºC, which provides adequate design margin above the LBE melting point of 129 ºC. The outlet temperature is constrained by the maximum allowable temperature for the structural material. Heat conduction to the back shield in the beam tube, natural convection, and radiation to the sodium pool are used for decay heat removal. These objectives and constraints are utilized to develop the target design.

The ability of the target structure to withstand the mechanical and thermal loads is determined by comparing the induced stresses to allowable stresses based on the APT supplemental structural design requirements, the international thermonuclear experimental reactor, and the ASME Code. The allowable stresses take into consideration the change in the mechanical properties due to the radiation exposure. The ANSYS general-purpose finite element code [5] was used with a two-dimensional axisymmetric finite element model for the target. The LBE hydrostatic pressure load and the thermal stresses caused by the temperature gradient in the target structure were used in the ANSYS analysis. The buckling capabilities of the structure were initially evaluated using the ASME code. Then, a nonlinear buckling analysis was performed using ANSYS code.

3. Design Description and Analyses

The proton beam has a total current of 8.33 mA distributed uniformly over a circular cross section. The beam radius is 8.14 cm with a current density of 40 $\mu A/cm^2$. The beam tube has 10-cm radius to accommodate the halo current. A hemi-spherical geometry is used for the target window, which is connected to the beam tube. The beam tube is enclosed inside two coaxial tubes to provide inlet and outlet manifolds for the LBE target coolant. The double function of the LBE as a target material and coolant does simplify the design. The radii of these tubes were adjusted to achieve the same average velocity in the inlet and the outlet manifolds. The outer manifold is used for the inlet flow for efficient beam window cooling. The edge of the inside tube between the inlet and the outlet flow is terminated with a rounded fairing to improve the flow stability. The fairing is tangent to the inlet side surface of the middle wall and extends into the outlet flow field. The geometrical details of the target design are shown in Figure 1. A guard tube is used to enclose the target. It provides a confined space to check and contain any LBE leakage. Also, this space provides a buffer between the subcritical assembly sodium pool and the LBE. Helium gas at low pressure is used to fill this space. HT-9 alloy is the structural material. The LBE oxygen concentration is maintained in the range of 10^{-6} to 10^{-4} at% to avoid corrosion concerns.

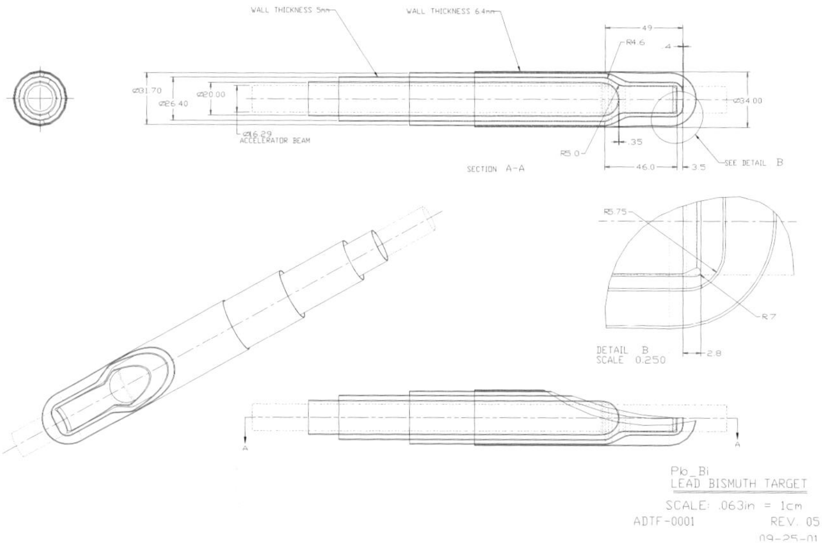

Figure 1. Lead-bismuth eutectic target design.

The beam tube enters the subcritical multiplier building horizontally above the subcritical multiplier. Then the beam is bended 90° to reach the subcritical multiplier. The vertical section of the beam tube is ~14.1 m after the last bending magnet. The coolant manifolds have a vertical length of about 10.1 m before changing direction to connect horizontally with the external section of the LBE loop. Pressurized helium gas is used to heat the target tubes before the target is filled with the LBE material. Also, helium is utilized to drain the LBE using a small vertical tube(s) of ~1-cm diameter, which reaches the target bottom section. In the target replacement procedure, the LBE is drained before the target tubes are disconnected for removal. The overhead crane is used to pull the empty target structure inside a target replacement cask.

Inside the guard tube, chemical and pressure sensors are used to check for Na or LBE leakage to shut down the target operation. This early warning avoids the possibility of mixing the two fluids, which reduces the maintenance down time and improves the safety performance. The beam tube vacuum is also monitored to detect any LBE leakage through the beam window.

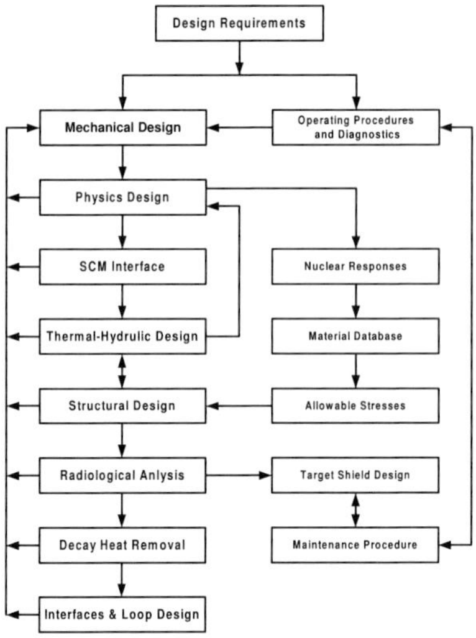

Figure 2. Target design methodology.

The target design analyses were iterated using the design methodology shown in Figure 2. A sample of the results is discussed to highlight the performance parameters of the target design. MCNPX computer code was utilized to calculate spatial neutron distribution shown in Figure 3 and the energy spectrum of the spallation neutrons, the energy deposition in the target material displayed in Figure 4, and the nuclear responses in the beam window given in Table 1. In this target design, the proton beam generates a total of 10.3 neutrons per proton from which 7.8 neutrons are utilized for the SCM. In the beam window, the neutrons are responsible for 69% of the atomic displacement and the protons are generating more than 96% of the gas production rate. The neutron distribution peaks at ~12 cm from the upper surface of the lead-bismuth material while the high-energy neutrons (above 20 MeV) peak at ~14.5 cm. The peak to the average is 1.33 and the peak to the minimum is 4.11. The peak energy deposition value is 796 W/cm^3 at 1.75 cm from the LBE surface.

The thermal hydraulic characteristics of the evaluated target geometries were simulated using the commercially available CFD code STAR-CD and the temperature distribution of the target structure was transferred to the structural analyses. The simulation uses a uniform inlet velocity of 2 m/s and an inlet temperature of 220 °C. Sufficient inlet and outlet manifold length is included to insure fully developed velocity and temperature profiles. Velocity and temperature profiles are shown in Figure 5. The peak structural temperature is 501 °C and the maximum LBE interface temperature is 340 °C.

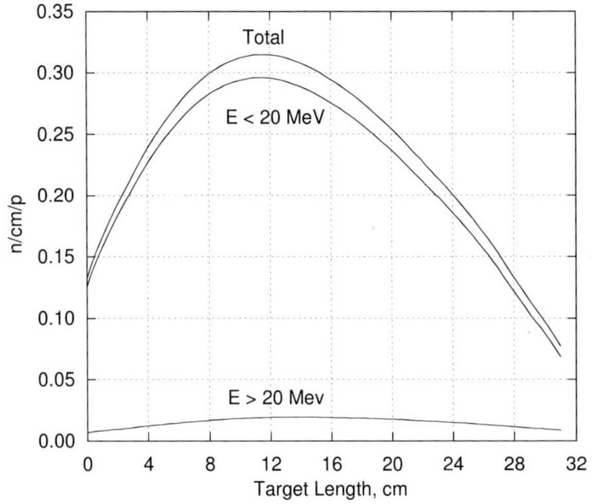

Figure 3. Spallation neutron distribution along the beam axis at the target boundary.

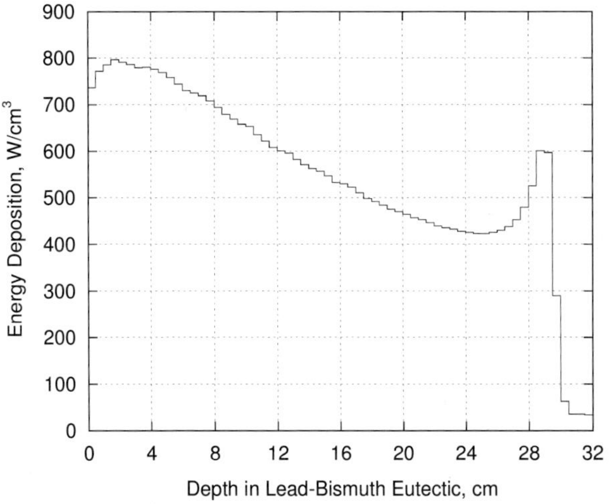

Figure 4. LBE axial energy deposition.

The ability of the target to withstand the mechanical and the thermal loads is determined by comparing the induced to allowable stresses. A stress analysis was performed to develop a beam window configuration, which would satisfy the stress and the buckling criteria. The ANSYS general-purpose finite element code was used with a two-dimensional axisymmetric finite element model for the target. The LBE hydrostatic pressure load and the thermal stresses caused by the temperature gradient in the target structure were used. The buckling capabilities of the structure were initially evaluated using the ASME code. Then, a nonlinear buckling analysis was performed using ANSYS code. Table 2 shows the calculated stresses during operation versus the allowable stresses for HT-9 with 72 dpa [6].

Radiological analyses were performed to define the spallation products. These products define the radiological toxicity and the decay heat source from LBE as function of the time after shutdown. The design analyses utilized the decay heat source to define the design performance during abnormal conditions with respect to the maximum allowed temperature for the structural material. Also, the dose rate from the gamma rays of the LBE spallation products was

Table 1. Beam window nuclear responses.

Energy deposition, W/cm^3	766.49
Atomic displacement, dpa/y	
Neutrons	46.2
Protons	21.1
Total	67.3
Helium production, appm/fpy	
Low energy neutrons $\leq$ 20 MeV	5.7
High energy neutrons > 20 MeV	50.2
Protons	1437.3
Total	1493.2
Hydrogen production, appm/fpy	
Low energy neutrons $\leq$ 20 MeV	6.3
High energy neutrons > 20 MeV	1010.1
Protons	26753.1
Total	27769.5

Table 2. Calculated stresses and HT-9 allowable stresses for the 3.5 mm spherical beam window.

Stress Component	Allowable stress	Calculated Stress (MPa)	HT-9 Allowable (MPa)	Temperature (°C)
P_M	S_m	23.6	181*	277
P_L+P_B	$K_{eff}\ S_m$	24.3	--**	277
P_L+Q_L	S_e	56.9	181	417
P_L+P_B+Q+F	S_{d1}	423.5	--***	500
P_L+P_B+Q	S_{d2}	374.2	386	500

* he allowable is calculated at the maximum temperature (417 C) to be on the conservative side since the S_m value is not available at 277 C ** Larger or equal to Sm *** Larger than S_{d2}

(a)

(b)

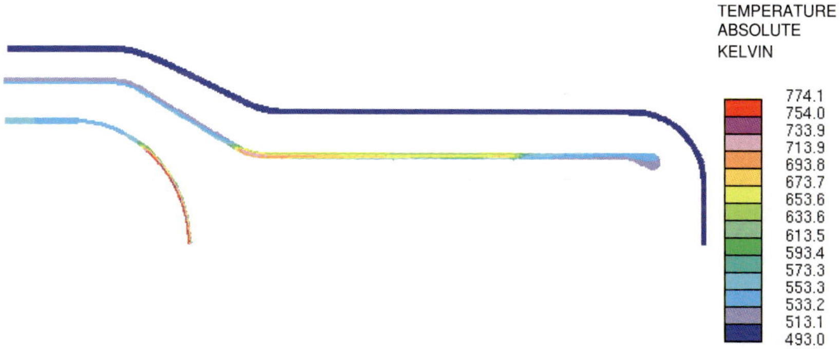

(c)

Figure 5. Lead Bismuth Eutectic Target Contour Plots Showing (a) Fluid Velocity, (b) Fluid Temperature, and (c) Structural Temperature Profiles calculated to define the required input for calculating the appropriate time and the shielding requirements for maintaining the target system. The loss of flow analysis show that the management of the decay heat does not require an active engineering system for the decay heat removal [6].

Conclusions

A target design methodology has been developed and successfully utilized for the LBE target of the subcritical multiplier station of the accelerator-driven test facility. In the design process, design objectives and engineering constraints were defined and satisfied. The target design has a coaxial geometrical configuration and HT-9 structural material, which achieves the ADTF design goals.

Acknowledgements

The submitted manuscript has been created by UChicago Argonne, LLC, Operator of Argonne National Laboratory ("Argonne"). Argonne, a U.S. Department of Energy Office of Science laboratory, is operated under Contract No. DE-AC02-06CH11357. The U.S. Government retains for itself, and others acting on its behalf, a paid-up, nonexclusive, irrevocable worldwide license in said article to reproduce, prepare derivative works, distribute copies to the public, and perform publicly and display publicly, by or on behalf of the Government.

References

1. L. S. Waters, Editor, MCNPX User's Manual – Version 2.1.5, TPO-E83-UG-X-00001, Revision 0, Los Alamos Report, November 14, 1999 and Version 2.4.j.
2. Star-CD Version 3.1B, CD-Adapco Group.
3. Supplemental Structural Design Requirements, PPO-P00-G-PRD-X-00006, Rev. A.
4. S. Majumdar and P. Smith, Treatment of irradiation effects in structural design criteria for fusion reactors, Fusion Engineering and Design 41, 1998.
5. ANSYS Version 5.7, ANSYS, Inc., 2000.
6. Yousry Gohar, Phillip Finck, Albert Hanson, Joe Herceg, Maria Koploy, Lubomir Krajtl, Panto Mijatovic, W. David Pointer, James Saiveau, Tanju Sofu, and Michael Todosow, "Lead- Bismuth Target Design for the Subcritical Multiplier (SCM) of the Accelerator Driven Test Facility (ADTF)," Argonne National Laboratory Report, ANL/TD/02-01, 2002.

DESIGN CONSIDERATIONS FOR ACCELERATOR TRANSMUTATION OF WASTE SYSTEM

WON SIK YANG

Argonne National Laboratory, Nuclear Engineering Division
Argonne, Illinois 60439, USA

This paper briefly discusses the design objectives of accelerator transmutation of waste systems and important design parameters and constraints to be considered.

1. Design Objectives of ATW

The main objective of the accelerator transmutation of waste (ATW) system is to reduce the mass, toxicity and longevity of the high-level waste to be disposed in the repository. This can be achieved by separating transuranic (TRU) elements from the spent fuel and transmuting them in the subcritical multiplier of ATW.

One practical measure for the performance of the ATW system fueled with a uranium-free metallic dispersion fuel is the fraction of the initial TRU inventory that is not transmuted and lost to the waste stream. This fractional loss of the initial TRU inventory can be represented as

$$l_w = (1 - B_d) f / [(1 - B_d) f + B_d] , \qquad (1)$$

where B_d is the fractional TRU discharge burnup and f is the fraction of TRU lost in recycle/re-fabrication [1]. Therefore, in order to minimize this fractional loss, it is necessary to maximize the fractional discharge burnup and minimize the fractional loss in recycle/re-fabrication.

The fission power produced by the subcritical multiplier varies with static reactivity ρ as

$$P_{fission} \propto SI_s / (-\rho) , \qquad (2)$$

where ρ is related to the effective multiplication factor k (k<1) as ρ=1-1/k, S is the spallation neutron source, and I_s is the source importance factor [1]. As TRU fuel is depleted over an irradiation cycle, k decreases and ρ becomes more negative. As a result, absent compensating measures, the fission power declines with fuel depletion, making it difficult to design an economic heat removal system and reducing the generation of electric power. In order to minimize the resulting needs for increasing accelerator power and/or introducing an excess reactivity and active reactivity control, therefore, it is necessary to minimize the reactivity loss over an operating cycle, which is somewhat contradictory to the objective of maximizing discharge burnup.

2. Design Considerations

2.1. *Subcriticality and Source Importance*

The subcriticality of the ATW multiplier is an important design parameter. As the subcriticality level increases, the beam-current to generate a desired system power increases, and the margin to the critical reactor performance regime increases. Thus, the subcriticality level needs to be optimized by trade-off between minimizing the beam current and maximizing the margin to criticality.

As shown in Eq. (2), for a given spallation source intensity (i.e., beam current), the fission power is proportional to the source importance factor that depends on the space and energy distributions of source neutrons. Thus, the spallation target and subcritical multiplier need to be arranged in such a way to maximize the source importance.

2.2. *Fuel Discharge Burnup*

Discharge burnup is proportional to the average power density and the fuel residence time, and inversely proportional to the fuel volume fraction and the TRU fraction in fuel. Thus, the discharge burnup can be maximized by designing for the maximum power density and fuel residence time and the minimum fuel volume fraction. However, these design parameters are interrelated and limited by various design constraints. The following points need to be considered in determining these design parameters.

The TRU fraction in fuel is determined such that the desired subcriticality level is achieved for the selected multiplier configuration and fuel management scheme, and the maximum TRU fraction is constrained by the irradiation performance of selected fuel form. The fuel residence time is typically constrained by the peak fast fluence limit for the structural material to ensure the fuel pin integrity. The peak linear power is constrained by the need to limit peak fuel and/or cladding temperatures, and the minimum fuel volume fraction required to satisfy the specified constraint on peak linear power increases as the power density increases. The power density and coolant volume fraction are also interrelated for adequate cooling, and for a specified coolant velocity, the minimum coolant volume fraction increases as the power density increases.

2.3. *Burnup Reactivity Loss*

The decline in multiplier fission power over an irradiation cycle can be mitigated in three ways. The first option is the gradual addition of reactivity, e.g., by continuous replacement of depleted fuel with fresh fuel or by withdrawal of control rods, but this adds to system complexity/cost and creates a potential accident initiator.

The second choice is to increase of the neutron source strength by gradually increasing beam power, but this requires an accelerator that is "overdesigned" for the lower TRU depletion state early in the irradiation cycle and creates a potential for source increase accidents.

Finally, the source importance factor can be increased, e.g., by reducing the fraction of source neutrons lost by leakage or through capture in the target, but this option would likely be similar in terms of cost/complexity as control rods and also introduces the possibility of accidental increases in source importance.

Irrespective of the method used to compensate for the reactivity decline, there are strong economic and safety incentives to minimize the decline itself. Therefore, the design parameters discussed in Sec. 2.2 should be optimized by trade-off between maximizing the discharge burnup and minimizing the burnup reactivity loss.

2.4. *Buffer Design*

The spallation neutron target and subcritical multiplier are coupled through a buffer region to enhance the diffusion of spallation neutrons to the multiplier and to reduce the damage to fuel by high-energy neutrons.

The target and buffer designs should be performed in conjunction with the subcritical multiplier design. For example, by designing target and buffer for a higher source importance, the desired multiplication factor at the beginning of cycle can be achieved with a reduced amount of fuel. The reduced fuel loading would increase the discharge burnup for a fixed fuel residence time. This indicates that the target position and the buffer thickness need to be determined such that the source importance is maximized. At the same time, the peak damage to fuel needs to be reduced to increase the fuel residence time and hence the discharge burnup. Therefore, the target position and buffer thickness need to be optimized based on a compromise between these competing objectives.

It should be also noted that the burnup reactivity loss decreases with reduction in buffer thickness, but the reduced buffer thickness results in a large increase in the fluence to burnup ratio, implying a much lower discharge burnup for a given fluence limit [2]. In addition, the high-energy neutron and proton damages are much more severe at reduced buffer thickness. Thus, there exists a clear design trade-off between increased accelerator size/cost (thick buffer) and increased fuel processing demand (reduced burnup with thin buffer).

2.5. *Irradiation Damage to Fuel Cladding*

For investigating the irradiation damage to fuel, the atomic displacement, hydrogen production, and helium production rates should be evaluated at the buffer and multiplier interface where the peak fast fluence occurs.

Atomic displacement rate of structural material is mainly induced by low-energy (< 20 MeV) neutrons, and high-energy neutron and proton induced damages are negligible. On the other hand, high-energy (>20 MeV) neutron contribution to hydrogen production is significantly larger than low-energy neutron contribution. High-energy neutron contribution to helium production is comparable to low-energy neutron contribution. Proton induced hydrogen and helium production rates are generally much smaller than neutron induced ones, but become non-negligible for thin buffer.

It is necessary to investigate the combined effects of various irradiation damage components and to devise the design criteria for the irradiation damage of structural materials using actual irradiation data.

3. Concluding Remarks

Physics of subcritical multiplier are not much different from those of critical reactors, and various subcritical multipliers can be designed, depending on the mission such as actinide transmutation, power production, material irradiation test, etc. However, in order to compete with alternative solutions, economical competitiveness is crucial, and it is necessary to develop unique physical or technical advantages.

In general, the optimum design is mainly determined by the imposed constraints, rather than the governing equations. Therefore, to eliminate unnecessarily conservative design margins, realistic design criteria based on experimental data need to be developed.

References

1. W. S. Yang and H. Khalil, *Nuclear Technology*, **135**, 162-182 (2001).
2. W. S. Yang, L. Mercatali, T. A. Taiwo, and R. N. Hill, "Effects of Buffer Thickness on ATW Blanket Performance," *Proc. of the ANS Topical Meeting on Nuclear Applications in the New Millennium*, Reno, Nevada, November 11-15, 2001.

JAPAN ADS PROGRAM

TOSHINOBU SASA

J-PARC Center, Japan Atomic Energy Agency
Tokai-mura, Ibaraki-ken, 319-1195, Japan

The Japan Atomic Energy Agency (JAEA) has been proceeding with the research and development on accelerator-driven system (ADS) for the transmutation of long-lived radioactive nuclides. The ADS proposed by JAEA is a lead-bismuth eutectic (LBE) cooled fast subcritical core with 800 MW_{th}. Various activities were conducted to investigate the feasibility of the ADS from viewpoints of the accelerator, LBE handling technology, minor actinide bearing fuel and subcritical core design. The design study and discussion of effective application on the Transmutation Experimental Facility (TEF) was also continued under a framework of J-PARC (Japan Proton Accelerator Research Complex) project.

1. R&D on Linear Proton Accelerator (LINAC)

The proton accelerator for the Accelerator-driven System (ADS) requires high intensity more than tens MW, high acceleration efficiency around 20 to 30 percent, and high reliability. A superconducting proton LINAC (SC-LINAC) is considered as a primary candidate for ADS. As for the R&D on the SC-LINAC, a prototype 9-cell cryomodule gives fairly good electric field by liquid-helium cooling around 4.2°K to 2.1°K. Further engineering tests are underway.

Japan Atomic Energy Agency (JAEA) precedes operation of high-intensity proton accelerators within the framework of J-PARC (Japan Proton Accelerator Research Complex) project with High Energy Accelerator Research Organization (KEK). The proton beam was accelerated up to 3 GeV by the connection of LINAC with 180 MeV and send to rapid cycle synchrotron. Proton beam is accelerated to 3 GeV-120kW and introduced to mercury spallation target. Operation record of J-PARC LINAC will be used to evaluate the reliability of high power proton accelerator.

2. R&D on Lead-Bismuth Technology for Spallation Target and Core Coolant

Lead-Bismuth eutectic (LBE) is proposed as the prime candidate that works both as the spallation target and as the core coolant of the ADS. It is well known that in high temperature range above 500°C, LBE becomes corrosive liquid against structural material. JAEA continues the implementation of a static corrosion test and a loop corrosion test in parallel.

Corrosion tests in pots were conducted to elucidate corrosion behavior of various steels in LBE for 3000 hours under the condition of low oxygen concentration at 450°C and 550°C, respectively[1)]. Significant corrosion was not observed at 450°C for ferritic/martensitic steels, F82H, Mod.9Cr-1Mo steel, 410SS, 430SS except 2.25Cr-1Mo steel. LBE penetration into steels and dissolution of elements into LBE were severe at 550°C even for ferritic/martensitic steels. Typical dissolution attack occurred for pure iron both at 550°C without surface Fe_3O_4 and at 450°C with a thin Fe_3O_4 film. Ferritization due to dissolution of Ni and Cr, and LBE penetration were recognized for austenitic stainless steels, 316SS and 14Cr-16Ni-2Mo steel at both temperatures of 450°C and 550°C. The phenomena were mitigated for 18Cr-20Ni-5Si steel. In some cases oxide films could not be a corrosion barrier in LBE.

3. R&D on Subcritical Design and Technology

The ADS proposed by JAEA is an $800MW_{th}$, minor actinide (MA) Nitride fuel, LBE cooled, tank-type subcritical core operated at $k_{eff} < 0.97$. The plant image is shown in Figure 1. The various analyses were performed to improve the reference design of the LBE cooled ADS[2)-4)].

Figure 1. Conceptual bird-eye view of JAEA-proposed ADS (1.5GeV-20MW beam, $800MW_{th}$).

Because of the subcritical core configuration, the power peaking is rather large than the typical critical fast reactors. To apply the LBE coolant with ductless fuel subassembly (no flow rate control) and to obtain more efficient transmutation of MA, it is important to flatten the neutron flux in the subcritical core. A survey analyses to flatten the heat power in the core region at each cycle was performed to decrease the maximum temperature at the surface of the fuel pin. By adjustment of the ZrN inert matrix ratio for four fuel zones at each beginning of the cycle (BOC), the subcriticality of the BOC can be equalized and the maximum temperature at the outer surface of the fuel pin was decreased from 577°C to 493°C.

For the design of the beam window, the parametric survey for the buckling failure was carried out to investigate allowable ranges for the thickness and the outer radius. The methods to increase the margin for the buckling pressure were also discussed through this survey. From the analyses, it was shown the decrease of the heat power density caused by the proton beam injection is the effective method to improve the allowable buckling pressure. The changes of the window thickness and outer radius also have the possibility to improve the buckling pressure if the thermal stress can be reduced. At the same time, it is necessary to study the safety factor for the buckling failure of the beam window to obtain more reliable design.

The impact to introduce Partitioning-Transmutation (P-T) technology to LWR high level waste disposal and effective time for predisposal storage was studied[5)]. It is noted that the introduction of P-T processes can bring an appreciable reduction in waste management size.

4. Transmutation Experimental Facility

To study the basic characteristics of the ADS and the transmutation technology from viewpoints of the reactor physics and the spallation target engineering, JAEA plans to build the Transmutation Experimental Facility (TEF) under a framework of J-PARC Project. TEF consists of two buildings: the Transmutation Physics Experimental Facility (TEF-P) and the ADS Target Test Facility (TEF-T) as shown in Figure 2[6)]. TEF-P is a zero-power critical facility where a low power proton beam is available to study the neutronic characteristics of the transmutation systems and the controllability of the ADS. TEF-T is a material irradiation facility which can accept a maximum 200kW-600MeV proton beam into the LBE spallation target.

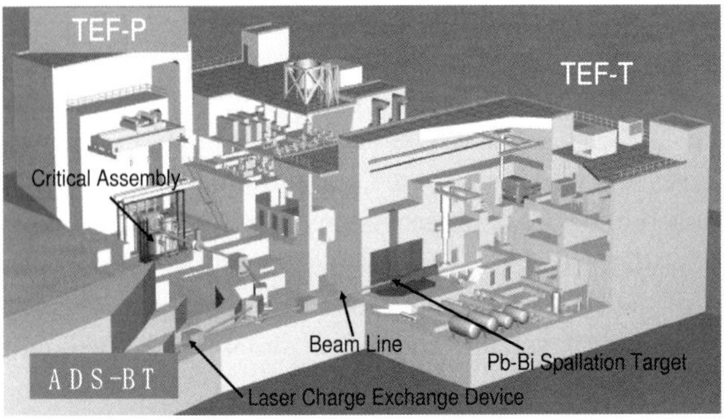

Figure 2. Transmutation Experimental Facility proposed within the framework of J-PARC project.

As for TEF-P, the engineering feasibility to handle MA-bearing fuel was studied. Although the fuel includes several percent of MA, remote handling system is required to storage, core loading and transportation between storage and reactor room. The sequence to use the MA-bearing fuel in TEF-P was specified with several remote handling devices.

5. International Collaboration

The J-PARC team called for the Preliminary Letters of Intent (Pre-LOI) for experiments using TEF in the beginning of 2006[6]. More than hundred persons including 7 foreign organization/universities are propose 37 research themes. It is helpful for us to realize the facility construction and to establish the worldwide collaboration scheme between J-PARC and the anticipated outside users. The collection of Pre-LOI continues in the immediate future.

From the 2003, Asia ADS Network initiative was launched to enhance the information exchange concerning ADS and innovative nuclear reactor systems such as LBE cooled fast reactor. Seven meetings were held annually at Japan, Korea and China. From sixth meeting, Indian activities were introduced in the discussion. Further collaborations are expected with other countries.

References

1. Y. Kurata, M. Futakawa, and S. Saito, *J. Nucl. Mat.* **373**, pp.164-178 (2008).
2. K. Tsujimoto et al., *Nucl. Tech.* **161**, No. 3, pp.315-328 (2008).
3. T. Sugawara, et al., *Proc. Global2007* (CD-ROM, 2007).
4. T. Sasa and H. Oigawa, *Proc. AccApp07* (CD-ROM, 2007).
5. K. Nishihara et al., *J. Nucl. Sci. Technol.,* **45**, No.1, pp.84-97 (2008).
6. J-PARC web site: http://j-parc.jp/index-e.html

OVERVIEW OF MEMBERS STATES' AND IAEA ACTIVITIES IN THE FIELD OF ACCELERATOR DRIVEN SYSTEMS (ADS)

ALEXANDER STANCULESCU

Department of Nuclear Energy, International Atomic Energy Agency, Wagramerstrasse 5, 1400 Vienna, Austria

Plutonium recycle in fast reactors, as well as utilization/transmutation of minor actinides and long-lived fission products in hybrid reactor systems (e.g. accelerator driven systems, ADS) offer promising nuclear fuel backend management options. Several R&D programs in various IAEA Member States are actively pursuing these options, along with the energy production and breeding mission of fast reactor systems. The paper presents an overview of the major national and international R&D programs in the area of ADS for transmutation of spent nuclear fuel. It also describes IAEA's ongoing and planned activities in this field.

1. Introduction

Based on a cumulative experience of almost 13000 reactor-years, nuclear power is a mature technology that makes a large contribution to worldwide energy supply. As of March 2009, there were 436 nuclear power plants operating in the world with a total net installed electrical capacity of 370 GW supplying slightly more than 15% of the world's electricity, and 44 nuclear power plants of 38 GW electrical capacity under construction [1]. In 2007, the nuclear share in electricity generation ranged from maxima of 76.9% (France), 64.4% (Lithuania), and 54.3% (Slovakia) to minima of 2.8% (Brazil), 2.5% (India), 2.3% (Pakistan), and 1.9% (China) [1].

There are four major challenges facing the long-term development of nuclear energy as a part of the world's energy mix: improvement of the economic competitiveness, meeting increasingly stringent safety requirements, adhering to the criteria of sustainable development, and public acceptability. While not involving the large quantities of gaseous products and toxic solid wastes associated with fossil fuels, radioactive waste disposal is also today's dominant public acceptance issue. One of the primary reasons that are cited is the long life of many of the radioisotopes generated from fission. This concern has led to increased R&D efforts to develop a technology aimed at reducing the amount of long-lived radioactive waste through transmutation in fission reactors or accelerator driven systems (ADS). In recent years, in various countries and at an international level, more and more studies have been carried out on advanced and innovative waste management strategies (i.e. actinide separation and utilization/elimination).

This paper summarizes the major programs worldwide, as well as IAEA's activities (implemented within the framework of the "Project on Technology Advances in Fast Reactors and Accelerator Driven Systems" [2]) on utilization of plutonium and minor actinides.

2. Advanced Nuclear Fuel Cycles

Presently, industrial maturity can be claimed for two fuel cycle strategies, i.e. the "Once Through Fuel Cycle" (OTC), and the "Reprocessing Fuel Cycle" (RFC) in which plutonium and very limited uranium quantities are being recycled. Taking Yucca Mountain (63000 t heavy metal capacity) as reference repository, the present worldwide accumulated spent fuel and high level waste inventories would require two repositories. For the USA alone (OTC strategy), assuming a life time extension of the present nuclear reactors to 60 years, and no new reactors, the capacity of Yucca Mountain will be exceeded by ~2050. Given the strong public opposition to the construction of geologic repositories, it is understandable that over the last decade or so, in various countries and at an international level, more and more studies have been carried out on advanced and innovative nuclear fuel cycle backend strategies [including Partitioning and Transmutation (P&T)] aiming at reducing the amount of long-lived radioactive material through utilization and transmutation in fission reactors or ADS.

3. Status of ADS Research and Technology Development

P&T is a complex technology that implies the availability of advanced reprocessing plants, facilities for fuel fabrication of trans-uranics, and irradiation facilities beyond the presently existing nuclear reactors. In principle, partitioning consists in extending the current reprocessing techniques: in addition to uranium, plutonium and ^{129}I, also minor actinides (neptunium, americium and curium), and, possibly, also some long-lived fission products (^{99}Tc, ^{93}Zr, ^{135}Cs, ^{107}Pd and ^{79}Se) would be extracted from the liquid high level waste.

Transmutation requires fully new fuel fabrication plants and irradiation technologies and their implementation on an industrial scale. New concepts, i.e. dedicated fast reactors, and sub-critical systems (ADS and even fusion/fission hybrids) have been proposed as transmutation devices.

ADS rely on the availability of a hard neutron source produced by the spallation process induced by a high-energy proton beam impinging on a heavy nuclide target. While ADS present an attractive potential [specifically, improvement of the dynamics behavior (and hence safety characteristics), and the enhanced flexibility linked to an improved neutron balance gained from the availability of an external neutron source], various specific economics and technological issues remain to be solved (e.g. capital costs, additionasl energy consumption, accelerator reliability, various physics and material science issues linked, e.g. to nuclear data, fuel performance, heavy liquid thermal hydraulics, etc). The major ongoing R&D programs in Europe and Asia, aiming at substantiating the potential of ADS and studying their role in innovative reactor and fuel cycle strategies, are summarized below.

3.1. *European Unions*

In the European Union, research and technology development in the area of ADS is part of European Atomic Energy Community's 6th (2002 – 2006) and 7th (2007 – 2011) Framework Programs (FPs) [4,5]. The objective of these activities is to investigate ways and means of reducing the amount and/or the hazard of the nuclear waste. The activities include research and technology development in partitioning and transmutation with the expressed goal of developing pilot facilities. In FP6 (to be continued in FP7), the focal point for ADS research and technology development was the EUROTRANS project (funded with a total of 45 million Euro, of which the EC contributed 23 million). EUROTRANS had two main objectives: firstly, perform preliminary design studies of experimental ADS (XT-ADS) in the thermal power range 50 – 100 MW, and, secondly, perform the conceptual design of a modular European Transmutation Demonstrator (ETD) in the thermal power range of several hundred MW. The mission of the XT-ADS is to demonstrate the feasibility of transmutation with ADS. Both the spallation target material and the coolant are lead-bismuth eutectic. The subcritical core is fuelled with mixed uranium-plutonium oxide, and the accelerator is a 600 MeV / 2.5 mA LINAC. The Belgian SCK•CEN Research Centre at Mol has aligned its R&D activities aiming at the design of the full scale ADS demonstrator MYRRHA with the XT-ADS efforts. All major possible design cliff edges are being (will be in FP7) investigated, the key accelerator components demonstrated, the spallation target thermal hydraulics design completed, and, last but not least, the experimental coupling of a sub-critical core with a neutron source realized. The latter (the GUINEVERE experiment) will also provide an experimental facility allowing physics experiments and technological research under conditions representative for XT-ADS. GUINEVERE, currently being erected at SCK•CEN Mol, couples a deuteron accelerator functioning in continuous, pulsed, and beam trip modes with a Ti^3H target producing 14.1 MeV neutrons, surrounded by a fast sub-critical metallic uranium fuelled core in a lead matrix.

3.2. *Belarus*

The Joint Institute for Power and Nuclear Research – SOSNY (JIPNR-SOSNY) of the National Academy of Sciences of Belarus is implementing the experimental program YALINA with the objective of studying the physics of ADS and investigating the feasibility of minor actinides and long-lived fission product transmutation in fast spectrum sub-critical facilities [6]. The research activities pursued by the YALINA group include the development of experimental techniques for monitoring sub-criticality levels, the investigation of spatial kinetics of sub-critical systems driven by external neutron sources, the measurement of transmutation reaction rates, and investigations of maintenance and operation characteristics of sub-critical systems driven by external neutron sources. The YALINA experimental programme contributes, on the one hand side, to the EUROTRANS project, and, on the other, aims at converting the HEU loaded fuel zones of the sub-critical YALINA-BOOSTER configuration to LEU loaded zones without penalizing its functionality [7].

3.3. *India*

In addition to the rationale for ADS development mentioned in Section 2, India is pursuing an ADS programme in view of its potential to enhance breeding of fissile material in conjunction with the utilization of thorium fuel in nuclear power systems [8]. The programme includes research in the areas of high power proton accelerator technology, spallation target systems, heavy liquid metal thermal hydraulics, and sub-critical core design. In the field of accelerator technology, the Indian research programme addresses the challenges at both the low (due to high currents) and high (superconducting RF cavities) energy end. R&D related to the spallation target includes thermal hydraulics experiments on a lead-bismuth eutectic loop with proton heating of the beam window simulated by a plasma torch and electron beam. Extensive validation and qualification of Computational Fluid Dynamics codes is being done based on these experiments. This loop is also being used for corrosion testing. The core design studies look at various sub-critical systems in view of thorium fuel utilization. The Indian program includes also an experimental program coupling a deuterium-tritium neutron source with a sub-critical water cooled, natural uranium fuelled reactor.

3.4. *Japan*

R&D on ADS is pursued by the Japan Atomic Energy Agency (JAEA) with the objective of transmuting long-lived radioactive nuclides [9]. JAEA proposes an 800 MWth lead-bismuth eutectic cooled ADS. Various activities are conducted to investigate the feasibility of the ADS from the viewpoints of accelerator technology, lead-bismuth technology, and sub-critical core design. Within the framework of the J-PARC (Japan Proton Accelerator Research Complex) project, JAEA is pursuing the TEF (Transmutation Experimental Facility) design studies. Specifically, handling of minor actinides bearing fuel studies were performed, and a remote handling system was designed for both the transportation of the fuel assemblies and for the core storage. Corrosion tests for various structural material candidates were performed in low oxygen concentration condition at 450°C and at 550°C.

4. IAEA Activities

IAEA activities include information exchange and collaborative R&D. In collaboration with the International Centre for Theoretical Physics (ICTP), the IAEA convened an "Advanced Workshop on Model Codes for Spallation Reactions" [10], and the "Workshop on Nuclear Reaction Data for Advanced Reactor Technologies". IAEA's Department of Nuclear Energy and Department of Nuclear Sciences and Applications are organizing in collaboration with ICTP the School on "Physics, Technology and Applications of Innovative Fast Neutron Systems" (Trieste, 9 – 20 November 2009). Two major international conferences related to the scope of this paper are organized by the IAEA in 2009: the "Topical Meeting on Nuclear Research Applications and Utilization of Accelerators, AccApp '09" (in collaboration with the ANS, Vienna, 4 – 8 May 2009); and the conference on "Fast Reactors and Related Fuel Cycles: Challenges and Opportunities (FR09)", organized by IAEA's Department of Nuclear Energy in cooperation with various Japanese and international organizations, and hosted by JAEA (Kyoto, 7 – 11 December 2009).

With regard to collaborative R&D, the IAEA has completed the Coordinated Research Project (CRP) on "Studies of Advanced Reactor Technology Options for Effective Incineration of Radioactive Waste" [11], and has an ongoing CRP (2005-2009) on "Analytical and Experimental Benchmark Analyses of Accelerator Driven Systems (ADS)" [12].

References

1. INTERNATIONAL ATOMIC ENERGY AGENCY, Power Reactors Information System (PRIS), http://www.iaea.org/programmes/a2/index.html.
2. INTERNATIONAL ATOMIC ENERGY AGENCY, Project on Technology Advances in Fast Reactors and Accelerator Driven Systems, http://www.iaea.org/inisnkm/nkm/aws/fnss/index.html.
3. OECD NUCLEAR ENERGY AGENCY, Accelerator-driven Systems (ADS) and Fast Reactors (FR) in Advanced Nuclear Fuel Cycles: a Comparative Study, Paris, 2002.
4. V.P. Bhatnagar, G. Van Goethem, Overview of EU Activities in Partitioning and Transmutation Research in the EURATOM 6th and 7th Framework Programmes, OECD/NEA Actinide and Fission Product Partitioning and Transmutation 9th Information Exchange Meeting, Nîmes, 25-29 Sept 2006.
5. V.P. Bhatnagar, Overview of EU Research Activities in Transmutation and Innovative Reactor Systems within the EURATOM Framework Programmes, Topical Meeting on Nuclear Research Applications and Utilization of Accelerators, AccApp '09, Vienna, May 2009.
6. A. Kiyavitskaya, V. Burnos, et al., Application of High Intensity Neutron Generator for the Investigations of ADS Neutronics, Topical Meeting on Nuclear Research Applications and Utilization of Accelerators, AccApp '09, Vienna, May 2009.
7. Y. Gohar et al., YALINA Booster Conversion Project, Topical Meeting on Nuclear Research Applications and Utilization of Accelerators, AccApp '09, Vienna, May 2009.
8. P.K. Nema et al., ADS Programme and Associated Developments in India; Roadmap and Status of Activities, Topical Meeting on Nuclear Research Applications and Utilization of Accelerators, AccApp '09, Vienna, May 2009.
9. H. Takei, N. Ouchi, et al., Research and Development Programme on ADS in JAEA, Topical Meeting on Nuclear Research Applications and Utilization of Accelerators, AccApp '09, Vienna, May 2009.
10. Joint ICTP-IAEA Advanced Workshop on Model Codes for Spallation Reactions, International Nuclear Data Committee Report, INDC(NDS)-0530, August 2008.
11. INTERNATIONAL ATOMIC ENERGY AGENCY, Advanced Reactor Technology Options for Utilization and Transmutation of Actinides in Spent Nuclear Fuel, IAEA-TECDOC-1626, 2009 (in print).
12. A. Abánades, G. Aliberti, et al., IAEA Coordinated Research Project (CRP) on "Analytical and experimental benchmark analyses of accelerator driven systems". PHYSOR 2008, Interlaken, September 2008, FP305.pdf.

LINAC FOR ADS APPLICATION - ACCELERATOR TECHNOLOGIES*

ROBERT W. GARNETT and RICHARD L. SHEFFIELD
Accelerator Operations and Technology Division, Los Alamos National Laboratory
Los Alamos, New Mexico 87545, USA

Significant high-current, high-intensity accelerator research and development have been done in the recent past in the US, centered primarily at Los Alamos National Laboratory. These efforts have included designs for the Accelerator Production of Tritium Project (APT), Accelerator Transmutation of Waste (ATW), and Accelerator Driven Systems (ADS), as well as many others. A 6.7-MeV, 100-mA, CW proton demonstration accelerator was operated successfully as a proof-of-principle for the APT Project that also showed promise as the front-end of a GWth-class ADS driver. This past work and some specific design principles that were developed to optimize linac designs for ADS and other high-intensity applications will be discussed briefly.

1. ADS Linac Requirements

The Accelerator-Driven Systems (ADS) application requires a hard neutron spectrum that can be produced by a spallation process from a proton beam. The required power of an ADS linac is determined by the design of the subcritical multiplier. The beam energy and beam current of the linac are determined by the required neutron yield and profile of the power density deposited by the proton beam in the spallation target where a uniform profile as a function of depth into the target is ideal to reduce excessive local heating in the target [1]. Recently, the trend is towards higher criticality of the subcritical multiplier and lower accelerator beam currents. The optimal proton energy for ADS is near 1500MeV.

The engineering challenges for an ADS system lie in the safe, controlled coupling of an accelerator to a subcritical reactor through a spallation target. System control and safe operation will demand the understanding and resolution of the potentially complex behavior of this coupled accelerator/target/reactor system with the key to success being how well the ADS system will be integrated in terms of design and operation.

---

* This work is supported by the U. S. Department of Energy Contract W-7405-ENG-36.

An aspect of the integrated operation of an ADS system is that of handling accelerator faults and beam interrupts that can impact the lifetime of the target/reactor system. Recent efforts have focused on improving performance of the driver linac beyond what is typical for operating machines today, emphasizing improved equipment protection and high availability, but not acceptable interrupt-free operation. Progress is still needed in the design of the linac, as well as improvements towards a more interrupt-tolerant target/reactor system in order for ADS to become viable. Arriving at credible criteria for accelerator trip rates and modifying the target/reactor to have a reasonable fault acceptance may be the greatest challenges ahead.

2. The Advantages of Superconducting (SC) RF Technology

The present trend is for new high-power proton-linac projects to use SC accelerator technology almost exclusively. The APT project at Los Alamos in the 1990s established the necessary technology base required for high-power CW SC proton-linac technology by demonstrating the feasibility and advantages of using SC elliptical cavities at high beam velocities [2, 3]. Since about 1995, all new high-power proton-linac projects include SC accelerator sections. Additionally, cavity developments such as the multi-spoke resonator are allowing application of SC technology to lower beam velocities.

The use of SC cavities for ADS is of interest because of the significant thermal issues with room-temperature structures, the larger SC-cavity apertures, and the lower operating costs (improved electrical efficiency). These characteristics also allow operational flexibility not available with room-temperature structures. An example of this is the ability to independently control the phase and amplitude of cavities which allows operations in the event of a cavity failure by rapid retuning, as well as others.

3. Accelerator Design Approach for ADS

Historically, as in all projects, accelerator systems have been designed to meet specific technical performance requirements as well as balancing costs. Less attention has been paid to optimizing reliability, availability, or maintainability (RAM). However for ADS to be successful, an integrated systems design approach is warranted that uses "lessons-learned" design principles that lead to inherent fault tolerance as well as uses formal RAM and Failure Mode and Effects (FEM) analyses [4]. This may require a iterative process to optimize a design for technical performance, cost, and RAM and FEM. Significant improvements in the implementation of appropriate real-time monitoring systems (diagnostics and controls) may also be necessary.

Much has been learned in recent years about high-intensity linac beam physics that can contribute to more reliable designs for ADS. Tools now exist to produce designs that focus on optimizing beam physics by avoiding beam envelope instabilities and resonances, achieve high beam capture, provide nearly current-insensitive focusing lattices and transitions, allow efficient acceleration, and improve inherent insensitivity to faults. This approach has been used in varying degrees for many Los Alamos projects such as Accelerator Production of Tritium, Accelerator Transmutation of Waste, ADS applications, and SNS.

4. APT and the Low-Energy Demonstration Accelerator (LEDA)

Figure 1 shows the layout of the APT SC accelerator design completed at Los Alamos in 1999 [5]. This concept along with several technology developments (a high-power 1-MW CW RF coupler, multi-gap spoke resonator cavities, and advanced RF power and controls technology) provided the design basis for an ADS system. The design included the use of three sections of multi-cell elliptical cavities to efficiently accelerate a 100-mA CW proton beam to a 1-GeV final energy.

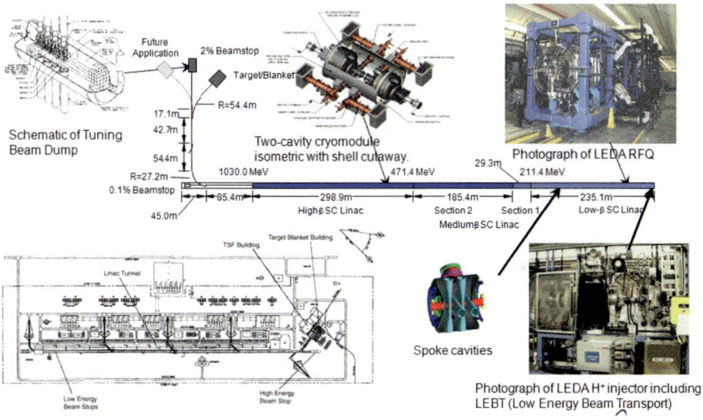

Figure 1. Layout of the 1-GeV, 100-mA (100 MW) APT linac and target concept developed at Los Alamos. It provides the design basis for an ADS system.

A 75-keV H^+ injector followed by a normal conducting 6.7-MeV 350-MHz RFQ provided the beam generation for the APT. This section of the linac, the Low-Energy Demonstration Accelerator (LEDA) was built and successfully operated at Los Alamos as a stand-alone accelerator from 1999-2001. It demonstrated the feasibility of making a two-order-of-magnitude jump in average beam current. This demonstrated technology serves as the basis for APT, ADS, and other high-power accelerator uses. The LEDA Facility was proposed as an ADS high-power accelerator test bed in 2003 [1], but was never realized.

Another important consideration for high-power beams is beam halo. An experimental study of the beam halo in a high-current 6.7-MeV proton beam propagating through a 52-quadrupole periodic-focusing channel was done at LEDA. Beam emittances and beam widths were obtained from measured profiles over a wide dynamic range for comparisons with maximum emittance-growth predictions of a free-energy model and maximum halo-amplitude predictions of a particle-core model. The experimental results supported both models and the present theoretical picture of halo formation [6].

5. Advanced SC Accelerator Design Concepts for ADS

Several advanced designs based on APT were developed at Los Alamos during 2002-2004 for an ADS driver linac. All designs assumed use of the 6.7-MeV LEDA RFQ as an injector to the SC linac. The final design iteration resulted in a linac design capable of accelerating a 20-mA CW beam to 600 MeV using only 350-MHz SC multi-spoke resonator cavities operating at 4°K [7]. Replacing the 2°K, 700-MHz SC elliptical cavity sections with spoke resonators has several advantages including reduced cryoplant operating cost and an improved real-estate accelerating gradient due to the longer active lengths of the 350-MHz cavities. Figure 2 shows schematic layouts for the initial baseline design and the advanced design concept. The use of 5-gap spoke resonators was also proposed.

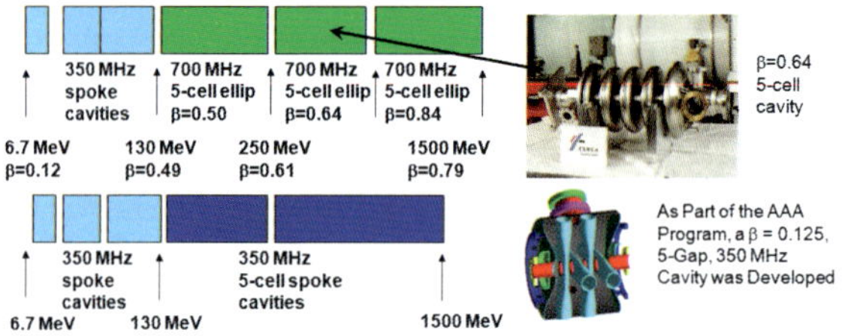

Figure 2. Schematic layouts of advanced SC linac concepts for ADS.

References

1. R. W. Garnett and R. L. Sheffield, *AccApp'03 Proceedings*, 752 (2003).
2. R. W. Garnett and T. P. Wangler, *Linac'96 Proceedings*, 707 (1996).
3. R. W. Garnett and A. J. Jason, *HPPA4 Proceedings,* 57 (2004).
4. C. M. Piaszczyk, *PAC'99 Proceedings*, 1465 (1999).
5. J. Tooker and G. Lawrence, *PAC'99 Proceedings*, 3567 (1999).
6. J. Qiang, P. L. Colestock, D. Gilpatrick, H. V. Smith, T. P. Wangler, M. E. Schulze, *Phys. Rev. ST Accel. Beams* 5, 124201 (2002); C. K. Allen, K. C. D. Chan, P. L. Colestock, K. R. Crandall, R. W. Garnett, J. D. Gilpatrick, W. Lysenko, J. Qiang, J. D. Schneider, M. E. Schulze, R. L. Sheffield, H. V. Smith, T. P. Wangler, *Phys. Rev. Lett. 89, 214802* (2002)
7. R. W. Garnett, F. L. Krawczyk, and T. P. Wangler, *HPPA4 Proceedings,* 235 (2004).

SRF LINACS AND ACCELERATOR DRIVEN SUB-CRITICAL SYSTEMS — SUMMARY WORKING GROUPS 3 & 4

J. DELAYEN
Old Dominion University, Norfolk, Virginia 23529
TJNAF, Newport News, Virginia 23606

Y. GOHAR
Argonne National Laboratory, Argonne, Illinois 60439

R. RAJA
Fermilab, Batavia, Illinois 60510

A. STANCULESCU
IAEA, A-1400, Vienna, Austria

The summary of working groups WG3 and WG4 SRF linacs and Accelerator driven subcritical systems (ADS) is given here.

1. Introduction

Most of the energy consumed today comes from oil (37%), coal (25%) and natural gas (23%), with nuclear energy making up only 6% of the total. Thus 85% of the energy production is from non-renewable sources that are a leading cause of global warming. Currently the world emits 7 billion tons of carbon into the atmosphere every year. This figure is expected to double by the year 2055 if nothing is done to change the mix of energy production. If we increase nuclear energy production to offset just one seventh of the rate of increase in carbon emission, using solar, wind, geothermal and other technologies to offset the remainder, we will have to double the number of nuclear reactors by the year 2055.

Conventional nuclear reactors use less than 0.7% of the mined Uranium. After a single pass through a nuclear reactor, the U.S stores away the waste. This waste has to be stored for geological periods of time (>100,000 years) to decay away the residual radio-activity in the long-lived actinide waste. The residual fission energy in the nuclear waste remains largely untapped, some of it is wasted as the decay processes heat up the storage facility over geological periods of time. There are two known ways to solve the waste problem and to utilize more sources of fuel such as Thorium 232 and Uranium 238. Both involve bombarding the fuel mix with fast neutrons. The first method uses fast critical reactors that produce fast neutrons using an un-moderated core of plutonium. The second uses a proton accelerator that directs a beam of protons (10mA current, 1GeV proton energy, 10MegaWatts of power) at a spallation target made of high atomic weight (e.g lead, lead/bismuth eutectic, tungsten, Uranium). The neutrons thus produced are directed at the fuel (or fuel/waste mix), that is operated well away from criticality (typical criticality factors k range from 0.95-0.98). This second method is collectively called Accelerator Driven Sub-Critical Systems (ADS). One of the main reasons such systems have not been built since the method was proposed by Rubbia et al in the early 1990's is that the accelerator of requisite power was deemed too expensive to build. The advent of super-conducting radio frequency (SRF) cavities has made building such accelerators easier and more cost effective and the use of SRF for ADS systems will be the subject of this paper.

2. ADS Efforts Worldwide

The workshop heard reports on the status of ADS research in Europe (EUROTRANS /MYRRHA), Japan (JAEA/JAERI), China (CSNS) and India (BARC, RRCAT). The MYRRHA effort in Belgium plans to have an experimental demonstration of an ADS system (SRF Linac and reactor) by 2020. The Japanese design efforts are well advanced and the Chinese spallation neutron source (CSNS, 2016 completion) plans to upgrade its accelerator to facilitate an ADS effort. In India, there is vigorous research into advanced reactors, with fast breeder reactors being in phase 2 and ADS is expected to be part of phase 3.

In the U.S, the ADS effort suffered from a 1996 National Research Council study, which was predicated on assumptions that no longer hold. The study considered costs associated with building a 250mA, 1.6 GeV 1MV/m room temperature accelerator that was shown to be too expensive. The advent of SRF has made the conclusions of this study subject to re-examination. In 2004, DOE-NE dropped ADS research in favor of solely utilizing fast reactors. It can be shown that ADS is capable of deep burning of minor actinide waste, since it is less dependent on the abundance of delayed neutrons for its safe operation. ADS is also effective in utilizing Thorium to breed U-233 fuel. Given these considerations and the intense activity abroad into ADS research, it is perhaps appropriate to restart ADS research in the U.S.

3. Feasibility of Building a 10MW Proton SRF Linac in the 1-2GeV Energy Range

The overall conclusion of WG3 was that there are no show-stoppers in building such a machine. Most of the hardware required to make a CW linac of the requisite power exists already. Vintage designs (LANL, 1999) were shown for building a 1.5 GeV SRF CW linac of adequate power. A general comment was that the optimized RF frequency should be significantly lower than 1GigaHz below 2 GeV proton energies. Designs were shown that used 350MHZ rf below 130MeV and 700MHz rf below 1.5 GeV. This was mainly motivated by the desire for larger apertures to reduce beam losses. The main challenges in a final design may well be the fraction of halo in the high-power proton beam and the associated beam losses and trips. These issues can be studied and controlled at a low energy front-end to the linac that delivers CW high current (30mA) and few tens of MeV energy. The critical accelerator physics related to beam quality is below 50 MeV. The Fermilab project HINS can play an important role here.

A critical need in studying the halo problem is the development of better diagnostics to measure the halo. Better simulation tools to predict and study the halo problem are also called for, where we include all the associated accelerator errors (both static and dynamic).

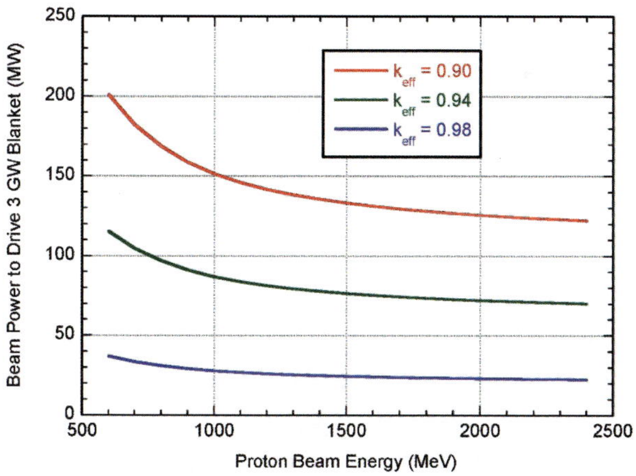

Figure 1. Shows the dependence of the beam power in MW needed to produce 3 GWth ADS reactor as a function of proton energy and the effective criticality factor keff. The beam power needed for other thermal outputs can be obtained by scaling the above curves for the appropriate keff factor.

4. Reactor Beam Requirements

ADS neutron multiplication varies as $1/(1-k_{eff})$, where k_{eff} is the effective criticality factor. Figure 1 shows the beam power in MW needed to power a reactor of 3GW (thermal) energy output as a function of beam proton energy (LANL study). It can be seen that all the curves flatten out above 1 GeV proton energy. However, the reactor designers prefer beam energies in the 1-2 GeV range and not higher in order to keep the length of the target as small as possible., to reduce the structural material damage from high energy neutrons and to reduce the required shielding volume.

5. Tolerable Beam Trip Rates

A Japanese study (JAEA) showed that the target window can tolerate 10^5 beam trips of <1 sec per 2 years. The reactor vessel can tolerate 10^4 trips per 40 years of duration 1sec to 5 minutes. The system availability requirements will not permit downtimes of > 5 mins once per week. A European study has even more stringent requirements on beam trips (>1sec to a few mins) of less than 3 per year. These requirements are motivated to try and minimize the thermal stresses on the fuel rods as well as the stability of power output from the reactor. The thermal stress requirement may be relaxed in future with progress in fuel structure and the stability requirement may be ameliorated by the development of energy storage devices such as superconducting coils or pumping water uphill.

6. Potential of Project-X to Contribute to ADS Research

The design of Project-X has evolved to an extent that ICD-2 design provides for a CW SRF linac to 2 GeV. This linac can be designed in such a way as to permit intensity upgrade to 10MW (5mA current at 2 GeV), should the U.S decide that ADS research is a national priority. The 10MW beam can be dumped on to a target station that can be used for muon collider/neutrino factory pion production. This technology can then be transferred to a partner laboratory that has nuclear reactor expertise (Argonne, ORNL, BARC) and a full demonstration system can be put together. No attempt should be made to require super-stable beams at Fermilab short of trying to limit halos. This work can be done for the final demonstration machine. Table 1 shows the power requirements of various possible uses of Project-X beams. The total requested equivalent power budget at 2 GeV excluding ADS is (as of AHIPA09) is 10.417MW. So having an ADS class machine will help power the rest of the physics. Project-X can come online at 2MW with an upgrade path to 10MW.

Project-X user	Power need	Equivalent Power at 2 GeV
Muon Collider/Neutrino factory	4 MW each @ 8 GeV	2 MW
ADS	--	10MW
Main Injector Neutrinos	2 MW @ 60GeV	0.067MW
Low Energy Neutrinos	1 MW @ 8GeV	0.25MW
Rare Kaon Decays K^+ ,K^0_L decays		2+2 = 4MW
Rare Muon decays mu2E,g-2, muon EDM		0.2+0.4+2= 2.6MW
Material Science and Nuclear science		1.5 MW

Table 1. Shows the power budget of Project-X for various activities discussed at AHIPA09. Rare Kaon decays (K^+ and K^0_L) each demand 2 MW at 2 GeV, yielding a demand for 4 MW. Similarly for rare muon physics.

Working Group 5

Nuclear and Material Science and Material Production

Conveners

Eric Pitcher(Los Alamos)
Jerry Nolen(Argonne)
Harold Kirk(Brookhaven)

PRODUCTION OF ACTINIUM-225 VIA HIGH ENERGY PROTON INDUCED SPALLATION OF THORIUM-232

JAMES HARVEY

NorthStar Medical Radioisotopes, LLC, Madison, WI

JERRY A. NOLEN*

Argonne National Laboratory, Argonne, Illinois 60439

THOMAS KROC†

Fermi National Accelerator Laboratory, Batavia, Illinois 60510

ITACIL GOMES

I.C. Gomes Consulting, Naperville, Illinois 60565

E. PHILIP. HORWITZ

PG Research Foundation, Lisle, Illinois 60532

DANIEL R. MCALISTER

PG Research Foundation, Lisle, Illinois 60532

The science of cancer research is currently expanding its use of alpha particle emitting radioisotopes. Coupled with the discovery and proliferation of molecular species that seek out and attach to tumors, new therapy and diagnostics are being developed to enhance the treatment of cancer and other diseases. This latest technology is commonly referred to as Alpha Immunotherapy (AIT). Actinium-225/Bismuth-213 is a parent/ daughter alpha-emitting radioisotope pair that is highly sought after because of the potential for treating numerous diseases and its ability to be chemically compatible with many known and widely used carrier molecules (such as monoclonal antibodies and proteins/peptides). The object of this effort is to refine the simulations for producing actinium-225 at proton beam energies of 400 MeV and above up to about 8 GeV. Once completed, the simulations will be experimentally verified using 400 MeV and 8 GeV protons available at Fermi National Accelerator Laboratory. Targets will be processed at

* Argonne National Laboratory work is supported by the U.S. Department of Energy, Office of Nuclear Physics, under Contract No. DE-AC02-06CH11357.

† Fermi National Accelerator Laboratory is operated by Fermi Research Alliance, LLC under Contract No. DE-AC02-07CH11359 with the U. S. Department of Energy.

New effort to be funded by DOE Office of Science

Argonne National Laboratory to separate and purify the actinium-225 that will subsequently be transferred to NorthStar laboratory facilities for product quality testing and comparison to the product quality of ORNL produced actinium-225, which is currently the industry standard. The test irradiations at FNAL will produce 1-20 mCi per day which is more than sufficient for quantitative evaluation of the proposed production process.

1. Background

The science of cancer research is currently expanding its use of alpha particle emitting radioisotopes. Coupled with the discovery and proliferation of molecular species that seek out and attach to tumors, new therapy and diagnostics are being developed to enhance the treatment of cancer and other diseases. Some of these molecules have the ability to be conjugated with a radioisotope whereby the radioactivity can be delivered directly to the cells to be treated. Some of today's therapies involve high-energy beta emitters. Research is showing promise that an even more effective therapy can be developed with alpha emitters - with far less side effects and patient discomfort. This latest technology is commonly referred to as Alpha Immunotherapy (AIT). Scientists are looking for ways to minimize side effects by using radioisotopes that produce alpha radiation, which has the benefit of a short destructive path and minimizes damage to adjacent healthy tissue. Depending on the isotope, another desirable feature is that the carrier molecules have far more success of reaching the tumor sites without being destroyed in transit. Many common cancer-fighting drugs today use beta emitters, which are comparatively hundreds of times more destructive to surrounding good tissue.

Currently, research with bismuth-213, and its parent actinium-225, includes clinical trials for therapies for acute myeloid leukemia (AML). Scientist at the Albert Einstein College of Medicine described at a recent the American Association for the Advancement of Science (AAAS) Annual Meeting work in infectious diseases including HIV. The HIV research has currently completed several mouse trials and is expected to proceed to primate trial in 2010. If these efforts are successful and FDA provides approval to begin a Phase 1 human trial, the need beginning in 2011 is expected to be at least 50+mCi of actinium-225 per shipment with 6-12+ shipments needed annually. This one trial could outstrip the entire available US supply. Failure to provide this amount of actinium-225 will severally hamper progress of getting a potential treatment for HIV to patients in need. Other results have been published demonstrating the potential application of bismuth-213 as a therapeutic for prostate, gastric, breast, pancreatic, and ovarian cancer along with Hodgkin's lymphoma, and bone marrow transplant.

The current worldwide supply of actinium-225 can be summarized as follows:

> *ORNL:* Currently produces the bulk of the supply of actinium-225 from a single 150mCi source of actinium-225's parent, thorium-229 that is milked ~ 6 times per year.
> *ITU:* ITU has a small amount of thorium-229 supplied by ORNL a number of years ago and produces ~350mCi of actinium-225 annually. ITU does not market or sell this actinium-225 but provides it only to their "collaborators".
> *Russia:* There have been various attempts to ship small amounts (few mCi) of actinium-225 from Russia but the actual sources are largely unknown and highly sporadic in availability.

More than half of the ORNL supply is contracted for on an annual basis, leaving less than 200mCi available for others.

Several routes have been proposed for potential actinium-225 supply resolution. The most immediate way for supply to be expanded is to expand production at ORNL from the current thorium-229 material. While this material is limited, and an increase in production is likely limited to at best a 2X increase in the amount of actinium-225 to market, this would be a small but nevertheless useful expansion of the actinium-225 supply for the near term.

Unfortunately, all remaining DOE controlled thorium-229 is currently part of the uranium-233 stocks held at either INL or ORNL. About 45% of the available Th229 is tied up in the LWBR fuel pellets. DOE has proceeded with disposal of this source at NTS. Sections of the core began shipping in 2008 and a portion of the drummed material has also been dispositioned for disposal. The permanent loss of this source represents the permanent loss of about 45,000mCi of annual actinium-225 production. The remaining thorium-229 is part of the high enrichment uranium-233 housed at ORNL in building 3019. Recovery of this source of thorium-229 would allow for production of about 60,000 mCi of actinium-225 annually. Moving forward though is complicated by Congressional action in 2005 that removed DOE's flexibility to consider alternative uses, however temporary, of this uranium-233. If relief from the Congressional actions of 2005 cannot be obtained and down-blending proceeds as scheduled, the permanent loss of this source represents the permanent loss of up to 60,000mCi of annual actinium-225 production.

The Figure 1 stresses the importance of actinium-225 supply expansion and clearly indicates that loss of access to DOE controlled thorium-229 sources means supply expansion can only occur with new production technologies.[1] It

[1] USES FOR 233U, Forsberg, C.W., Chemical Technology Division, Oak Ridge National Laboratory, Briefing for U.S. Department of Energy, Oak Ridge, Tennessee, March 22, 2000

also clearly shows that even with access to DOE thorium-229 sources, the approval of multiple treatments using actinium-225, or its daughter bismuth-213, could require alternative production methodologies regardless.

New processes for supply expansion have not been fully developed, currently have only been demonstrated on a minute scale, and currently do not produce any commercially available quantities. Further, all alternative processes being considered rely on either an accelerator approach (producing radium-225 or actinium-225 directly) or a high-flux reactor to produce thorium-229 and all require use of radium-226 as the target material (the reactor approach could potentially use thorium-228 as target material).

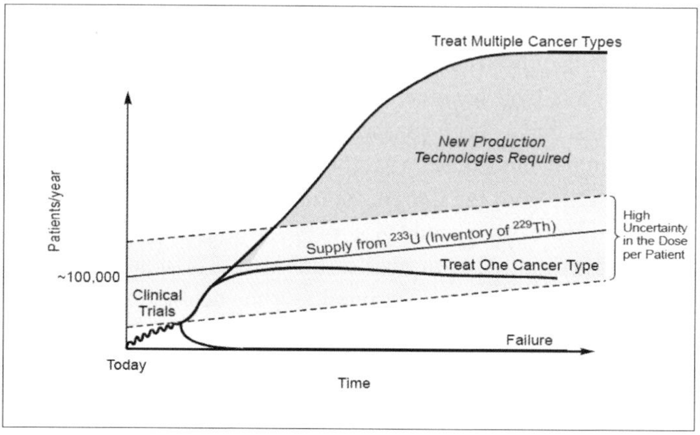

Figure 1. Relative need of actinium-225 with time.

Use of radium-226 or thorium-228 in each of these alternative approaches comes complete with a plethora of nuclear and safety issues complicating its use. The route proposed herein utilizes high energy protons to produce actinium-225 via spallation of a thorium-232 target. The resulting irradiate target is processed using chemistry developed as part of NorthStar's CRADA at INL to extract micro quantities of actinium-225 from macro quantities of thorium.

2. Description of Spallation Process

As part of previous R&D efforts carried out at Argonne National Laboratory recently in support of the proposed US FRIB facility, it was shown that a very effective production mechanism for actinium-225 is spallation of thorium-232 by high energy proton beams. These studies used the Los Alamos CEM nuclear model within the MCNPX simulation code. The high yields are predicted for proton beam energies of 200 MeV and above. For a given proton beam power the yields for a given target thickness, ~50 g/cm2, are highest at 200 MeV. For a given beam current they increase slightly with energy to about 400 MeV and then decrease slightly for beam energies in the several GeV region. Spallation of heavy elements is a common reaction mechanism for the production of short-lived isotopes for nuclear physics research; this is the mechanism used for so-called ISOL-type radioactive beam facilities, such as ISOLDE at CERN and ISAC at TRIUMF. This mechanism is not commonly used for the production of relatively long-lived medical isotopes such as actinium-225, but is being proposed in this effort as a potential solution to the supply issue surrounding actinium-225.

The base-line simulation for the production rate of actinium-225 by this reaction mechanism is 8E12 atoms per second at 200 MeV proton beam energy with 50 g/cm2 thorium target and 100 kW beam power. An irradiation of one actinium-225 half-life (10 days) produces ~10 Ci of actinium-225. Hence, this mechanism has the potential to completely solve the supply problem of actinium-225 and its daughter bismuth-213, made available as a daughter product of actinium-225 decay.

The effective production cross section extracted from these preliminary simulations is about 20 mb at 200 MeV. The present simulations give a slightly larger cross section than the 14 +/- 3 mb measured at 340 MeV[2]. This result is probably due to secondary production reactions in the relatively thick target of these preliminary simulations. More detailed simulations accounting for all such effects will be carried out during the planning stages of the proposed irradiations at FNAL. For the present planning purposes, described below, a cross section of 20 mb is used for the potential irradiations at 400 MeV and 8 mb at 8 GeV[3].

3. Description of the Irradiation Process

Two options are being considered for irradiating the thorium targets. The first option uses the 400 MeV proton (H-) beam at the end of the FNAL LINAC. The second option uses the 8 GeV proton beam in the booster synchrotron.

400 MeV Option: In normal operation, the initial 10 μsec of every beam pulse that goes to the Booster is cut off and sent to the 400 MeV beam dump. An irradiation sample placed there can receive from 5×10^{16} p/hr to 1×10^{17} p/hr which would result in a production rate of 1-2 mCi 225Ac per day of irradiation per mm of thorium sample thickness (10-20 mCi per day for a 1-cm thick sample). This option involves designing, constructing, and installing an experimental cross with a vacuum lock in the 400 MeV beamline immediately upstream of the 400 MeV dump. Once the experimental cross has been installed, samples can be quickly

[2] M. Linder and R.N. Osborne, Phys. Rev. 103 (1956) 378

[3] B.D. Pate and A.M. Poskanzer, Phys. Rev. 123 (1961) 647

interchanged with minimal interruption to LINAC operations by using access opportunities resulting from downtimes of other accelerator systems in the FNAL complex.

8 GeV Option: The FNAL Booster synchrotron has an 8 GeV beam dump in which the thorium sample could be placed. In this case, no vacuum components would be required. Beam rates would be between 1 and 5 $\times 10^{15}$ p/hr which would result in 1-5 mCi 225Ac for a 10-day irradiation of a 1-cm thick thorium sample.

The test irradiations at FNAL will produce 1-20 mCi per day which is more than sufficient for quantitative evaluation of the proposed production process.

4. Chemical Separation of Actinium from Thorium Target

Post irradiation, the thorium-232 target will contain a plethora of radioisotopes as a result of the spallation reaction and thus will require a robust chemical separation to isolate the actinium isotopes. After a sufficient cool-down period of about 10-days, the residual target activity will be reduced significantly and is expected to be handled in a shielded glove box. The only actinium isotopes present will be actinium-225 and actinium-227 in an activity ratio expected to be about 1000:1. The 1.2-day actinium-226 will be decayed at least 3 orders of magnitude by processing time.

5. Evaluation of Product Quality

Once the chemical separations have been completed, the final actinium product will be evaluated for both chemical and radiochemical purity. Since the current standard in the US is the only source of actinium-225 supplied by ORNL, actinium-225 will be purchased from ORNL and the product produced via the method being developed here will be tested relative to the ORNL material. Further, since actinium-225's daughter, bismuth-213 is used on clinical research, it is imperative the bismuth-213 also be evaluated.

Actinium and bismuth product evaluations will consist of the following:

- Alpha spectroscopy of both actinium and bismuth products to ascertain presence of unwanted alpha emitters,
- Gamma spectroscopy of both actinium and bismuth products to ascertain presence of unwanted beta/gamma emitters,
- Decay/half-life studies of both actinium and bismuth products over multiple half-lives,
- ICP analysis of the actinium and bismuth products for metals content, and
- Labeling efficacy of the bismuth-213 using an accepted mAb.

All results will be presented in a side-by-side comparison to the accepted product from ORNL.

SEARCH FOR THE ELECTRIC DIPOLE MOMENT OF RADIUM-225

ROY J. HOLT,[1] ZHENG-TIAN LU,[1,2] PETER MUELLER[1]

[1]*Physics Division, Argonne National Laboratory, 9700 S. Cass Ave., Argonne, IL 60439*
[2]*Department of Physics and Fermi Institute, University of Chicago, Chicago, IL 60637*

The nuclear Schiff moment and its resulting atomic electric dipole moment (EDM) are signatures of time-reversal and parity violation. They represent an important window onto physics beyond the Standard Model. We are developing a next generation experiment to search for the Schiff moment and EDM of ^{225}Ra ($t_{1/2}$ = 15 d) based on laser-cooled and –trapped radium atoms. Due to octupole deformation of the nucleus, ^{225}Ra is predicted to be 2-3 orders of magnitude more sensitive to T-violating interactions than ^{199}Hg (stable), which currently sets the most stringent limits in the nuclear sector. At present, ^{225}Ra samples at the level of a few mCi (~ 10^{14} atoms) are available from the decay of the long-lived ^{229}Th in stock. A future ISOL facility driven by a high-intensity accelerator could deliver 4-5 orders of magnitude more ^{225}Ra. It holds the potential to further improve the EDM search sensitivity.

The quest for nonzero permanent electric dipole moments (EDMs) represents an outstanding opportunity to search for new physics beyond the Standard Model. The four general classes of EDM searches are the neutron, nuclei (either bare or embedded in diamagnetic atoms), the electron (in paramagnetic atoms or molecules), and the muon. Searches in these areas compete against each other for the first discovery of a nonzero EDM. At the same time, they are also complimentary to each other as each is sensitive to different classes of new physics. In order to fully understand the origin of a nonzero EDM, it is perhaps necessary to measure the EDMs in all these systems. Despite of the ever increasing search sensitivity, EDM measurements in the past five decades have yielded results consistent with zero. Nevertheless, the non-observation of EDM has provided important guidance to developing new theories beyond the Standard Model, and has complimented other studies on fundamental symmetries in both nuclear and high-energy physics.[1]

We describe the progress in a search for a nonzero EDM in the diamagnetic ^{225}Ra atom. Its atomic ground-level has J = 0, resulting in a null electronic magnetic moment. Its nuclear spin I = ½, and its nuclear magnetic moment is -

0.75 NM. It was Leonard Schiff who first pointed out that, in a neutral atom, the nuclear EDM is shielded by the atomic electrons, but the shielding is imperfect due to the difference in spatial distributions of charge and EDM over the nucleus. The remaining measurable quantity characterized by a radially weighted EDM of the nucleus is called the nuclear "Schiff moment". It induces a proportional and measurable atomic EDM. The conversion between these two moments can be calculated using atomic structure theories.

Sensitive measurements on EDMs of diamagnetic atoms have been performed on stable isotopes such as ^{199}Hg and ^{129}Xe. The most sensitive measurement to date was recently performed on ^{199}Hg, whose EDM was found to be less than 3×10^{-29} e-cm at the 90% confidence level.[2] This is four orders of magnitude above the approximate EDM value (10^{-33} e-cm) for ^{199}Hg according to the Standard Model. A promising way to further improve search sensitivity in the nuclear sector was pointed out by Victor Flambaum and co-workers,[3] who indicated that P-odd and T-odd effects in the ^{225}Ra atom should be greatly enhanced because of its nuclear octupole deformation. The octupole deformation leads to a collectivity effect as well as near parity doublets that enhance strongly the Schiff moment. Indeed, both collective and mean-field calculations have predicted that the EDM of ^{225}Ra is 2-3 orders of magnitude larger than that of ^{199}Hg.[4] There are additional experimental advantages that make ^{225}Ra an exceptional candidate for EDM searches: it has a reasonably long half-life (15 d) so that the experiment can be performed off-line; its nuclear spin is ½ so that electric quadrupole effects are eliminated; and, as has been demonstrated recently, radium atoms can be laser trapped and cooled.[5]

^{225}Ra is a decay product of the long-lived ^{229}Th ($t_{1/2}$ = 7300 yr). With a stock sample of ^{229}Th, ^{225}Ra is constantly generated, and can be chemically extracted on a regular basis. ^{225}Ra sources at the level of a few mCi (~ 10^{14} atoms) can be purchased from the DOE isotope program located at Oak Ridge National Laboratory. A future ISOL facility driven by a high-intensity accelerator could deliver 4-5 orders of magnitude more ^{225}Ra. It holds the potential to significantly improve the EDM search sensitivity.

Here we describe an experiment that exploits both the enhancement of EDM due to the nuclear octupole deformation of ^{225}Ra and the beneficial properties of optically trapped atoms. Optically trapped atoms exhibit a small velocity (~ mm/s), form a compact cloud (~ 1 mm), and have a long spin-coherence time (~ 100 s). Laser trapping and cooling of radium atoms on both ^{225}Ra and the long-lived ^{226}Ra ($t_{1/2}$ = 1600 yr, I = 0) have been realized at Argonne National Laboratory.[5] The experimental apparatus is shown in Figure 1. The ^{225}Ra atoms emerging from a highly collimated oven are first transversely cooled and then

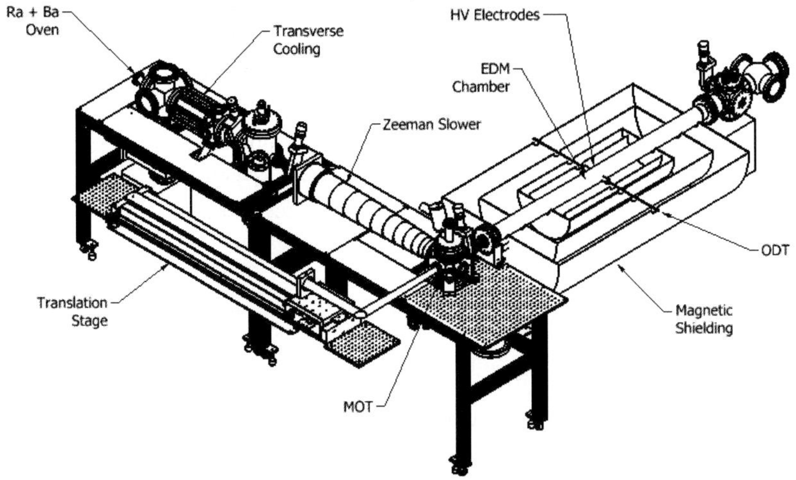

Figure 1. Schematic diagram of the setup for laser trapping and EDM measurement of ^{225}Ra. The overall length is approximately 2 m.

longitudinally slowed in a Zeeman slower before being trapped in a magneto-optical trap (MOT). The force for laser manipulation is generated by repeatedly exciting the atoms on the intercombination transition ($^1S_0 - {}^3P_1$) at 714 nm. Up to 15,000 radium atoms can be accumulated in the MOT with a trap lifetime of 5 s. Currently, the transfer of atoms from the MOT to an optical dipole trap (ODT) is under investigation. The ODT is formed by a focused beam of a 10 Watt fiber laser at the wavelength of 1.5 μm. The plan is to move the focal point of the laser beam to a neighboring vacuum chamber inside a 3-layer shields of magnetic field, and carry the cold atoms into a 2 mm gap between two parallel high-voltage electrodes where the EDM will be measured. During the development of these techniques and the apparatus, a number of key atomic transition energies and rates (lifetimes) relevant to controlling radium atoms have been measured.[6,7]

The statistical error on an EDM signal for the near-term phase of the experiment is projected to be 10^{-26} e-cm for 10^4 trapped atoms and a measurement time of 10 days. Due to the large enhancement factor of ^{225}Ra, the near-term goal should already be competitive with the current best limits set by the ^{199}Hg experiment in terms of sensitivity to T-violating interactions. In a second phase of the experiment, the aim is to increase the number of atoms to 10^6 and the measurement time to 100 days in order to achieve a limit of 3×10^{-28} e-cm. A scheme that makes use of the much stronger atomic transition ($^1S_0 - {}^1P_1$) is being evaluated as a means to increase the radium atoms in the trap by a factor of 100. Although additional repump lasers would be necessary, this scheme has already been demonstrated at KVI for trapping barium atoms.[8] In the future, a dramatic increase in the availability of ^{225}Ra made possible by an ISOL facility would enable further improvements in the search sensitivity. It is possible to eventually reach the Standard Model value. It is our hope that new physics would be discovered along the way with a nonzero EDM value.

This work is supported by the U.S. Department of Energy, Office of Nuclear Physics, under Contract No. DE-AC02-06CH11357.

References

1. P5 Report: The Particle Physics Roadmap (October 2006).
2. W.C. Griffith *et al.*, *Phys. Rev. Lett.* **102**, 101601 (2009).
3. V. Spevak, N. Auerbach, V.V. Flambaum, *Phys. Rev.* **C56**, 1357 (1997).
4. J. Engel *et al.*, *Phys. Rev.* **C68**, 025501 (2003).
5. J.R. Guest *et al.*, *Phys. Rev. Lett.* **98**, 093001 (2007).
6. N.D. Scielzo *et al.*, *Phys. Rev.* **A73**, 010501 (2006).
7. W.L. Trimble *et al.*, *Phys. Rev.* **A80**, 054501 (2009).
8. S. De *et al.*, *Phys. Rev.* **A**79, 041402(R) (2009).

SRF LINAC AND MATERIAL SCIENCE AND MEDICINE* — SUMMARY OF WORKING GROUP 5

J. NOLEN
Argonne National Laboratory, Argonne, Illinois 60439

E. PITCHER
Los Alamos National Laboratory, Los Alamos, New Mexico 87545

H. KIRK
Brookhaven National Laboratory, Upton, New York 11973

The summary of working group WG15 exploring topics in material science and medicine enabled by SRF technology is given here.

1. Introduction

One version of Project X makes available high CW proton beams at 2 GeV, as well as high power pulsed proton beams at 8 GeV.

These beams from the Project-X linac can be used for materials studies, fundamental measurements in neutron and nuclear physics, and medical isotope production. The irradiation capability of this beam could support a) closing of the nuclear fuel cycle, b) fusion reactor development and c) MW class accelerator target development. The beam can also be used for fundamental studies of radiation damage in a wide variety of materials, including metals, insulators, superconductors, and permanent magnets to understand the how radiation changes their bulk properties and micro-structure. These beams can also be used for precision nuclear and neutron physics related to the Standard Model and beyond (as discussed in paper [1]) and would also be able to provide certain important isotopes for nuclear medicine. For example, ^{225}Ac is urgently needed for clinical trials for cancer and HIV therapy but is now available only in very limited quantities.

---

* This work was partially supported by the U.S. Department of Energy, Office of Nuclear Physics, under Contract No. DE-AC02-06CH11357.

2. Materials studies:

- **Irradiation Capability to support the closing of the nuclear fuel cycle**
  - Fast neutron spectrum as in a fast reactor
    - requires ~10^{15} n/cm^2/s, ~1 MW beam power, 1 to 3 GeV beam energy and ~70% availability (6000 hours/year) to support fusion reactor deployment
- **Irradiation capability to support fusion reactor deployment**
  - Neutron damage rate >40 dpa/year
    - requires ~2 MW beam power, 1 to 3 GeV beam energy, and ~70% availability (6000 hours/year)
- **Irradiation capability to support MW-class accelerator targets**
  - Dose rate >40 dpa/year
    - requires ~1 MW beam power
- **Fundamental studies of radiation damage to understand how radiation changes bulk properties and micro-structure in materials**
  - Can study metals, insulators, superconductors, and permanent magnet materials
    - requires damage rates ~ 10–6 dpa/s with *in situ* "probing" capability

3. Nuclear Physics

- **Fundamental symmetry studies in nuclei and atoms:**
  - Intensities available at Project X will enable extensions to much higher sensitivity fundamental measurements such as search for an EDM in diatomic atoms (^{229}Ra) in a search for physics beyond the Standard Model.
  - Similarly precision measurements of parity non-conservation in atoms of isotopes of elements such as francium could greatly extend such studies presently being carried out in isotopes of lighter elements such as cesium.

4. Neutron Physics

- **Precision measurement of the neutron lifetime:** Precision measurement of the neutron lifetime can be used to determine the fundamental coupling of the up-quark to the down-quark with high accuracy. This determination combined with other precision measurements of quark couplings is sensitive to new families of quarks that may exist beyond the known three generations.

- **Search for the neutron electric dipole moment:** The search for an electric dipole moment of the neutron is broadly recognized as one of the most incisive probes of new physics today. The current search sensitivity of neutron EDMs tightly constrains the character of new physics beyond the Standard Model, and further sensitivity driven by Megawatt class proton beams can probe many new physic models.

5. Isotope Production

There are potentially very important isotopes for nuclear medicine, such as ^{225}Ac, that are currently available in very limited quantities. This particular isotope, for example, is the precursor of ^{213}Bi which is in clinical trials for cancer and HIV therapy. These trials are, however, severely limited due to the short supply of this isotope. Proton irradiation of ^{232}Th, however, is a prolific production mechanism of this isotope at beam energies of 200 MeV and above. Project X could provide this isotope in adequate quantities to support extensive clinical trails and possibly even for production quantities if the trials are successful, using only a small percentage of the CW beam available at 2 GeV.

Workshop on Applications of High Intensity Proton Accelerators

October 19-21, 2009
Fermi National Accelerator Laboratory, Batavia, IL, USA

List of Participants

Last Name	First Name	Institution
Abrams	Robert	Muons, Inc. E-mail: rabrams@fnal.gov
Adolphsen	Chris	SLAC National Accelerator Laboratory E-mail: star@slac.stanford.edu
Ahmed	Shahid	Illinois Institute of Technology E-mail: shahid@fnal.gov
Alexahin	Yuri	Fermi National Accelerator Laboratory E-mail: alexahin@fnal.gov
Ankenbrandt	Chuck	Muons, Inc. E-mail: ankenbra@fnal.gov
Bailey	James	Argonne National Laboratory E-mail: jbailey@anl.gov
Banerjee	Srikumar	Director, Bhabha Atomic Research Centre E-mail: sbanerji@barc.gov.in
Barlow	Roger	Manchester University E-mail: roger.barlow@manchester.ac.uk
Beard	Carl	ASTeC, STFC Daresbury Laboratory E-mail: carl.beard@stfc.ac.uk
Beard	Kevin	Muons, Inc. E-mail: beard@muonsinc.com
Bellantoni	Leo	Fermi National Accelerator Laboratory E-mail: bellanto@fnal.gov
Bellettini	Giorgio	Fermi National Accelerator Laboratory E-mail: giorgiob@fnal.gov

Bertsche	Kirk	SLAC National Accelerator Laboratory E-mail: bertsche@slac.stanford.edu
Bhandari	Rakesh K.	Government of India / Dept. of Atomic Energy E-mail: bhandari@veccal.ernet.in
Bhat	Chandra	Fermi National Accelerator Laboratory E-mail: cbhat@fnal.gov
Bhat	Pushpa	Fermi National Accelerator Laboratory E-mail: pushpa@fnal.gov
Bishai	Mary	Brookhaven National Laboratory E-mail: mbishai@bnl.gov
Brown	Bruce	Fermi National Accelerator Laboratory E-mail: bcbrown@fnal.gov
Bryman	Douglas	University of British Columbia E-mail: doug@triumf.ca
Cahalan	James	Argonne National Laboratory E-mail: jecahalan@anl.gov
Campbell	Myron	University of Michigan E-mail: myron@umich.edu
Cao	Yan	Argonne National Laboratory E-mail: ycao@anl.gov
Caretta	Ottone	Rutherford Appleton Laboratory E-mail: o.caretta@rl.ac.uk
Champion	Mark	Fermi National Accelerator Laboratory E-mail: champion@fnal.gov
Childress	Sam	Fermi National Accelerator Laboratory E-mail: childress@fnal.gov
Chou	Weiren	Fermi National Accelerator Laboratory E-mail: chou@fnal.gov
Christian	David	Fermi National Accelerator Laboratory E-mail: dcc@fnal.gov
Chupp	Tim	University of Michigan E-mail: chupp@umich.edu
Cooley	Lance	Fermi National Accelerator Laboratory E-mail: ldcooley@fnal.gov
Corlett	John	Lawrence Berkeley National Laboratory E-mail: jncorlett@lbl.gov
Cummings	Mary Anne	Muons, Inc. E-mail: macc@fnal.gov
Delayen	Jean	Thomas Jefferson National Accelerator Facility E-mail: delayen@jlab.org
Dhavale	Asavari	Bhabha Atomic Research Center E-mail: ajd@jlab.org

Dudnikov	Vadim	Muons, Inc E-mail: dvg43@yahoo.com
Efthymiopoulos	Ilias	CERN E-mail: ilias.Efthymiopoulos@cern.ch
Eichten	Estia	Fermi National Accelerator Laboratory E-mail: eichten@fnal.gov
Fancher	James	Consultant E-mail: j.fancher@ieee.org
Favale	Anthony	Advanced Energy Systems, Inc. E-mail: tony_favale@mail.aesys.net
Finley	David	Fermi National Accelerator Laboratory E-mail: finley@fnal.gov
Fischler	Mark	Fermi National Accelerator Laboratory E-mail: mf@fnal.gov
Fu	Shinian	Institute of High Energy Physics E-mail: fusn@ihep.ac.cn
Galambos	John	Oak Ridge National Laboratory E-mail: jdg@ornl.gov
Garnett	Robert	Los Alamos National Laboratory E-mail: rgarnett@lanl.gov
Garoby	Roland	CERN E-mail: roland.garoby@cern.ch
Garren	Al	Lawrence Berkeley National Laboratory (Retired) E-mail: al.garren@gmail.com
Geer	Steve	Fermi National Accelerator Laboratory E-mail: sgeer@fnal.gov
Gerig	Rod	Argonne National Laboratory E-mail: rod@aps.anl.gov
Ginsburg	Camille	Fermi National Accelerator Laboratory E-mail: ginsburg@fnal.gov
Glass	John	Able Electropolishing Company E-mail: jglass@ableep.com
Gohar	Yousry	Argonne National Laboratory E-mail: gohar@anl.gov
Gomes	Itacil	I.C.Gomes Consulting & Investment Inc E-mail: icgomes@icgomes.com
Gonzalez-Romero	Enrique	CIMEAT E-mail: enrique.gonzalez@ciemat.es
Group	Craig	Fermi National Accelerator Laboratory E-mail: group@fnal.gov
Grudzinski	Jim	Argonne National Laboratory E-mail: jjg@anl.gov

Gschwendtner	Edda	CERN E-mail: edda.gschwendtner@cern.ch
Harms	Elvin	Fermi National Accelerator Laboratory E-mail: harms@fnal.gov
Harvey	James	NorthStar Medical Radioisotopes, LLC E-mail: jharvey@northstarnm.com
Henderson	Stuart	Oak Ridge National Laboratory E-mail: shenderson@ornl.gov
Henning	Walter	Argonne National Laboratory E-mail: wfhenning@anl.gov
Holmes	Steve	Fermi National Accelerator Laboratory E-mail: holmes@fnal.gov
Hurh	Patrick	Fermi National Accelerator Laboratory E-mail: hurh@fnal.gov
Hylen	James	Fermi National Accelerator Laboratory E-mail: hylen@fnal.gov
Jaffe	David	Brookhaven National Laboratory E-mail: djaffe@bnl.gov
Jain	Supriya	University of Oklahoma E-mail: sjain@fnal.gov
Jansson	Andreas	Fermi National Accelerator Laboratory E-mail: jansson@fnal.gov
Jenner	Leo	Imperial College London E-mail: l.jenner@imperial.ac.uk
Johnson	David	Fermi National Accelerator Laboratory E-mail: dej@fnal.gov
Johnson	Rolland	Muons, Inc. E-mail: roljohn@aol.com
Johnstone	Carol	Fermi National Accelerator Laboratory E-mail: cjj@fnal.gov
Junquera	Tomas	Institut Physique Nucleaire Orsay E-mail: junquera@ipno.in2p3.fr
Kahn	Stephen	Muons, Inc. E-mail: kahn@fnal.gov
Kellogg	Roger	Argonne National Laboratory E-mail: rkellogg@anl.gov
Kephart	Robert	Fermi National Accelerator Laboratory E-mail: kephart@fnal.gov
Kerby	Jim	Fermi National Accelerator Laboratory E-mail: kerby@fnal.gov
Kharoua	Cyril	ESS-S E-mail: cyril.kharoua@esss.se

Kikuchi	Kenji	Ibaraki University E-mail: kikuchik@mx.ibaraki.ac.jp
Kim	Young-Kee	Fermi National Accelerator Laboratory E-mail: ykkim@fnal.gov
Kirk	Harold	Brookhaven National Laboratory E-mail: hkirk@bnl.gov
Kobayashi	Hitoshi	KEK E-mail: hitoshi.kobayashi@kek.jp
Kolomiets	Andrey	ITEP E-mail: kolomiets@itep.ru
Kothapalli	Raju	Avasarala Technologies Limited E-mail: knr@avasarala.com
Kronfeld	Andreas	Fermi National Accelerator Laboratory E-mail: ask@fnal.gov
Kuno	Yoshitaka	Osaka University E-mail: kuno@phys.sci.osaka-u.ac.jp
Kurennoy	Sergey	Los Alamos National Laboratory E-mail: kurennoy@lanl.gov
Kustom	Robert	Argonne National Laboratory E-mail: rlk@aps.anl.gov
Latina	Andrea	Fermi National Accelerator Laboratory E-mail: alatina@fnal.gov
Lebedev	Valeri	Fermi National Accelerator Laboratory E-mail: val@fnal.gov
Lebrun	Paul L. G.	Fermi National Accelerator Laboratory E-mail: lebrun@fnal.gov
Letchford	Alan	STFC/Rutherford Appleton Laboratory E-mail: alan.letchford@stfc.ac.uk
Letchford	Alan	STFC/Rutherford Appleton Laboratory E-mail: alan.letchford@stfc.ac.uk
Leveling	Anthony	Fermi National Accelerator Laboratory E-mail: leveilng@fnal.gov
Lewis	Jonathan	Fermi National Accelerator Laboratory E-mail: jdl@fnal.gov
Li	Derun	Lawrence Berkeley National Laboratory E-mail: dli@lbl.gov
McDonald	Kirk	Princeton University E-mail: kirkmcd@princeton.edu
Mcintosh	Peter	STFC Daresbury Laboratory E-mail: peter.mcintosh@stfc.ac.uk
McManamy	Thomas	Oak Ridge National Laboratory E-mail: tnm@ornl.gov

McVea	James	Thales Components Corp E-mail: jmcvea@tccus.com
Mehdiyev	Rashid	University of Texas, Austin E-mail: rmehdi@fnal.gov
Micklich	Brad	Argonne National Laboratory E-mail: bjmicklich@anl.gov
Mishra	Shekhar	Fermi National Accelerator Laboratory E-mail: mishra@fnal.gov
Mokhov	Nikolai	Fermi National Accelerator Laboratory E-mail: mokhov@fnal.gov
Momozaki	Yoichi	Argonne National Laboratory E-mail: momo@anl.gov
Moore	Craig	Fermi National Accelerator Laboratory E-mail: cmoore@fnal.gov
Morfin	Jorge G.	Fermi National Accelerator Laboratory E-mail: morfin@fnal.gov
Mukherjee	Aseet	Fermi National Accelerator Laboratory E-mail: aseet@fnal.gov
Myers	Timothy	Advanced Energy Systems, Inc. E-mail: tim_myers@mail.aesys.net
Nagaitsev	Sergei	Fermi National Accelerator Laboratory E-mail: nsergei@fnal.gov
Nassiri	Alireza	Argonne National Laboratory E-mail: nassiri@aps.anl.gov
Neubauer	Michael	Muons Inc. E-mail: mike@muonsinc.com
Neuffer	David	Fermi National Accelerator Laboratory E-mail: neuffer@fnal.gov
Noah	Etam	ESS-Scandinavia E-mail: etam.noah@cern.ch
Nolen	Jerry	Argonne National Laboratory E-mail: nolen@anl.gov
Oddone	Piermaria	Fermi National Accelerator Laboratory E-mail: pjoddone@fnal.gov
Ostroumov	Peter	Argonne National Laboratory E-mail: ostroumov@anl.gov
Padamsee	Hasan	Cornell University E-mail: lsp2@cornell.edu
Palmer	Robert	Brookhaven National Laboratory E-mail: palmer@bnl.gov
Pasquinelli	Ralph	Fermi National Accelerator Laboratory E-mail: pasquin@fnal.gov

Peggs	Steve	Brookhaven National Laboratory/ESS-S E-mail: peggs@bnl.gov
Piel	Dr. Christian	RI Research Instruments GmbH E-mail: christian.piel@research-instruments.de
Pitcher	Eric	Los Alamos National Laboratory E-mail: pitcher@lanl.gov
Pogorelov	Ilya	Tech-X Corporation E-mail: ilya@xcorp.com
Pointer	David	Argonne National Laboratory E-mail: david.pointer@anl.gov
Popovic	Milorad	Fermi National Accelerator Laboratory E-mail: popovic@fnal.gov
Prebys	Eric	Fermi National Accelerator Laboratory E-mail: prebys@fnal.gov
Prior	Christopher	STFC/RAL E-mail: chris.prior@stfc.ac.uk
Prost	Lionel	Fermi National Accelerator Laboratory E-mail: lprost@fnal.gov
Rahaman	Azizur	University of South Carolina E-mail: rahaman@fnal.gov
Raja	Rajendran	Fermi National Accelerator Laboratory E-mail: raja@fnal.gov
Raparia	Deepak	Brookhaven National Laboratory E-mail: raparia@bnl.gov
Reed	Claude	Argonne National Laboratory E-mail: cbreed@anl.gov
Revol	Jean-Pierre Charles	CERN E-mail: jean-pierre.revol@cern.ch
Rimmer	Robert	Thomas Jefferson National Accelerator Facility E-mail: rarimmer@jlab.org
Romanenko	Olexander	Fermi National Accelerator Laboratory E-mail: aroman@fnal.gov
Ross	Marc	Fermi National Accelerator Laboratory E-mail: mcrec@fnal.gov
Rowe	Allan	Fermi National Accelerator Laboratory E-mail: arowe@fnal.gov
Sahni	Vinod	Director, Raja Ramanna Centre for Advanced Technology E-mail: vcsahni@gmail.com
Samulyak	Roman	Stony Brook University/Brookhaven National Laboratory E-mail: rosamu@bnl.gov

Sandha	Rajeshwar	Raja Ramanna Centre for Advanced Technology E-mail: rssandha@rrcat.gov.in
Sasa	Toshinobu	Japan Atomic Energy Agency E-mail: sasa.toshinobu@jaea.go.jp
Scarpine	Vic	Fermi National Accelerator Laboratory E-mail: scarpine@fnal.gov
Schellman	Heidi	Northwestern University E-mail: h-schellman@northwestern.edu
Seidel	Mike	Paul Scherrer Institut E-mail: mike.seidel@psi.ch
Shemyakin	Alexander	Fermi National Accelerator Laboratory E-mail: shemyakin@fnal.gov
Shiltsev	Vladimir	Fermi National Accelerator Laboratory E-mail: shiltsev@fnal.gov
Shishlo	Andrei	Oak Ridge National Laboratory E-mail: shishlo@ornl.gov
Sinha	Ashok	Economics & Technology Associates International E-mail: aksinha@sbcglobal.net
Solyak	Nikolay	Fermi National Accelerator Laboratory E-mail: solyak@fnal.gov
Stanculescu	Alexander	IAEA, Austria E-mail: a.stanculescu@iaea.org
Stefanski	Ray	Fermi National Accelerator Laboratory E-mail: stefanski@fnal.gov
Stratakis	Diktys	Brookhaven National Laboratory E-mail: diktys@bnl.gov
Sun	Yin-E	Fermi National Accelerator Laboratory E-mail: yinesun@fnal.gov
Syphers	Michael	Fermi National Accelerator Laboratory E-mail: syphers@fnal.gov
Talamo	Alberto	Argonne National Laboratory E-mail: alby@anl.gov
Thangaraj	Charles	Fermi National Accelerator Laboratory E-mail: jtobin@fnal.gov
Tigner	Maury	Cornell University E-mail: mt52@cornell.edu
Tollestrup	Alvin	Fermi National Accelerator Laboratory E-mail: alvin@fnal.gov
Torun	Yagmur	Illinois Institute of Technology E-mail: torun@iit.edu

Tschirhart	Robert	Fermi National Accelerator Laboratory E-mail: tsch@fnal.gov
Van den Bosch	Joris	Los Alamos National Laboratory E-mail: jvdbosch@lanl.gov
Vincke	Heinz	CERN E-mail: heinz.vincke@cern.ch
Volk	James	Fermi National Accelerator Laboratory E-mail: volk@fnal.gov
Wagner	Robert	Fermi National Accelerator Laboratory E-mail: wagner@fnal.gov
Ward	Thomas	Techsource Inc. / DOE-NE AFC R&D E-mail: tward@techsource-inc.com
Warner	Arden	Fermi National Accelerator Laboratory E-mail: warner@fnal.gov
Webber	Robert	Fermi National Accelerator Laboratory E-mail: webber@fnal.gov
Weerts	Harry	Argonne National Laboratory E-mail: weerts@anl.gov
Wendt	Manfred	Fermi National Accelerator Laboratory E-mail: manfred@fnal.gov
Weng	Wu-Tsung	OHEP/DOE E-mail: bill.weng@science.doe.gov
Wester	William	Fermi National Accelerator Laboratory E-mail: wester@fnal.gov
White	Herman B.	Fermi National Accelerator Laboratory E-mail: hwhite@fnal.gov
Yang	Won Sik	Argonne National Laboratory E-mail: wyang@anl.gov
Yeh	G.P.	Fermi National Accelerator Laboratory E-mail: gpyeh@fnal.gov
Yoshikawa	Cary	Muons, Inc. E-mail: cyoshikawa1@gmail.com
Zhong	Zhaopeng	Argonne National Laboratory E-mail: zzhong@anl.gov
Zisman	Michael	Lawrence Berkeley National Laboratory E-mail: mszisman@lbl.gov
Zwaska	Robert	Fermi National Accelerator Laboratory E-mail: zwaska@fnal.gov

Author Index